Disaster Management and Human Health Risk VII

WIT*PRESS*

WIT Press publishes leading books in Science and Technology.
Visit our website for new and current list of titles.
www.witpress.com

WITeLibrary

Home of the Transactions of the Wessex Institute.
Papers published in this volume are archived in the WIT eLibrary in volume 207 of WIT Transactions on The Built Environment (ISSN 1743-3509).
The WIT eLibrary provides the international scientific community with immediate and permanent access to individual papers presented at WIT conferences.
Visit the WIT eLibrary at www.witpress.com/elibrary.

Seventh International Conference on Disaster Management and Human Health Risk: Reducing Risk, Improving Outcomes

Disaster Management 2021

Honorary Chairman

Massimo Guarascio
University of Rome, "La Sapienza", Italy

Conference Chairmen

Giorgio Passerini
Marche Polytechnic University, Italy
Member of WIT Board of Directors

Fabio Garzia
University of Rome "La Sapienza", Italy

Mara Lombardi
University of Rome "La Sapienza", Italy

International Scientific Advisory Committee

Sponsored by

WIT Transactions on the Built Environment

Organised by

Wessex Institute, UK
University of Rome "La Sapienza", Italy

WIT Transactions

Wessex Institute
Ashurst Lodge, Ashurst
Southampton SO40 7AA, UK

Editorial Board

Disaster Management and Human Health Risk VII

Reducing Risk, Improving Outcomes

Editors

Giorgio Passerini
Marche Polytechnic University, Italy
Member of WIT Board of Directors

Fabio Garzia
University of Rome "La Sapienza", Italy

Mara Lombardi
University of Rome "La Sapienza", Italy

WITPRESS Southampton, Boston

Editors:

Giorgio Passerini
Marche Polytechnic University, Italy
Member of WIT Board of Directors

Fabio Garzia
University of Rome "La Sapienza", Italy

Mara Lombardi
University of Rome "La Sapienza", Italy

Published by

WIT Press
Ashurst Lodge, Ashurst, Southampton, SO40 7AA, UK
Tel: 44 (0) 238 029 3223; Fax: 44 (0) 238 029 2853
E-Mail: witpress@witpress.com
http://www.witpress.com

For USA, Canada and Mexico

Computational Mechanics International Inc
25 Bridge Street, Billerica, MA 01821, USA
Tel: 978 667 5841; Fax: 978 667 7582
E-Mail: infousa@witpress.com
http://www.witpress.com

British Library Cataloguing-in-Publication Data

A Catalogue record for this book is available
from the British Library

ISBN: 978-1-78466-445-9
eISBN: 978-1-78466-446-6
ISSN: 1746-4498 (print)
ISSN: 1743-3509 (on-line)

The texts of the papers in this volume were set individually by the authors or under their supervision. Only minor corrections to the text may have been carried out by the publisher.

Preface

These Proceedings contain the revised versions of selected papers presented at Disaster Management 2021, the 7th International Conference on Disaster Management and Human Health Risk, held online, in November 2021. The conference was jointly organized by Wessex Institute (UK), University of Rome "La Sapienza" (Italy) and Marche Polytechnic University (Italy). The 7th International Conference on Disaster Management was reconvened online, due to the criticality of being able to travel in the period scheduled for the conference, following the success of the previous five meetings, held at Wessex Institute in the New Forest in 2009, the University of Central Florida in Orlando, USA in 2011, A Coruña, Spain in 2013, Istanbul Technical University, Turkey in 2015, Seville, Spain in 2017 and Ancona, Italy in 2019.

This series originated with the need for academia, stakeholders and field operators to exchange knowledge and experience required to handle the increasing risk of natural and man-made disasters. Recent major earthquakes, tsunamis, hurricanes, floods and other natural phenomena have resulted in huge losses in terms of human life and property destruction. A new range of human-made disasters have afflicted humanity in modern times; terrorist activities have been added to more classical disasters such as those due to the failure of industrial installations for instance. Finally, Global Warming and the related important alterations in the behavior of Nature, including the associated extreme events, represent the new spearhead of all such issues.

A full understanding of the nature of these global risks is important to be able to develop models, scenarios, and strategies to prepare for future events and plan effective responses in terms of disaster management. The cooperation between researchers, stakeholders and field operators is essential to achieve such knowledge and fully apply it. Disaster Management offers a great opportunity for gathering, presenting the findings of delegates in form of presentations and papers but also fostering active discussions.

The Editors wish to acknowledge the authors, the members of the Scientific Committee, the Referees, the Institutional Partners, who supported the Conference, and, in particular, Irene Moreno, of the Conference Secretariat, who did an outstanding job of coordination together with the WIT Press staff.

Finally, the Editors, the Chairmen, the Members of Scientific Committee and of the Conference Secretariat wish to remember the late Professor Carlos Brebbia, founder of Wessex Institute, who foresaw and established this series of meetings.

The Editors
November 2021

Contents

SECTION 1
PUBLIC HEALTH RISK

REDUCING COMMUNITY VULNERABILITY IN PACIFIC ISLANDS FORUM COUNTRIES: LESSONS FROM A PANDEMIC

VIVIENNE SAVERIMUTTU
School of Business, Western Sydney University, Australia

ABSTRACT

The Oceania region is frequently under threat to natural disasters. In the event of a category 5 cyclone, under the FRANZ Agreement, France, Australia and New Zealand stand ready to coordinate disaster reconnaissance and provide relief assistance when requested by Pacific Islands Forum countries. Australia and New Zealand also provide financial assistance to Pacific nations to support their economic development. In 2020, the declaration of a pandemic by the World Health Organization prompted immediate border closures in Pacific countries. As a result, the relentless spread of the COVID-19 virus, evidenced in other parts of the world, has not been replicated as yet among these countries. Some Islands are still entirely free of infections. The remoteness of the many islands and atolls that characterise this region does afford some protection from COVID-19. Nevertheless, these developing, mostly small, island nations have not escaped the pandemic unscathed. Border closures and travel restrictions have affected tourism, supply chains, the agricultural sector, and food security. Pacific Islanders are resilient and ready to face these new challenges. However, the vulnerabilities underlined by the emergence of this new threat have revealed a continued need for assistance from Australia and New Zealand. This discourse examines community-based solutions drawing from secondary data sources, academic literature and practical examples, especially from the rural communities of Pacific nations. The paper addresses enduring vulnerabilities emphasised by the pandemic with reference to Nobel Laureate Amartya Sen's concepts of "capabilities", "functionings" and "freedom". The study seeks to demonstrate that developing "capabilities" that build or strengthen existing social capital, strongly aligned to traditional roots in Pacific Islands Forum nations, could lead to reducing vulnerabilities by building resilience that increases coping and adaptive capacities
Keywords: capabilities, freedom, functionings, PIF nations, resilience, social capital, traditional roots, vulnerability.

1 INTRODUCTION

An unrestricted lethal flu, disseminated by global travel, was predicted in 2006 and became a reality in 2020 [1]. In that same year, the World Economic Forum (WEF), with its global risk perception survey, identified that extreme weather and other environmental risks would dominate the world over the next ten years, in terms of likelihood and impact [2]. Among the world's regions, Oceania, which includes the Pacific Islands Forum (PIF) nations, is the most vulnerable to extreme weather events such as cyclones, earthquakes, volcanic eruptions and other environmental hazards [3], [4]. When COVID-19 was declared a pandemic in 2020 during the cyclone season, this new threat added to the vulnerability of the PIF countries [5]. The leaders of the PIF nations as signatories invoked the Biketawa Declaration, a framework in place for mounting a coordinated response to any regional crisis [6]. This declaration, issued in 2000 at the Kiribati forum, emphasises the Pacific way of co-operation [7], [8]. Since COVID-19 was unknown prior to 2020, there was no pre-prepared plan on how to respond to a fast-spreading virus [6]. The measles epidemic with its tragic consequences led to a hard-line approach to COVID-19, in Samoa. Perhaps, Samoa's experience prompted other PIF nations to respond similarly, imposing border closures, travel restrictions and social distancing measures swiftly [9].

The first cases of COVID-19 infections were detected in Fiji and Papua New Guinea (PNG) in March 2020. On 2 April, category 1 tropical cyclone (TC) Harold struck the Solomon Islands (SI), then gained strength to move on to strike Vanuatu as a category 5 cyclone on 5 April. Fiji was next in its pathway, at category 4 level of severity on 7 April, and finally Tonga as a category 5 cyclone on 9 April. This was the tail end of the cyclone season in the Southwestern Pacific. In the aftermath of the cyclone, there was a trail of death and an enormous amount of destruction, especially in Fiji and Vanuatu [5]. Under the FRANZ agreement, the four PIF nations requested the assistance of Australia, New Zealand and France who have, in circumstances of extreme weather conditions, coordinated disaster reconnaissance and relief assistance in the Pacific, since 1992 [10]. Due to the severe restrictions that had been placed to stem the spread of the virus, foreign humanitarian workers had been asked to return to their countries. In the wake of the cyclone, to manage the recovery process, Fiji, had little choice but to recall the humanitarian aid workers who now had to undergo a process of adequate quarantining, while in Vanuatu and Tonga, the borders remained shut [5]. The introduction of the strict restrictions was successful in containing the COVID-19 spread in most of the PIF nations. However, in the aftermath of the cyclone, timely humanitarian assistance was not forthcoming for all those affected [5]. Infections in Fiji increased to 18 after the cyclone and since then, in 1.5 years, cumulative infections recorded number approximately 49,500 with 544 deaths reported. PNG has currently recorded approximately 18,100 infections and 207 deaths. Both countries are still reporting new cases [11].

The cumulative total of infections for the SI is at 20, all related to returning passengers from the Philippines [5]. The Republic of Marshall Islands reported four infections, Palau five, the Independent State of Samoa one, and Vanuatu three infections, while no deaths have been recorded so far from any of these nations. Currently there are no new cases reporting either. Apart from the restrictions and the generally smaller populations than in Fiji or PNG, the remoteness of the many islands and atolls that characterise this region do afford some protection from COVID-19. The Kingdom of Tonga, the Republic of Kiribati, the Federated States of Micronesia and the much smaller nations, the Republic of Nauru and the island of Tuvalu have reported no cases at all [11]. Nevertheless, these developing and mostly rural island nations, engaged in subsistence living have not escaped the pandemic unscathed [12]. Meanwhile, the threat of extreme weather events continues to intensify and increase with inadequate attention being paid to climate policy by the rest of the world, while affected nations continue to become increasingly vulnerable [13]. In 2019, at the Tuvalu Forum, the collective voice of the PIF was strongest when the Kainaki II Declaration for Urgent Climate Change Action was issued as its key advocacy instrument [14].

The majority of the Pacific nations' population are located in rural areas. They have coped for centuries with severe cyclones that are now increasingly ferocious [4]. Their vulnerability and resilience addressing the damage from these storms will continue to be a challenge. However, the weaknesses within these economies, which have been underlined by the emergence of this new threat COVID-19, have revealed a continued and perhaps increased need for assistance from Australia and New Zealand in the immediate future. This analysis will focus on seven independent PIF nations, namely Vanuatu, Tonga, SI, PNG, Fiji, Kiribati and Samoa, four of which were struck by Tropical Cyclone Harold while dealing with border closures and severe social distancing restrictions to combat the spread of COVID-19. The aim of this paper is to address the enduring vulnerabilities emphasised by the pandemic and consider a new approach on how to build resilience within this new context of dealing with a health crisis of this nature in the midst of an extreme weather event. This discussion will explore community-based solutions drawing from various data sources, academic literature and practical examples, especially from the rural communities of PIF nations. Referring to Nobel Laureate Amartya Sen's concepts of *capabilities, functionings* and *freedom,* the study seeks to demonstrate that strengthening or adding to "*capabilities*" that underlie existing social capital, which is strongly aligned to traditional roots, could lead to reducing the vulnerability of PIF nations during extreme weather events and other crises.

2 THE IMPACT OF COVID-19 ON SELECTED PIF COUNTRIES

Despite continuously facing extreme events, almost all of the PIF countries were underprepared. Their health systems are grossly underfunded and inadequate to cope with a transmission rate the magnitude of COVID-19. The expected goal of 4.5 health workers per 1,000 population is met by only 12 of the PIF nations. The global average per capita of health expenditure is US$ 1,000 per annum, with 10 of the PIF countries spending less than US$ 500 per annum [15]. Historically, influenza pandemics have had a devastating impact on the mortality rate in the Pacific Islands [16]. Thus, there was no alternative other than to shut down the borders to global air travel and prevent transmission of the virus to vulnerable populations [17]. However, keeping the virus at bay has come at a huge cost to these nations and the overall economic impact is likely to have a long-lasting impact, not just on individual nations but also on the region, and could result in a structural change for better or for worse.

2.1 International tourism and global air travel

Tourism is seen as a viable alternative livelihood in these Pacific nations and its collapse is having a devastating impact from which the industry may never recover. In Fiji, tourism accounts for 39% of GDP, and a third of the workforce either had their hours reduced or were laid off [18]. In Samoa and Vanuatu, with both nations heavily dependent on tourism, there have been no cruise ships from 2020, perhaps signalling the end of cruise ship tourism. In Samoa, 33 hotels and 131 of the 141 handicraft businesses were closed due to the restrictions [19]–[21]. In Tonga, tourism accounts for 25% of GDP and 21% of employment. As flight restrictions affected tourism, the Tongan Government established a national carrier to service its outer islands following the cessation of a private domestic carrier [22].

The national carriers of most PIF nations have been affected or will be affected due to the lack of traffic or restrictions in air travel. Air Niugini, after three years of losses, was just moving into profitable territory. Air Kiribati has invested heavily in two aircraft costing approximately half of the country's GDP. The carriers of Vanuatu and the SI are in a similar situation. The Samoan Government has cancelled the operating lease of its airline and Fiji Airways flights to Samoa [23]. Due to the remoteness of the islands and atolls, national carriers, heavily subsidised by the respective governments, are a necessity. Perhaps the most affected is Samoa, with no way out except with Air New Zealand and an underutilised newly rebuilt international airport. The region's most vulnerable and smallest nations are located in the Pacific. The mortality rate in Oceania during the 1918 influenza pandemic was 22% [24]. Thus, from a regional perspective given Oceania's vulnerability to global disease threats, the future of the airline industry in these nations seems grim.

2.2 Labour mobility, seasonal worker programs and remittances

Australia and New Zealand as PIF members have long supported the weaker economies of the PIF with mutually beneficial bilateral arrangements, allowing Pacific Islanders access to higher paying work opportunities in their more dynamic labour markets. Migrant and seasonal workers from the Pacific, through the Seasonal Workers Program (SWP) and the

Pacific Labour Scheme (PLS) in Australia, and the Recognised Seasonal Employer (RSE) scheme in New Zealand, generally find low-skilled but better paid employment in the agriculture sector. When COVID-19 emerged, many workers were stranded in Australia and New Zealand while others could not access their regular seasonal employment. These migrant and seasonal workers from the Pacific number around 25,000 in Australia and New Zealand, and remittances to the Pacific nations are an important source of income support [25] to their families.

In Fiji, remittances form the second largest source of foreign exchange [18] next to tourism and are expected to fall, whereas in Samoa, remittances [19] had already fallen to a low of 16% in 2019 and was expected to fall further in 2020. Tonga [22] is also extremely reliant on remittances, which account for 38% of GDP, while for the past 15 years, remittances were approximately 10% of GDP in Kiribati [26]. In PNG, the population was able to rely more on subsistence agriculture in comparison to other nations and thus it was possible to rely less on remittances after COVID-19 materialised [25]. PNG also has a large informal sector and, given its land mass, most urban householders maintain a "home garden" for cultivating staple food and "greens". There are many Papua New Guineans resident in Australia under the Pacific Labour Scheme. Of the migrant workers and families in Australia, the highest rate of unemployment for those aged 15 and over is among those from PNG, the SI, Tonga and Samoa [25], [27]. Both SI and Vanuatu experienced reduced remittances and rising unemployment [28], [29].

2.3 The agricultural sector supply chains and food security

Pacific countries were dependent on food imports prior to the onset of COVID-19. Additionally, rural production of staple foods looked to the urban areas and larger islands as sources of revenue. As markets closed in urban areas, the availability of fresh fruit, vegetables and other staples from rural areas decreased. Further, as international agricultural trade was blocked, flows of imported staples to urban areas were no longer available. With dwindling or no supplies of locally grown or imported staple foods in urban areas, most residents either returned of their own will or were encouraged to return to rural areas. In Vanuatu and in the Solomon Islands, urban residents, especially those working in the informal sector, were encouraged to move closer to food sources [30]. The increased demand for food in rural areas has led to increased production of staples and vegetables in home gardens. In Fiji and the Solomon Islands, the return of urban dwellers has increased agricultural production, but at the same time complaints have arisen regarding theft of land and unsustainable farming practices [31]. In Samoa and Tonga there are labour shortages as not all migrant workers were able to return home due to the border closures. The increase in home gardening is a positive impact in PNG and especially in the SI with favourable conditions for growth and the majority engaged in some form of agricultural production [25], [32]. Where migrants have returned, there is reduced access to limited resources like water, food and planting material. In Kiribati, although 42% is agricultural land, this nation is more dependent on fisheries, which make up approximately 10% of GDP [30].

2.4 The overall economic impact of COVID-19 and TC Harold in 2020

Fig. 1 [33] illustrates year-on-year changes in real GDP growth rates in 2019, and 2020 and GDP per capita PPP in international dollars in 2020, for the seven PIF nations. Real GDP growth rates increased in 2019 in all the PIF nations except Fiji, where the economy shrank by 1.3%, due to declining business confidence after an election in 2018 [34].

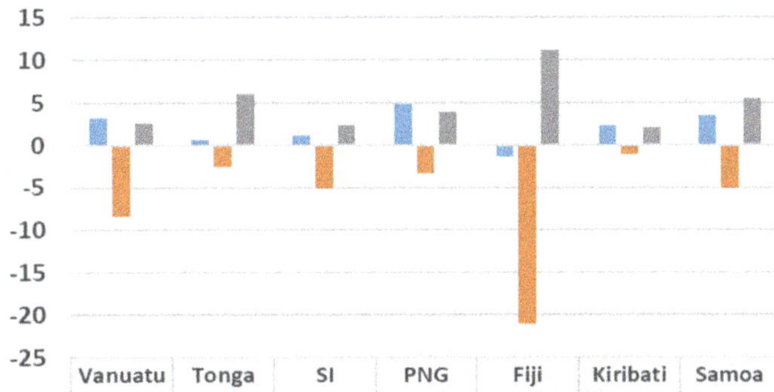

	Vanuatu	Tonga	SI	PNG	Fiji	Kiribati	Samoa
Real GDP Growth 2019 (% change yoy)	3.3	0.7	1.2	4.9	-1.3	2.3	3.5
Real GDP Growth 2020 (% change yoy)	-8.3	-2.5	-5	-3.3	-21	-1.1	-5
GDP per capita PPP 2020 (Int'l $ '000s)	2.6	6.1	2.4	3.9	11.2	2.1	5.5

Figure 1: Real GDP growth percentage change year on year and GDP per capita PPP International $ [33].

The Fijian economy shrank by 21%, the highest decline in 2020. Falling business confidence was exacerbated by spiralling infections, deaths, a lockdown stifling economic activity, and TC Harold. Economic activity declined by 8.3% in Vanuatu, and by 5% each in SI and Samoa. With no deaths and negligible infections in these three economies, the lockdown and TC Harold affected Vanuatu. Similarly, the lockdown combined with extensive floods damaging agricultural land, accounts for the decline in SI. In 2019, unlike in SI where the increase was marginal, the economies of both Vanuatu and Samoa expanded by 3.3% and 3.5% (Fig. 1). However, in 2019 the measles epidemic [9] caused a 6-week lockdown in November in Samoa. In 2020, Samoa was affected only by the border closures and restrictions that, by disrupting links to global value chains, caused job losses in the business sector as well as labour shortages in the agricultural sector [35]. The Tongan economy, though affected by the lockdown as well as by TC Harold at category 5 level of intensity, shrank by only 2.5% (Fig. 1).

PNG performed best in 2019, expanding by 4.9%. Further, the economic impact of the lockdown and cumulative infections and deaths are lower than in Fiji. The economy shrank by only 3.3%, in 2020, performing better than Fiji, Vanuatu, SI and Samoa. PNG has the advantage of both rural and urban residents having access to land and engaging in home gardening. Considering the size of PNG's population, the spread of the virus is not extensive (Fig. 1) [25]. In Kiribati, a remote small island nation, despite border closures, flights for humanitarian purposes have been allowed [26]. As a result of the lockdown, the Kiribati economy shrank by 1.1% (Fig. 1), but subsistence farmers have required assistance, which was provided as training to improve vegetable and fruit production, and local fruit tree cuttings to expand home gardens and produce for sale [36]. PPP in international dollars for all seven nations, reflects a comparison of the standards of living. Not surprisingly, the standard of living is highest in Fiji, the most developed among the seven. In Samoa and Tonga, the higher standards reflect the support received through remittances from kinfolk, which was experiencing a boom prior to 2020 [37].

3 VULNERABILTY, SOCIAL CAPITAL AND RESILIENCE

The Kainaki II declaration emphasises the vulnerability of PIF nations to climate change, which is high on their agenda for action. Except for Samoa, six of the selected nations rank among the top 20 countries in the world with the highest overall risk of experiencing extreme weather conditions (Table 1). From these six, except for Kiribati ranked at 22, the other five rank among the top 20 with highest exposure to risk. Overall risk, as defined [3], is the interaction of vulnerability and exposure to risk. Vulnerability is assessed as the mean of susceptibility, coping and adaptive capacity. Examples of susceptibility are the lack of infrastructure, food supply or other economic framework conditions, with a likelihood of suffering. Coping capacity indicates the ability to address the negative consequences of weak governance, poor health care systems and the lack of material security, while adaptive capacity refers to competence in initiating long term strategies for societal change, addressing challenges such as future natural hazards or climate change [3].

Table 1: Environmental risk, population distribution and land [3], [33].

Country	Risk ranking			2019 rural population % of total	2019 total population '000	Agri land '000 km^2	Total land '000 km^2
	Overall	Exposure	Vulnerability				
Vanuatu	1	1	46	74.6	299.9	1.9	12.2
Tonga	2	4	69	76.9	104.5	0.4	0.7
SI	5	9	38	75.6	669.8	1.2	28.0
PNG	8	18	13	86.8	8776.1	11.9	452.9
Fiji	15	14	88	43.3	890.0	4.3	18.3
Kiribati	18	22	44	45.2	117.6	0.3	0.8
Samoa	98	109	72	81.9	161.5	0.8	2.8

PNG, ranked 13 in the world for vulnerability, is at very high risk of susceptibility and lack of coping and adaptive capacity. Table 1 also displays specific characteristics of PNG that underlie its vulnerability as well as its capacity for resilience during the COVID-19 crisis. PNG is a large country in terms of land area with roughly 8.8 million population, 87% of whom reside in rural areas mainly engaged in subsistence farming. Village economies are slow due to poorly maintained infrastructure and transport networks lacking effective access to markets. Communication has improved enormously due to mobile phones, though many areas are affected by poor connectivity. Health care and other economic framework services are grossly underfunded. PNG's capacity for agricultural production is underutilised. SI, Kiribati and Vanuatu are among the top 50 nations in terms of vulnerability. 75% of the total population in SI and Vanuatu and 45% of the population in Kiribati are rural, mostly engaged in subsistence farming or fisheries (Table 1). SI is considered at very high risk for susceptibility and lack of coping capacity and at high risk for adaptive capacity. Both Vanuatu and Kiribati are considered at high risk for all three categories.

Kiribati is at high risk to sea level rise (SLR) and is likely to be one of the first islands to disappear. Fiji is considered at medium risk for vulnerability as a whole, with its comparatively higher standard of living (PPP$) almost twice that of Tonga and Samoa (Fig. 1). Tonga is at high risk for susceptibility and medium risk for lack of coping capacity, while Samoa is at medium risk for both. Both nations are categorised as high risk in terms of adaptive capacity and initiating policies addressing future environmental challenges [3].

Pacific countries are inhabited by traditional societies. Polynesians (Tonga and Samoa), Melanesians (Fiji, PNG, SI and Vanuatu) and Micronesians (Kiribati) depend on community and extended kinship and community-based networks that operate on trust, when exposed to situations that endanger life or cause vulnerabilities to surface. The dependence, and resulting collective action, increases social cohesion, leadership, decision-making and equal access to services that over time builds or enhances social capital. Social capital increases the ability of such communities to pursue resilient strategies that reduce vulnerability [39], [40]. Pacific nations, due to their low incomes, lack of technical skills and financial know-how, are often viewed as vulnerable when disaster strikes. Despite the presence of local community organisations, churches and NGOs, the international community has always rushed to provide humanitarian aid and technical assistance. Border closures prevented such offers of assistance, and each PIF nation has had to manage their crisis themselves. In the process, there have been many displays of resilience and changes in aspirations that indicate the possibility of a change in the status quo to the benefit of the region [41].

In SI and Fiji, a study of 13 communities indicates a resurgence of root crops vegetables and fruit production in home gardens, a return to traditional food systems, cultural values and safety networks such as the barter system, and what appears to be a move towards revitalising the agricultural sector [31]. The Fijian Government seems to validate this strategy of local food production, targeting urban residents with a home gardening package to move them away from imported primary products and towards growing their own food. The other initiative, the Farm support package of one million Fijian dollars, was released with a reminder of pre-colonial days and self-sufficiency in food [42]. The PNG Prime Minister too has indicated the possibility of increasing agricultural production for regional trade. PNG currently utilises only about 5% of its agricultural land [43]. A similar move to increase home gardening and the production of fruit was referred to before in relation to Kiribati, receiving assistance from the Food and Agricultural Organisation (FAO) [36]. Vanuatu has also seen an increase in local fishing in coastal areas for own consumption and sales to neighbouring villages [20].

When TC Harold struck Vanuatu amid the pandemic, due to the border closures and the absence of foreign humanitarian aid workers, ni-Vanuatu emergency workers and local volunteers armed with previous training and local knowledge went into action as first responders, paving the way for locally-led responses in the future [41]. Additionally, in PIF nations, chiefs in the villages are encouraging the organisation of community-based stockpiling and urging traditional means of stockpiling food [43]. These examples are indications of building resilience and strengthening capabilities that appear promising for meeting future challenges.

4 DISCUSSION AND CONCLUSION

4.1 Development, capabilities and freedom

Social capital is particularly important for nations that lack material security to improve the wellbeing of its population. Instead of focusing on income as a means to development and improvement in wellbeing, Amartya Sen suggests that communities should concentrate first on developing capabilities or the capacity to achieve, with freedom of choice as an important functioning, among others. Capability, is a real opportunity and results in the development of a particular "functioning" or achievement, e.g., food security [44]. The importance of freedom of choice is that individuals would be responsible for their own choices within a society even as they participate in collective action for the wellbeing of the community.

Focusing on capabilities allows for a range of possible choices, though assessing the outcome of wellbeing would require a measurement of the functioning and the capacity to achieve it [44], [45]. The possession of social capital is recognised as an important functioning in a community. Thus, by increasing the ability of closely knit communities to pursue resilient strategies, new capabilities are achieved, which in turn leads to enhancing social capital. Thus, social capital, functionings and capabilities are connected in a dynamic process [46] leading to increased wellbeing and development [44]. The capability approach is well established as a meaningful way of assessing wellbeing and underlies the basic insights of the Human Development Index (HDI). The HDI includes "life expectancy at birth, education (adult literacy and educational enrolment rates) and adjusted GDP per capita", which serves as a proxy for the material aspects of functionings and wellbeing [45]. The HDI ranks PNG as 7th and Kiribati as 4th among these seven nations. The other five nations follow the same order as for standard of living based on GDP per capita PPP International $ [48].

4.2 Conclusion

When food security was adversely affected by the lockdown, many urban residents turned to traditional food systems and subsistence food production reflecting their achieved capabilities and functionings, and away from dependence on imports. Arguably, subsequent inclusion in community organisations is likely to have directly addressed vulnerabilities and improved wellbeing during engagement in addressing the problem of food insecurity.

Previous discussion includes examples of resilience by PIF communities addressing the combined impact of COVID-19 and TC Harold. Such actions point to a generation of new capabilities and functionings in the process of addressing enduring vulnerabilities from the past, and strengthening coping and adaptive capacities. PIF nations are rich in terms of access to land governed by customary tenure, which is approximately 80% of total land, a huge advantage to PIF nations if they choose to pursue a strategy for regional food security, harnessing the majority's capabilities and functionings in agriculture. The government could provide marketing opportunities as a public good for income generation in rural areas. Further, cyclones are the most frequent natural disaster in this region [4] and Vanuatu has shown the way for the other nations to follow suit as first responders. There is still need for international humanitarian assistance but this example augurs well for the future.

The GHS Index 2019 [47] "is a comprehensive benchmark of health security and related capabilities" that concentrates on countries "capacities, and existence of functional, tested, proven capabilities for stopping outbreaks" of epidemics and pandemics at the source. Indicators included also assess factors that can hinder or enhance that capability. Scores range from 0 to 100 as best i.e., less risk, and are divided into three tiers, 0.0–33.3 as low, 33.4–66.6 as moderate and 66.7–100 as high scores. The Index is available for 195 countries and being the first of its kind the jury is still out in terms of its limitations.

Finally, Fig. 2 illustrates indicators of resilience, risk and vulnerability in the PIF nations from the GHS Index. Assessment of socio-economic resilience, includes literacy, gender equality, poverty levels, public confidence in governments and local media reporting. Environmental risks involve urbanisation, land use and natural disaster risk. Assessment of public health vulnerabilities are based on access to quality health care, potable water and sanitation, and public health care spending levels.

Except for PNG, with the lowest score for socio-economic resilience, and Kiribati, both of which fall within the moderate tier, the other five PIF nations are in the top tier. Samoa scored the highest, reflecting its success with the measles epidemic. Interestingly, Vanuatu

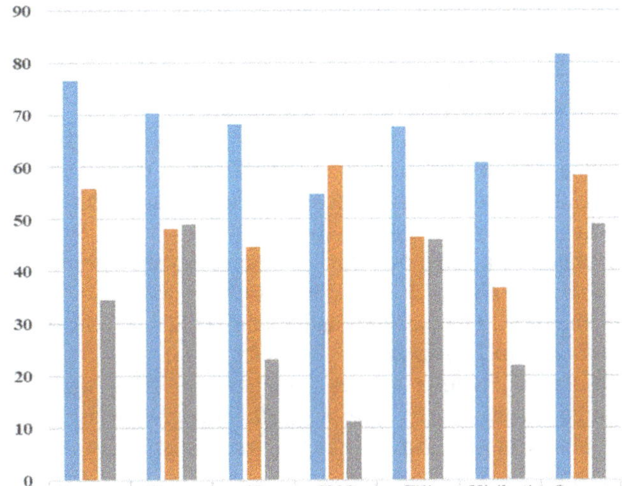

	Vanuatu	Tonga	SI	PNG	Fiji	Kiribati	Samoa
Socio-Economic Resilience	76.6	70.2	68.1	54.9	67.7	60.7	81.5
Environmental Risks	55.8	48.2	44.6	60.3	46.6	36.7	58.3
Public Health Vulnerabilities	34.7	49	23.2	11.2	46.1	21.9	49.1

Figure 2: Socio-economic resilience, environmental risk, public health vulnerabilities [47].

was next highest at 76.6, and this was reflected in the handling of the twin disasters in 2020, by the ni-Vanuatu. Only PNG and Kiribati fall below the global average of 66.1 [47].

Environmental risks, for all the PIF nations are classified as of moderate risk. In comparison with the World Risk Report, the GHS Index indicator includes land use and urbanisation in its assessment. Kiribati and Fiji with more than 50% urbanisation may partially explain the differences in ranking. PNG, Samoa and Vanuatu with large rural populations are above the global average of 52.9. Public health is a vulnerability that requires urgent attention and justifies the lockdown in PIF nations. Samoa, Tonga and Fiji are classified as moderate and the rest as low, but only marginally so

To conclude, considering the resilience exhibited by PIF nations during the COVID-19 crisis and their capacity for building social capital, the high levels of socio-economic resilience are not surprising, and achieving aspirations of the Kainaki II declaration on climate change in the future look promising for PIF nations.

REFERENCES

[1] World Economic Forum, The Global Risks Report 2021 Insight Report, 16th ed. http://www3.weforum.org/docs/WEF_The_Global_Risks_Report_2021.pdf. Accessed on: 6 Aug. 2021.

[2] World Economic Forum, The Global Risks Report 2020 Insight Report, 15th ed. http://www3.weforum.org/docs/WEF_Global_Risk_Report_2020.pdf. Accessed on: 6 Aug. 2021.

[3] World Risk Report, 2020. WorldRiskReport-2020.pdf. Accessed on: 6 Aug. 2021.

[4] Diamond, H.J., Lorrey, A.M. & Renwick, J.A., A southwest Pacific tropical cyclone climatology and linkages to the El Nino-Southern oscillation. *Journal of Climate*, **26**, pp. 3–25, 2013. http://dx.doi.org/10.1175/JCLI-D-12-00077.1.

[5] Ober, K. & Bakumenko, S., Refugees international: Issue brief – A new vulnerability: COVID-19 and tropical cyclone Harald create the perfect storm. https://reliefweb.int/report/vanuatu/issue-brief-new-vulnerability-covid-19-and-tropical-cyclone-harold-create-perfect. Accessed on: 6 Sep. 2021.

[6] The Permanent Mission of Tuvalu to the UN, PIF agrees to establish a Pacific humanitarian pathway on COVID-19. https://www.un.int/tuvalu/fr/news/pacific-islands-forum-agrees-establish-pacific-humanitarian-pathway-covid-19. Accessed on: 6 Sep. 2021.

[7] Ivarature, H., Regionalism: Performance and promise. *Politics, Development and Security in Oceania*, **7**, p. 179, 2013.

[8] Shibuya, E., The problems and potential of the Pacific Islands forum. *The Asia-Pacific: A Region in Transition*, pp. 102–115, 2004.

[9] Iosefa, V., Samoa applies lessons from 2019 measles epidemic in response to coronavirus. *The Strategist*, 2020. https://www.aspistrategist.org.au/samoa-applies-lessons-from-2019-measles-epidemic-in-response-to-coronavirus/. Accessed on: 3 Sep. 2021.

[10] The FRANZ arrangement. https://www.mfat.govt.nz/assets/Aid-Prog-docs/NZDRP-docs/Franz-Arrangement-Brochure.pdf. Accessed on: 23 Aug. 2021.

[11] World Health Organization, Coronavirus (COVID-19) dashboard. https://covid19.who.int/. Accessed on: 1–30 Sep. 2021.

[12] Department of Foreign Affairs and Trade (DFAT), Pacific partnerships for recovery: Australian Government Development Response Plan. https://www.dfat.gov.au/publications/aid/partnerships-recovery-australias-covid-19-development-response. Accessed on: 20 Sep. 2021.

[13] Nand, M.M. & Bardsley, D.K., Climate change loss and damage policy implications for Pacific Island countries. *Local Environment*, **25**(9), pp. 725–740, 2020.

[14] Lowy Institute, The Kainakii declaration is a signal of Pacific strength (Dame Meg Taylor). *The Interpreter*. https://www.lowyinstitute.org/the-interpreter/kainaki-ii-declaration-signal-pacific-strength. Accessed on: 9 Sep. 2021.

[15] World Health Organization, Strengthening Pacific health systems. https://www.who.int/westernpacific/activities/strengthening-pacific-health-systems. Accessed on: 10 Sep. 2021.

[16] NZ History, The 1918 influenza pandemic, Influenza in Samoa, p. 10. https://nzhistory.govt.nz/culture/1918-influenza-pandemic/samoa. Accessed on: 24 Sep. 2021.

[17] Leal Filho, W., Lütz, J.M., Sattler, D.N. & Nunn, P.D., Coronavirus: COVID-19 transmission in Pacific small island developing states. *International Journal of Environmental Research and Public Health*, **17**(15), p. 5409, 2020.

[18] Department of Foreign Affairs and Trade (DFAT), Fiji partnerships for recovery, Australian Government Development Response Plan. https://www.dfat.gov.au/sites/default/files/covid-response-plan-fiji.pdf. Accessed on: 24 Sep. 2021.

[19] Department of Foreign Affairs and Trade (DFAT), Samoa partnerships for recovery, Australian Government Development Response Plan. https://www.dfat.gov.au/sites/default/files/covid-response-plan-samoa.pdf. Accessed on: 24 Sep. 2021.

[20] Steenbergen, D.J., Neihapi, P.T., Koran, D., Sami, A., Malverus, V., Ephraim, R. & Andrew, N., COVID-19 restrictions amidst cyclones and volcanoes: A rapid assessment of early impacts on livelihoods and food security in coastal communities in Vanuatu. *Marine Policy*, **121**, 104199, 2020.

[21] Connell, J. & Taulealo, T., Island tourism and COVID-19 in Vanuatu and Samoa: An unfolding crisis. um.edu.mt. Accessed on: 24 Sep. 2021.

[22] Department of Foreign Affairs and Trade (DFAT), Tonga partnerships for recovery, Australian Government Development Response Plan. https://www.dfat.gov.au/sites/default/files/covid-response-plan-tonga.pdf. Accessed on: 24 Sep. 2021.

[23] Pryke, J., Can Pacific Airlines pull out of the dive? Lowy Institute. https://www.lowyinstitute.org/the-interpreter/can-pacific-airlines-pull-out-dive. Accessed on: 20 Sep. 2021.

[24] Craig, A.T., Heywood, A.E. & Hall, J., Risk of COVID-19 importation to the Pacific Islands through global air travel. *Epidemiology and Infection*, **148**, 2020.

[25] World Bank, Pacific labour mobility, migration and remittances in times of COVID-19: Interim Report. https://documents.worldbank.org/en/publication/documents-reports/documentdetail/430961606712129708/pacific-labor-mobility-migration-and-remittances-in-times-of-covid-19-interim-report. Accessed on: 20 Sep. 2021.

[26] Department of Foreign Affairs and Trade (DFAT), Kiribati partnerships for recovery, Australian Government Development Response Plan. https://www.dfat.gov.au/sites/default/files/covid-response-plan-kiribati.pdf. Accessed on: 20 Sep. 2021.

[27] Department of Foreign Affairs and Trade (DFAT), PNG partnerships for recovery, Australian Government Development Response Plan. https://www.dfat.gov.au/sites/default/files/covid-response-plan-papua-new-guinea.pdf. Accessed on: 20 Sep. 2021.

[28] Department of Foreign Affairs and Trade (DFAT), Solomon Islands partnerships for recovery, Australian Government Development Response Plan. https://www.dfat.gov.au/sites/default/files/covid-response-plan-solomon-islands.pdf. Accessed on: 20 Sep. 2021.

[29] Department of Foreign Affairs and Trade (DFAT), Vanuatu partnerships for recovery, Australian Government Development Response Plan. https://www.dfat.gov.au/sites/default/files/covid-response-plan-vanuatu.pdf .Accessed on: 20 Sep. 2021.

[30] ACIAR, COVID-19 and food systems in the Indo Pacific: An assessment of vulnerabilities, impacts and opportunities for action, Chapter 5, 2020. https://www.aciar.gov.au/publication/covid-19-and-food-systems. Accessed on: 26 Sep. 2021.

[31] Iese, V., Wairiu, M., Hickey, G.M., Ugalde, D., Salili, D.H., Walenenea Jr, J., Tabe, T., Keremama, M., Teva, C., Navunicagi, O. & Fesaitu, J., Impacts of COVID-19 on agriculture and food systems in Pacific Island countries (PICs): Evidence from communities in Fiji and Solomon Islands. *Agricultural Systems*, **190**, 103099, 2021.

[32] Farrell, P., Thow, A.M., Wate, J.T., Nonga, N., Vatucawaqa, P., Brewer, T., Sharp, M.K., Farmery, A., Trevena, H., Reeve, E. & Eriksson, H., COVID-19 and Pacific food system resilience: Opportunities to build a robust response. *Food Security*, **12**(4), pp. 783–791, 2020.

[33] World Bank, World development indicators data bank. https://databank.worldbank.org/source/world-development-indicators. Accessed on: 20 Sep. 2021.

[34] Gounder, N., Fiji economic survey: Low growth the new normal? *Asia and the Pacific Policy Studies*, **7**(2), pp. 145–157, 2020.

[35] United Nations, COVID-19 socio-economic response plan, 2020. https://unsdg.un.org/resources/covid-19-socio-economic-response-plan-samoa. Accessed on: 20 Sep. 2021.

[36] FAO, Kiribati strengthens agriculture in the face of COVID-19, Asia Pacific Regional Office. http://www.fao.org/asiapacific/news/detail-events/en/c/1364849/. Accessed on: 20 Sep. 2021.

[37] Connell, J. & Brown, R.P., The remittances of migrant Tongan and Samoan nurses from Australia. *Human Resources for Health*, **2**(1), pp. 1–21, 2004.

[38] Saverimuttu, V. & Cochran, A.K., Addressing rural–urban disparities: A case of government service delivery in Papua New Guinea. *International Journal of Business and Economic Development (IJBED)*, **6**(2), 2018.

[39] Narayan, D. & Cassidy, M.F., A dimensional approach to measuring social capital: Development and validation of a social capital inventory. *Current Sociology*, **49**(2), pp. 59–102, 2001.

[40] Malherbe, W., Sauer, W. & Aswani, S., Social capital reduces vulnerability in rural coastal communities of Solomon Islands. *Ocean and Coastal Management*, **15191**, 105186, 2020.

[41] Kenni, L. & Wijewickrama, E., Vanuatu: A real test for local emergency response. https://www.lowyinstitute.org/the-interpreter/vanuatu-takes-emergency-response-amid-covid-19. Accessed on: 2 Oct. 2021.

[42] Randin, G., COVID-19 and food security in Fiji: The reinforcement of subsistence farming practices in rural and urban areas. *Oceania*, **90**, pp. 89–95, 2020.

[43] Piesse, M., Covid-19 and food security in the Pacific: Considerable challenges with a 2020 opportunity to "build back better". https://www.futuredirections.org.au/topics/global-food-water-crises/. Accessed on: 1 Oct. 2021.

[44] Sen, A., The concept of development. *Handbook of Development Economics*, **1**, pp. 9–26, 1988.

[45] Robeyns, I., The capability approach in practice. *Journal of Political Philosophy*, **14**(3), pp. 351–376, 2006.

[46] Migheli, M., Capabilities and functionings: The role of social capital for accessing new capabilities. *Review of Political Economy*, **23**(1), pp. 133–142, 2011.

[47] The Global Health Security (GHS) Index 2019. https://www.ghsindex.org. Accessed on: 20 Sep. 2021.

[48] United Nations Development Program. http://hdr.undp.org/en/content/latest-human-development-index-ranking. Accessed on: 9 Oct. 2021.

EFFECTIVENESS OF GOVERNMENT RESPONSES AND THE ROLE OF NATIONAL CULTURE IN ADDRESSING THE COVID-19 PANDEMIC: A GLOBAL PERSPECTIVE

WALID BAKRY[1], PETER JOHN KAVALAMTHARA[2], SAJAN CYRIL[3] & YIYANG LIU[1]
[1]School of Business, Western Sydney University, Australia
[2]UNSW Business School, University of New South Wales, Australia
[3]Australian Institute of Higher Education, Australia

ABSTRACT

The colossal proportion of the COVID-19 disaster, with approximately 213 million confirmed cumulative cases and 4.5 million deaths at the end of August 2021, is of historical significance that is unparalleled. The pandemic has had an impact on the physical, mental and economic welfare of human beings in every corner of the world. To contain the crisis, governments have implemented a set of policies, aimed at directing human behaviour and improving healthcare outcomes almost from the outset, with varying levels of success. The battle against the virus continues. This study utilises data on growth in daily COVID-19 confirmed cases, from 56 countries which were most affected by the pandemic and investigates the effectiveness of several policies implemented by governments. The investigation covers the period, 24 January 2020 to 25 June 2021. The policies analysed in detail constitute 13 containment and health measures, incorporated in the Oxford COVID-19 Government Response Tracker. The research is based on the premise that the contexts in which policies are implemented have a bearing on the outcomes. The contexts are classified under economic status, national culture, the level of human development and population density. We find that these contexts are relevant in determining the success of alternative policy prescriptions. Overall, the containment and health measures are effective in curtailing the growth in COVID-19 cases to varying degrees across all economic contexts of countries. Containment measures are least effective in lower-middle income countries. Three of the six dimensions of Hofstede's national culture classifications tend to be positively associated with infection rates and the other three negatively associated. Moreover, national culture has a more prominent impact on the effectiveness of actions to reduce the growth in COVID-19 cases in lower-middle income countries. It is important to improve effectiveness of government policy responses to combat the pandemic by tailoring them to country-specific contexts. Our findings contribute to this customisation.

Keywords: COVID-19, pandemic, OxCGRT, containment and health policies, economic context, Hofstede, national culture.

1 INTRODUCTION

The COVID-19 pandemic, which is still a threat to human lives in almost every part of the world, is in its second year. By the end of August 2021, human beings had experienced approximately 213 million cumulative cases of infections and 4.5 million deaths. At the same time the world is also beginning to experience some hope, as the effectiveness of the various government responses to the pandemic are now becoming evident [1]. During February 2020, a joint mission consisting of 25 national and international experts from China, Germany, Japan, Korea, Nigeria, Russia, Singapore and USA led by the World Health Organization began assessing transmission dynamics and control measures. The objective was to enhance understanding, share knowledge and generate recommendations surrounding COVID-19 outbreak [2]. Government responses differed among countries, some choosing a hard lockdown affecting people's mobility and economic performance of the country while others were less restrictive. A ban on public gatherings, travel restrictions and social distancing were the first among certain initiatives undertaken to manage and prevent the rapid transmission

www.witpress.com, ISSN 1743-3509 (on-line)
doi:10.2495/DMAN210021

of the virus [3], [4]. Closing business and imposing bans on gathering have been found effective in reducing COVID-19 transmissions [5]. Strict lockdown and other social distancing policies at a huge cost to the economy during the early days of the pandemic has helped China in drastically reducing infection rate [6].

Oxford COVID-19 Government Response Tracker reports the overall level of government responses to the pandemic in different countries [1]. Fig. 1 shows time varying correlations between growth in total world daily confirmed COVID-19 cases and growth in the worldwide daily average Government Response Index from quarter 1 2020 to quarter 3 2021. Overall, the direction of the relationship between the two metrics was mixed in 2020 but was predominantly negative in the first three quarters of 2021. Therefore, on a global basis, it seems there is an improvement in the effectiveness of government interventions so far in 2021. Countries seem to be getting better at controlling the infections based on lessons learnt in managing the crisis in 2020.

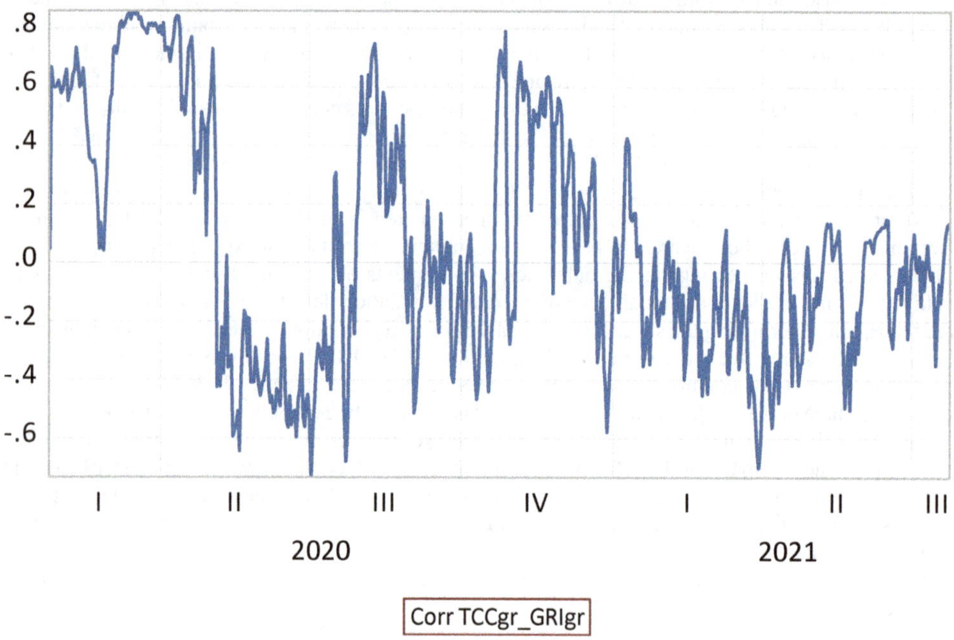

Figure 1: Time varying correlations (dynamic conditional correlations-GJRGARCH) between growth in total world daily confirmed COVID-19 cases and growth in world average daily Government Response Index [1], [14].

This study presumes that the efficacy of various government containment and health policies and policies to provide income support to citizens in controlling spread of the pandemic may be influenced by economic context, cultural context, level of human development and population density of the individual countries. Less developed countries seem to struggle to respond aggressively to the pandemic as compared to richer countries [7]. To examine social behaviour, scholars have used the framework of Hofstede's cultural dimensions, which are power distance, individualism vs. collectivism, masculinity vs. femininity, uncertainty avoidance, long term orientation vs. short term normative orientation and indulgence vs. restraint [8]. For example, long term orientation vs. short term and indulgence vs. restraint seem to have a significant effect on social distancing [6]. A Vietnamese study suggests high power distance, collectivism and uncertainty avoidance help to reduce the spread of COVID-19 [9]. There are mixed results on the effect of population density on the infections. Population density did not affect COVID-19 spread in US and China [10]. However, another study reports a positive relationship between the two variables in both high and low population countries [11]. There is evidence suggesting a positive association between Human Development Index and COVID-19 infections [12], [13].

This paper seeks explanations for differences in effectiveness of government policies to control COVID-19 infections. Our findings reveal important differences in the effectiveness of different policies. A country's economic context, dominant traits of national culture, population density and level of human development critically influence the effectiveness of the policies. The rest of this paper proceeds as follows: Section 2 describes the data. Section 3 presents the research methodology and Section 4 reports the empirical results. Finally, Section 5 concludes the study.

2 DATA DESCRIPTION

This study first analysed daily data for the top 70 countries, ranked by number of COVID-19 confirmed cases, for the period from 24 January 2020 to 25 June 2021. Countries with missing data for any of the variables included in this study were excluded. The final dataset used comprises of 56 countries, which account for 90.67% of cumulative COVID-19 total confirmed cases in the world as of 25 June 2021. These countries are listed in Table 1. The study employs data from four databases. Data relating to COVID-19 policies from Oxford COVID-19 Government Response Tracker (OxCGRT) [1], daily COVID-19 confirmed cases data from Refinitiv DataStream database [14], national culture data from Hofstede Insights [15] and Human Development Index, GNP per capita and population density data from Our World in Data database [16].

Table 1: Sample information. This table displays 56 countries most impacted by COVID-19 infections used in this study. The countries are classified into three groups based on their gross national income (GNI) per capita in U.S. dollars [17]. Our sample did not have any lower income countries.

Panel A. High income countries		Panel B. Upper-middle income countries		Panel C. Lower-middle income countries
Austria	Lithuania	Argentina	Malaysia	Bangladesh
Belgium	Netherlands	Azerbaijan	Mexico	Bolivia
Canada	Poland	Belarus	Paraguay	Egypt
Chile	Portugal	Brazil	Peru	India
Croatia	Saudi Arabia	Bulgaria	Romania	Indonesia
Czechia	Slovakia	Colombia	Russia	Iran
Denmark	Spain	Dominican Republic	Serbia	Morocco
France	Sweden	Georgia	South Africa	Pakistan
Germany	Switzerland	Iraq	Turkey	Philippines
Greece	United Kingdom	Jordan		Ukraine
Hungary	United States	Kazakhstan		
Ireland	Uruguay	Lebanon		
Japan				

3 RESEARCH METHODOLOGY

To investigate the impact of various government intervention policies on COVID-19 confirmed cases, the study uses the following common static pooled panel ordinary least squares (OLS) model:

$$TCP_{it} = \beta_0 + \beta_1 GCHIgr_{it} + \textstyle\sum_{c=1}^{C} \beta_c X_{c,i} + d_i + d_t + \varepsilon_{it}. \qquad (1)$$

where i and t refer to country and time respectively. TCP_{it} is the dependent variable representing the daily total COVID-19 confirmed cases as a percentage of population for country i and time t; β_0 is a constant term; $GCHIgr_{it}$ is the growth rate in Government Containment and Health Index which reflects governments' containment and health measures to fight COVID-19 spread; β_1 and β_c are coefficients of the variables under study. $X_{c,i}$ is a set of country-specific explanatory variables (including country's Human Development Index, population density and all six Hofstede cultural dimension indices); d_i is a time-invariant unobserved firm-specific effects of the error term; d_t is the firm-invariant unobserved time-specific effects of the error term; and ε_{it} is the error term. A detailed description of the dependent variable and the eight independent variables in eqn (1) is presented in Table 2.

The study re-estimates eqn (1) by replacing $GCHIgr_{it}$ with the growth in the Government Economic Support Index ($GESIgr_{it}$). Estimating these two general models will show if any of these indices influence the number of COVID-19 confirmed cases over time.

The next step is to test the impact of each of the component policies represented by the government policy indices which are identified in the previous step as significantly impacting the number of COVID-19 confirmed cases. This test will be done for the sample of all countries first and then for set of countries classified based on their economic context (that is, high income countries, upper-middle income countries, and lower-middle income countries) [17].

The study uses pooled panel OLS regression model with a robust standard error clustered by country and day (eqn (2)) to assess the relationship between various government intervention policies and COVID-19 confirmed cases:

$$\mathrm{TCP}_{it} = \beta_0 + \textstyle\sum_{j=1}^{J} \beta_j Y_{j,i,t} + \sum_{c=1}^{C} \beta_c X_{c,i} + \varepsilon_{it}. \qquad (2)$$

where i and t refer to country and time respectively. The dependent variable is the daily total COVID-19 confirmed cases as a percentage of population for country i and time t. $Y_{j,i,t}$ is a set of individual country-specific government intervention variables constituting either one or both of the government policy indices (that is, Government Containment and Health Index and Government Economic Support Index) which are found to be significantly related to daily total COVID-19 confirmed cases in the previous step. β_j and β_c are coefficients of the variables under study. $X_{c,i}$ is a set of country-specific explanatory variables as described in eqn (1). Table 2 contains detailed descriptions of these variables. The model corrects for potential cross-sectional and serial correlation by clustering the standard errors by country and time.

4 RESULTS AND DISCUSSION

Table 3 reports the results of the pooled panel OLS regression analysis (eqn (1)) of the effectiveness of government containment and health measures (Model 1) and the government economic support measures (Model 2) in controlling the spread of COVID-19 for the full set

Table 2: Variables and definitions [1], [14]–[16].

Variable	Definition
TCP	Daily total COVID-19 confirmed cases for country *i* at time *t* which is measured as the daily cumulative number of COVID-19 confirmed cases divided by country *i* total population
GCHIgr	Growth in Government Containment and Health Index for country *i* between time t and time *t–1*
GESIgr	Growth in Government Economic Support Index for country *i* between time t and time *t–1*
C1	Growth in the level of restrictions on school closures for country *i* at time *t*
C2	Growth in the level of restrictions on workplaces closures for country *i* at time *t*
C3	Growth in the level of restrictions on public events for country *i* at time *t*
C4	Growth in the level of restrictions on cut-off size of public gathering for country *i* at time *t*
C5	Growth in the level of restriction on public transport closures for country *i* at time *t*
C6	Growth in the level of restrictions on "stay-at-home" requirements for country *i* at time *t*
C7	Growth in the level of restriction on internal travel between regions/cities for country *i* at time *t*
C8	Growth in the level of restrictions on international travel for country *i* at time *t*
H1	Growth in the level of public information campaigns for country *i* at time *t*
H2	Growth in the level of government policy on testing policy for country *i* at time *t*
H3	Growth in the level of government policy on contact tracing for country *i* at time *t*
H4	Growth in the level of government policy on the use of facial coverings outside home for country *i* at time *t*
H5	Growth in the level of government policy on protecting elderly people in long term care facilities and/or the community and home setting for country *i* at time *t*
HDI	Human development index of country *i*
ln(PD)	the natural logarithm of the population density of country *i*
PDI	Level of power distance for country *i*
IDV	Level of individualism versus collectivism for country *i*
MAS	Level of masculinity versus femininity for country *i*
UAI	Level of uncertainty avoidance index for country *i*
LTO	Level of long-term orientation versus short-term normative orientation for country *i*
IVR	Level of indulgence versus restraint for country *i*

of 56 countries included in the study. Tightening of the containment and health measures has a significant and negative relationship with the growth in daily COVID-19 confirmed cases, emphasising the effectiveness of these two sets of measures. However, an increase in government economic support measures does not affect the growth in daily COVID-19 confirmed cases, although they may have other community benefits. Therefore, rest of this study will not include government economic support measures.

Table 3: Impact of government measures (containment and health measures – Model 1 and economic support measures – Model 2) on daily total COVID-19 confirmed cases over the period 22/01/2020–25/06/2021. The numbers in the parentheses are the robust standard errors. *, **, and *** indicate significance at the 10%, 5%, and 1% levels, respectively.

	Model 1	Model 2
Variable	TCP	TCP
GCHIgr	–0.0026** (0.0011)	
GESIgr		–0.0004 (0.0009)
HDI	0.4970*** (0.0019)	0.4971*** (0.0019)
Ln(PD)	–0.8026*** (0.0031)	–0.8022*** (0.0029)
PDI	0.1238*** (0.0007)	0.1238*** (0.0007)
IDV	–0.1106*** (0.0002)	–0.1107*** (0.0002)
MAS	0.0598*** (0.0002)	0.0597*** (0.0002)
UAI	–0.1032*** (0.0005)	–0.1033*** (0.0005)
LTO	–0.0431*** (0.0006)	–0.0431*** (0.0006)
IVR	–0.0278*** (0.0013)	–0.0279*** (0.0013)
Constant	–3.6305*** (0.9460)	–3.647221*** (0.9554)
Country effect	Yes	Yes
Day effect	Yes	Yes
Observations	27,403	27,403

The study now explores the impact of component policies of the Government Containment and Health Index on growth in daily COVID-19 confirmed cases. Table 4 shows the results of this analysis by estimating eqn (2) using the two-way cluster pooled panel OLS regression. The Government Containment and Health Index comprises eight measures of containment actions (denoted by *C1* to *C8*) and five measures of health-related actions (denoted by *H1* to *H5*). This analysis is undertaken first for the full set of 56 countries included in the study. Then the study explores whether the findings will hold if we segment our sample into high income countries, upper-middle income countries and lower-middle

Table 4: The effect of government containment and health-related intervention policies on the COVID-19 total confirmed cases in aggregate and in high, upper-middle and lower-middle income countries. The numbers in the parentheses are the robust standard errors. *, **, and *** indicate significance at the 10%, 5%, and 1% levels, respectively.

	All countries	High income countries	Upper-middle income countries	Lower-middle income countries
Variables	TCP	TCP	TCP	TCP
C1	−0.2319*** (0.0784)	−0.2759** (0.1177)	−0.2433*** (0.0825)	−0.1065** (0.0499)
C2	−0.1494 (0.0753)	−0.0908 (0.1007)	−0.1663** (0.0714)	−0.0591 (0.0513)
C3	−0.2466** (0.1029)	−0.2826* (0.1577)	−0.3563*** (0.1156)	−0.0346 (0.0835)
C4	−0.1874*** (0.0550)	−0.3278*** (0.0984)	−0.1121** (0.0553)	−0.0393 (0.0502)
C5	0.0427 (0.0914)	0.0148 (0.2244)	−0.1117 (0.0851)	0.0016 (0.0482)
C6	−0.0232 (0.0634)	−0.2699** (0.1289)	−0.0213 (0.0673)	0.0998 (0.0764)
C7	−0.0542 (0.0647)	−0.0942 (0.1316)	−0.0317 (0.0758)	−0.0632** (0.0275)
C8	−0.1953*** (0.0589)	−0.2219** (0.1001)	−0.1190 (0.0991)	−0.0650** (0.0268)
H1	−0.0748 (0.1462)	0.0404 (0.4057)	−0.1955 (0.1660)	−0.1063* (0.0631)
H2	−0.1610 (0.1089)	−0.4684*** (0.1544)	−0.0452 (0.0895)	0.1095 (0.0704)
H3	−0.3856*** (0.1219)	0.0499 (0.2623)	−0.5818*** (0.1576)	−0.2446** (0.0990)
H4	−0.2542*** (0.0789)	−0.6505*** (0.1847)	−0.2554*** (0.0595)	−0.0644 (0.0409)
H5	−0.2173*** (0.0583)	−0.4588*** (0.1113)	−0.0478 (0.0712)	−0.1225 (0.1008)
HDI	0.0512*** (0.0199)	0.0449 (0.0777)	0.0621** (0.0299)	0.0127 (0.0189)
Ln(PD)	0.1062 (0.1163)	0.2517 (0.1984)	0.1458 (0.1425)	0.0470** (0.0607)
PDI	0.0002 (0.0076)	0.0053 (0.0118)	−0.0234*** (0.0082)	−0.0135* (0.0071)
IDV	0.0188** (0.0094)	0.0253* (0.0142)	0.0090 (0.0116)	0.0237*** (0.0060)
MAS	−0.0139* (0.0081)	−0.0197*** (0.0073)	−0.0143 (0.0106)	0.0078 (0.0236)
UAI	0.0171** (0.0068)	−0.0018 (0.0102)	0.0222*** (0.0084)	0.0360*** (0.0090)
LTO	−0.0076 (0.0079)	−0.0033 (0.0143)	−0.0232*** (0.0080)	−0.0148*** (0.0045)
IVR	0.0001 (0.0092)	−0.0147 (0.0181)	−0.0125 (0.0091)	0.0287*** (0.0041)
Constant	−3.6600** (1.5517)	5.8323 (6.6531)	−1.5368 (2.8746)	−4.1153* (2.3601)
Observations	27,403	12,350	10,160	4,893
Prob > F	0.0000	0.0000	0.0000	0.0000

income countries. The rationale for this segmentation includes differences in economic conditions, healthcare system capabilities, educational levels and national cultures of countries belonging to these three categories.

4.1 Effect of containment measures

The results presented in Table 4 show that there are major differences among the three categories of countries in how the growth in daily COVID-19 confirmed cases responds to different government containment actions. Four containment measures – level of restrictions relating to school closures, conducting public events, the size of public gathering and international travel – have a significant negative relationship with the growth in daily COVID-19 confirmed cases when data are pooled together for all countries. However, the level of restrictions relating to school closures is the only containment measure which decreases daily COVID-19 confirmed cases across all the three country classifications. Restrictions on conducting public events and on the size of public gathering are effective in decreasing daily COVID-19 confirmed cases in high income countries and upper-middle income countries. Restrictions on international travel are an effective mitigant against increase in daily COVID-19 confirmed cases in high income countries and lower-middle income countries. Restrictions relating to public transport closures is the only containment measure which does not have a relationship with daily COVID-19 confirmed cases either in the pooled sample or any of the country categories. Although not observed in the pooled sample of all country categories, the following containment measures decrease growth in infections in specific country categories – stay at home restrictions in high income countries, restrictions on workplace closures in upper-middle income countries and restrictions on travel between regions and cities in lower-middle income countries.

It is a key finding from this study that the effectiveness of containment measures is related to the economic context of the countries. In Table 4, we note that in lower-middle income countries, only three containment measures have a significant negative relationship with the growth in daily COVID-19 confirmed cases whereas corresponding numbers for high income countries and upper-middle income countries are five and four respectively. Limited economic resources to implement the containment measures, weak governance and poor compliance regimes are possible explanations for the lower effectiveness of containment measures in lower-middle income countries [3], [18].

4.2 Effect of health measures

Results in Table 4 also show that there are key differences among the three categories of countries in how the five health-related measures impact growth in daily COVID-19 confirmed cases. When data are pooled together for all countries, three component policies – level of contact tracing, level of restrictions on the use of facial coverings outside home and level of protection for elderly people in long term care facilities and/or the community and home setting – help to slow the growth in infections. However, none of these three measures has any significant effect on infections in any of the three country classifications. Level of government policy on testing, level of requirements for the use of facial coverings outside home and level of protection for elderly people help to decrease the infections in high income countries. Contact tracing and the use of facial coverings outside home are helpful in curtailing infections in upper-middle income countries. Public information campaigns and contact tracing are the only health measures that seem to suppress infections in the lower-middle income countries.

4.3 Effect of national culture

The study finds that national culture of countries, measured by the six Hofstede cultural dimensions, has a major relationship with the spread of COVID-19. The direction of the relationships varies depending on the individual dimension. Table 5 summaries the inferences from Table 4 relating to these relationships.

Table 5: Relationship between growth in daily COVID-19 confirmed cases and six Hofstede cultural dimensions in aggregate and in high income, upper-middle income and lower-middle income countries.

Hofstede cultural dimension index	Nature of relationship with growth in daily COVID-19 confirmed cases	Applicable country categories
Individualism versus collectivism	Positive	All countries High income countries Lower-middle income countries
Uncertainty avoidance	Positive	All countries Upper-middle income countries Lower-middle income countries
Masculinity versus femininity	Negative	All countries High income countries
Power distance	Negative	Upper-middle income countries Lower-middle income countries
Long-term versus short-term normative orientation	Negative	Upper-middle income countries Lower-middle income countries
Indulgence versus restraint	Positive	Lower-middle income countries

Table 5 shows that three Hofstede cultural dimensions (individualism versus collectivism, uncertainty avoidance, and indulgence versus restraint) are positively related to growth in daily COVID-19 confirmed cases, whereas the other three (masculinity versus femininity, power distance, and long-term versus short-term normative orientation) are negatively related. Effectiveness of response to a pandemic requires collective effort of the communities and it should not be surprising that in individualism-dominant countries where people's self-image is defined by "I" rather than "we", it is harder to control COVID-19 infections. Countries with high uncertainty avoidance display more resistance to change accepted norms of behaviour. By not quickly making necessary changes to behaviour such as social distancing, these communities may struggle to contain COVID-19 infections. High scores on indulgence versus restraint denote a society which allows free gratification of natural human drives to enjoy life. Such social disposition can be expected to impede efforts to curtail spread of COVID-19 infections as evident from the positive relationship. Societies where the dominant cultural trait is masculinity are more assertive and achievement focussed and are not consensus oriented. Crisis events such as the pandemic call for decisive actions and therefore, the negative relationship shown in Table 5 for masculinity versus femininity index seems justified. High power distance in a country setting means people accept hierarchical order readily and such acceptance can be helpful when major changes in social behaviour are to be implemented rapidly during the course of a pandemic. Societies which have high long-term versus short-term normative orientation do not hold on to time-honoured traditions and nor do they view social changes with suspicion. Such a cultural orientation facilitates implementation of big changes necessary to bring down infection rates.

An important finding from our research is that national culture has a greater influence in lower-middle income countries in determining the effectiveness of measures to control the spread of COVID-19. Table 4 shows that five of the six Hofstede cultural dimensions have significant relationships with the growth in daily COVID-19 confirmed cases in lower-middle income countries, whereas corresponding numbers for high income countries and upper-middle income countries are three each. It seems that relatively weak economic strength, health infrastructure, governance structure and technological capability of the lower-middle income countries provide national culture a profound role in determining the success of these countries in their fight against the pandemic [19], [20].

4.4 Effect of other contextual variables

The study now examines results in Table 4 for the relationship of growth in daily COVID-19 confirmed cases with Human Development Index and population density in the countries. Human Development Index is positively related to growth in daily COVID-19 confirmed cases when data are pooled together for all countries and in the category of upper-middle income countries. This positive association is consistent with literature [12], [13], [21] and it may be attributed to higher risk of COVID-19 infection of populations in upper-middle income countries due to their longer life expectancy. The absence of such association in high income countries, despite their longer life expectancy, may be ascribed to better healthcare infrastructure in these countries. In addition to shorter life expectancy, under-reporting of COVID-19 cases due to limited capacity for testing may explain the absence of positive association between Human Development Index and growth in daily COVID-19 confirmed cases in lower-middle income countries [12]. In lower-middle income countries we find a positive relationship between growth in daily COVID-19 confirmed cases and population density. Weak healthcare infrastructure and less rigorous compliance with COVID-19 control measures may explain this association.

5 CONCLUSION

This study investigates effectiveness of country responses in controlling the spread of COVID-19 pandemic. Government responses are first grouped into Containment and Health Index and Economic Support Index. The investigation is based on the belief that the effectiveness of these responses is conditioned by the context in which they are implemented. The study, therefore, explores their effectiveness separately for high income countries, upper-middle income countries and lower-middle income countries. In addition to the economic context defined by the above country categories, we further explore whether other contextual variables, namely national culture, level of human development, and population density influence the effectiveness of the measures to combat the spread of COVID-19.

The research shows that, overall, the set of containment and health measures are effective in curtailing the growth of daily COVID-19 confirmed cases, but economic support measures are not effective. However, there are major differences in the usefulness of containment and health measures depending on the economic context of the countries. This calls for tailoring the responses of governments to the economic context of individual countries. For example, containment measures are less successful in lower-middle income countries compared to the other two categories of countries. Given the resource constraints and weaknesses in governance and compliance regimes in lower-middle income countries, it may be worthwhile for policy makers in these countries to customise their actions based on the findings presented in this paper.

National cultural traits have varying degrees of influence in supporting or hindering successful outcomes of the measures undertaken to decrease COVID-19 infections. The finding that low levels of individualism versus collectivism, uncertainty avoidance, and indulgence versus restraint and high levels of masculinity versus femininity, power distance, and long-term versus short-term normative orientation contribute to decrease in infections is useful to government policy makers and health management experts in designing suitable responses to fight the pandemic. Far greater influence of national culture in shaping successful policy outcomes in lower-middle income countries points to devoting special attention to cultural sensitivities when designing and implementing strategies to manage the pandemic in these countries.

COVID-19 pandemic has imposed unprecedented human and economic costs on almost all countries. Developing effective containment and health policies is of paramount importance to manage the disaster. This research contributes to existing literature in helping to develop these policies. It shows that it is imperative to tailor the policy mix of countries to achieve optimal outcomes in managing the pandemic. Since the fight against the pandemic is seriously constrained by resource availability, especially in lower-middle income countries, prioritising implementation of effective policies is critical. Our research assists in this prioritisation by taking into account the economic, cultural and human development contexts of individual countries. A topic for future research is the effect of COVID-19 vaccinations on infection and fatality rates and how the contextual variables used in this study affect these relationships.

REFERENCES

[1] Oxford COVID-19 Government Response Tracker. https://www.bsg.ox.ac.uk/research/research-projects/covid-19-government-response-tracker. Accessed on: 12 Jun. 2021.

[2] Report of the WHO-China Joint Mission on Coronavirus Disease 2019 (COVID-19). https://www.who.int/docs/default-source/coronaviruse/who-china-joint-mission-on-covid-19-final-report.pdf?sfvrsn=fce87f4e_2. Accessed on: 15 Jul. 2021.

[3] Zaremba, A., Kizys, R., Aharon, D.Y. & Demir, E., Infected markets: Novel coronavirus, government interventions, and stock return volatility around the globe. *Finance Research Letters*, 2020.

[4] Huynh, T.L., Does culture matter social distancing under the COVID-19 pandemic? *Safety Science*, 2020.

[5] Brauner J.M. et al., Inferring the effectiveness of government interventions against COVID-19. *Science (American Association for the Advancement of Science)*, 2021.

[6] Wang, Y., Government policies, national culture and social distancing during the first wave of the COVID-19 pandemic: International evidence. *Safety Science*, 2021.

[7] Chisadza, C., Clance, M. & Gupta, R., Government effectiveness and the COVID-19 pandemic. *Sustainability* (Basel, Switzerland), 2021.

[8] Hofstede, G., Dimensionalizing cultures: The Hofstede model in context. *Online Readings in Psychology and Culture*, 2011.

[9] Pham, N.C., Li, Y., Schapsis, C., Hossain, T., Pham, H.H., Fischer, D. & Yang, J., Intrinsic cultural factors that helped Vietnam overcome the COVID-19 pandemic compared with other countries. *Asia Pacific Journal of Health Management*, 2020.

[10] Sun, Z., Zhang, H., Yang, Y., Wan, H. & Wang, Y., Impacts of geographic factors and population density on the COVID-19 spreading under the lockdown policies of China. *Science of The Total Environment*, 2020.
[11] Velasco, J.M., Tseng, W.C. & Chang, C.L., Factors affecting the case and deaths of COVID-19 victims. *International Journal of Environmental Research and Pacific Health,* 2021.
[12] António, N., Rita, P. & Saraiva, P., COVID-19: Worldwide profiles during the first 250 days. *Applied Sciences*, 2021.
[13] Liu, J.H., Majority world successes and European and American failure to contain COVID-19: Cultural collectivism and global leadership. *Asian Journal of Social Psychology*, 2021.
[14] DataStream database. *Refinitiv Datastream.* Available online from a subscription service. Accessed on: 12 Jun. 2021.
[15] Hofstede Insights. https://www.hofstede-insights.com/. Accessed on: 12 Jun. 2021.
[16] Our World in Data database. https://github.com/owid/covid-19-data/tree/master/public/data. Accessed on: 12 Jun. 2021.
[17] New World Bank, Country classifications by income level: 2020–2021. https://blogs.worldbank.org/opendata/new-world-bank-country-classifications-income-level-2020-2021. Accessed on: 12 Jun. 2021.
[18] Engelhardt, N., Krause, M., Neukirchen, D. & Posch, P.N., Trust and stock market volatility during the COVID-19 crisis. *Finance Research Letters*, 2021.
[19] Islam, S.N., Cheng, H.J., Helgason, K.S., Hunt, N., Kawamura, H. & Lafleur, M., Variations in COVID strategies: Determinants and lessons. United Nations, 2020.
[20] Gupta, M., Shoja, A. & Mikalef, P., Toward the understanding of national culture in the success of non-pharmaceutical technological interventions in mitigating COVID-19 pandemic. *Annals of Operations Research,* 2021.
[21] Khazaei, Z., Goodarzi, E., Naemi, H., Hasanpour-Dehkordi, A. & Naghibzadeh-Tahami, A., COVID-19 pandemic in the world and its relation to Human Development Index: A global study. *Archives of Clinical Infectious Diseases,* 2020.

BUILDING RESILIENT COALITIONS: HEALTH AND MEDICAL RESPONSE THROUGH COVID-19 CONCURRENT EVENTS

LORI UPTON[1], ADAM LEE[1] & MAGDALENA ANNA DENHAM[2]
[1]SouthEast Texas Regional Advisory Council, USA
[2]Department of Security Studies, Sam Houston State University, USA

ABSTRACT

This study examines the role of the non-profit organization SouthEast Regional Advisory Council (SETRAC) in capacity-building among healthcare institutions in the largest medical center in the world in the Houston, Texas region. Specifically, using conceptual resilience model accessed by organizations as they respond to crisis or disruption, we assess the linking dimension of SETRAC in healthcare delivery during COVID-19 concurrent disasters through supporting and coordinating services to healthcare institutions amid the pandemic. Specifically, the assessment comprises two critical events in the region concurrent to the COVID-19 pandemic – Hurricane Laura in fall 2020 and Winter Storm Uri in spring 2021. Our inquiry adopts the classical content analysis suitable for analyzing large textual data of 802 open-ended after-action report (AAR) surveys conducted among healthcare system entities in the region of which 403 collected post Hurricane Laura and 399 AARs collected after the Winter Storm Uri. Our findings center on discussion of three categories of capabilities: (a) established capabilities; (b) developing capabilities; and (c) unmet capabilities. The purpose of the study is to illuminate the role of a non-profit coalition in enhancing and sustaining a decentralized network of healthcare providers in the management of increasingly complex and compounded disasters and to identify best practices for coordination and collaboration across a range of concurrent hazards and threats.

Keywords: COVID-19 concurrent crises, healthcare coalitions, healthcare disaster management, SETRAC, resiliency, crisis communications, healthcare systems, Hurricane Laura, Winter Storm Uri.

1 INTRODUCTION

The COVID-19 pandemic constitutes a paradigm shift in the approach to hazards and threats especially within the traditional cycle of emergency management. Specifically, natural hazard-triggered crises or threat-triggered crises tend to be characterized by distinct geographic locations, have clearly defined duration, and tend to fall within specific phases of response and recovery activities. Meanwhile, the protracted nature of the pandemic and its persistence, as well as its global dispersement present unique challenges to emergency management and require adaptations beyond accepted norms and existing structures [1]. Consequently, in healthcare and public health sectors, systems have had to perform under longitudinal stress [2]. Moreover, it has been difficult to discern concrete phases of response and recovery because of inherent uncertainty of the pandemic crisis, potential multiple waves as well as emerging new strands of the virus such as the UK, New York, Brazil, and Indian variants [3], all making the trajectory of the disaster quite non-linear. Indeed, "Transitioning from pandemic response to recovery in a spiral fashion: there is a high probability of the second and even the third wave of a pandemic if there is no vaccine or immunization, each new wave pushing the disaster risk reduction cycle from the recovery back to response phase" [1, p. 3]. Concurrently, the pandemic has ushered second and third-order harms affecting many sectors of the economy and society [4]. Most critically, the temporal persistence of the pandemic has created an opportunity for multiple intersections with cyclical weather events such as storms, earthquakes, wildfires, or other concurrent or systemic crises thus

WIT Transactions on The Built Environment, Vol 207, © 2021 WIT Press
www.witpress.com, ISSN 1743-3509 (on-line)
doi:10.2495/DMAN210031

compounding the complexity of required disaster management practices [2]. Markedly, meteorological hazards such as droughts or floods or geophysical hazards such as earthquakes occur commonly and cause disasters every year worldwide [5]. Admittedly, in 2020 more than 70 countries registered flood events coinciding with COVID-19 cases present in those countries [6]. The phenomenon of the overlap and compounding of one crisis with another has been referred to in research as concurrent crises [7] while others might refer to it as multi-hazard crises [8].

One of the earliest concurrent crises studied during COVID-19 was 21 March 2020, 5.3 Richter scale magnitude earthquake in the capital of Croatia, Zagreb [9]. The compounding disaster happened within the first month of COVID-19 response in the city causing structural damages to include critical hospital infrastructure. The affected zone comprised 199 health facility buildings, to include 42 healthcare centers, 115 hospitals and clinics, 20 health institutes and 22 pharmacies. Of those, 39 suffered moderate damage and eight were heavily damaged [10]. Consequently, hospital evacuations were necessary to include those from COVID-19 designated hospitals, which at that time provided care to 22 COVID-19 patients, the majority of whom were treated in the intensive care unit [9]. The earthquake response stressed already overextended healthcare workers, caregivers and other staff exposed to the infected and sick, placing them at a much higher risk. It is noteworthy that Croatia was under lockdown procedures at the time of the earthquake. Thus, mobilizing special damage assessment teams was difficult as many volunteer engineers living outside of Zagreb could not be mobilized for days [11]. Meanwhile dozens of residential homes were destroyed, and hundreds of individuals sought temporary housing. There were reports of 15,000 to 20,000 people moving out of the downtown area and a shelter created at a student dormitory housed approximately 500 displaced [10]. Even though the impact of the concurrent crisis was relatively localized and even though the authorities attempted to strictly enforce lockdown measures to prevent human social congregate behavior post disaster, infection rates within the incubation period after the earthquake revealed upward trends that could be attributed to temporary breakdown of social distancing practices [7]. Indeed researchers [10] concurred that in the initial hours post disaster hundreds of individuals were in the streets despite required lockdowns.

Crises concurrent to COVID 19 have been especially devastating to nations with rapid spread of COVID-19 such as Bangladesh because of (a) population density; (b) limited healthcare facilities; (c) sheltering camps housing more than a million refugees from Myanmar; and (d) migratory labor population traffic from COVID-19 affected countries; those factors contributed to the pandemic itself becoming a humanitarian crisis on an unprecedented scale [12]. Thus, the advent in May of 2020 of Cyclone Amphan, the most powerful cyclone in the region for two decades, represented in its timing, a threat to cyclone shelters that could become potential hotbeds of the spread of COVID-19. In Bangladesh cyclone shelters, community buildings, and schools conventionally used as evacuation centers during cyclonic activity, had at the time functioned as quarantine facilities and many served for quarantine processes of immigrant populations. The concurrent crises created a perfect storm of compatibility of two missions – that of sheltering vulnerable populations from the cyclone and that of protecting the public from COVID-19. According to research [3], international shelter space standards require 3.5 m^2 per individual; however, social distancing elevates that space to about 6 m^2 during a pandemic reducing overall sheltering capacity by 50%. This reduces the capacity of any existing evacuation center to almost 50%. Ultimately, in Bangladesh, rapid, ad hoc risk assessments were conducted and allowed for triaging shelters into half occupancy in areas of severe risk and full occupancy in those at a less serious risk of infection [13]. Overall, 12,078 cyclone shelters with 2.4 million evacuees were established in Bangladesh coastal areas affected by Amphan [14]. The unprecedented mobilization by humanitarian and governmental organizations admittedly limited deaths from the cyclone to 80, thus achieving the first mission. This included sheltering "in the densely crowded camps in Cox' Bazar, which shelter some 900,000 Rohingya refugees" [15, p. 1]. However, COVID-19 cases spiked in Amphan-affected regions underscoring challenges of weighing and prioritizing risks from concurrent disasters [16]. Indeed, based on accounts of community-based women's organization in Bangladesh called *Prerona* and whose members were supporting cyclone shelters' efforts "maintaining physical distancing was a major challenge in a densely occupied shelter. There were around 100–120 people sheltered in 5–7 rooms. In one cyclone shelter around 15–20 people stayed in one room, with 4–5 families who knew each other very well not wanting to be separated" [14, para. 9].

Beyond emerging research on concurrent crises in international contexts, US researchers stress the need to study compound crises because "for a long time, researchers and policy makers often treated disasters 'as though they were discrete events'" Peek said. "Our moment is obviously teaching us how important it is to think about people and systems when there are multiple disasters that are unfolding simultaneously" [17, para. 31]. Clearly, operational environments in contexts of multi-hazard scenarios are characterized by unique stressors that test the overall resiliency of response systems. For example, in context of COVID-19 and social distancing, intersecting hazards that necessitate implementation of evacuation procedures require adaptations from conventional models to dispersed evacuations or generation of alternative non-congregate sheltering arrangements such as hotels. Thus, because concurrent crises represent unique opportunities for researchers to study response systems pushed beyond their conventional capabilities and activities, the purpose of this study is to assess healthcare and public health system capabilities and challenges encountered during COVID-19 response at the intersection of two COVID-19 concurrent crises (fall 2020 and spring 2021) occurring in Texas and affecting the largest medical complex in the world. The Hurricane Laura that hit the US Gulf Coast in August 2020 was the strongest US storm on record in 2020, landing as category 4 with winds of 241 km/h (150 mph) [18]. While its most catastrophic impact was recorded in Louisiana where Hurricane Laura's intensity was the strongest storm recorded since 1851, Texas counties of Jefferson, Orange, Hardin, Tyler, Jasper, and Newton suffered significant damages [19]; moreover, Hurricane Laura caused 186,000 power outages in the state [20] and prompted evacuations of at least half million residents in Texas and Louisiana before the landfall. For example, Galveston County, Texas ordered mandatory evacuations of low-lying areas [21]; after the storm at least 5,000 individuals were in Texas shelters [22]. Whereas hurricanes represent a cyclical occurrence in Texas, the second concurrent COVID-19 crisis, Winter Storm Uri in February 2021 brought catastrophic freezing temperatures not recorded in the state since 1899 [23]. Winter Storm Uri left approximately 10 million Texans without power for several days causing near collapse of the independent Texas power grid, mass relocations, people seeking shelter at warming stations, and catastrophic damages to residential utilities' infrastructure due to lack of winterization. The so called "ice-pocalypse" claimed lives of at least 111 people due to low temperatures or carbon monoxide poisoning [24]. Both COVID-19 concurrent crises stressed already overextended capabilities of healthcare systems and thus constitute appropriate grounding for the current study.

2 HEALTHCARE COALITIONS

When assessing crisis management in the healthcare system in the United States, it is important to note that traditionally the system is ill-equipped to manage disaster preparedness and response. This is due to its dispersed, decentralized nature of various types of healthcare facilities, operating mostly at their daily capacities, and mostly in private sector. Conversely, crisis preparedness and response has been associated with management by public agencies at local, state, and federal levels [25]. Case in point, the American Hospital Association listed 6,090 hospitals operating in 2021 of which more than half (2,946) were non-for-profit, non-governmental community hospitals, 1,233 investor-owned (private) community hospitals [26]. Others fell under local, state, and federal government oversight. Those numbers do not include all other types of institutions that vary in type, funding, ownership, and size such as long-term care facilities for example. In such a landscape, the concept of healthcare coalitions (HCCs) defined as “a formal collaboration among hospitals, public health departments, emergency management and response agencies, and possibly other types of healthcare entities in a community that are organized to prepare for and respond to mass casualty and catastrophic health events” [25, para. 18] has been gaining in popularity, relevance, and governmental support. Clearly, HCCs inherently can build a more resilient community by bringing together the dispersed healthcare community, emergency management organizations, public health officials, first responder agencies, and elected jurisdictional officials to participate in joint trainings and exercises, which in turn results in a more coordinated and cohesive response. In current practice, the US Department Health and Human Service (DHHS) Assistant Secretary for Preparedness and Response (ASPR), a position created under the *Pandemic and All Hazards Preparedness Act* post-Katrina, provides federal funding to states to develop HCCs that focus on healthcare preparedness and response [26]. By 2020, “the HHS ASPR Hospital Preparedness Program (HPP) reported 85% of hospitals nationwide participate in HCCs along with 82% of local health departments, 56% of emergency management organizations, and 27% of emergency management services” [27, p. 3]. Even though researchers have recognized the value of coalitions in securing access to resources and political influence [28], [29], and even though the HCC concept appears to have been endorsed nationally, there is paucity of research focusing on healthcare coalitions for emergency management and healthcare resiliency-building specifically. Our study aims to bridge that gap by exploring the role of an HCC in Southeast Texas in building healthcare system resilience during COVID-19 concurrent crises.

The HCC concept of collaborative governance differs from models among countries with socialized medicine. During any disaster that would affect the health and medical area in a country with socialized medicine, the preparedness and response is the responsibility of the government. When a country has privatized medicine, the event or incident starts and ends at the local level, leaving the responsibility of preparedness and response to the individual facilities, cities, and counties. State and federal assistance are provided only when local resources have been or will be exceeded, and a formal request for assistance has been made by the Jurisdiction Having Authority (i.e., Mayor, Governor). Thus, instituting Regional Advisory Councils (RACs) and Medical Operation Centers (MOCs) as central coordination hubs of information, facility status, available resources, critical resource needs, and the singular collective voice for the individual organizations in the health and medical arena, provides a robust regional perspective of the critical healthcare needs and status and allows individual facilities the leverage needed to address these concerns.

During a disaster that affects the health and medical infrastructure in Southeast Texas, the Catastrophic Medical Operations Center (CMOC) is activated and staffed by the SouthEast Texas Regional Advisory Council (SETRAC), a non-profit organization. The Regional Healthcare Preparedness Coalition (RHPC) of which SETRAC is an example, serves as the governing body and adopts a representative form of governance. Due to the sheer size of the region, comparable to the size of Ireland with nearly twice the population, SETRAC divides the 25-county region into five corridors: North, South, East, West, and Downtown (Fig. 1).

The corridors are based on geographic locations and common risks and hazards. Each corridor has an elected official (Corridor Chair) who in turn represents the interests of all healthcare agencies in that corridor on the RHPC governing board. Each Corridor meets every other month to address preparedness or response-related issues. The RHPC Board provides the strategic vision and overall priorities of preparedness and response (trainings, regional plans, exercises) for the Corridors to implement at the local level. The CMOC serves as the resource and coordination hub for the 25-county area by (a) collecting/disseminating information; (b) planning for the next operational period from a regional perspective; (c) coordinating with other stakeholders outside of health and medical sectors; (d) providing resources; (e) anticipating resource needs; (f) providing logistical support; and (g) centralizing the coordination of patient movement and repatriation. An example of this was the COVID-19 response and the concurrent disasters within.

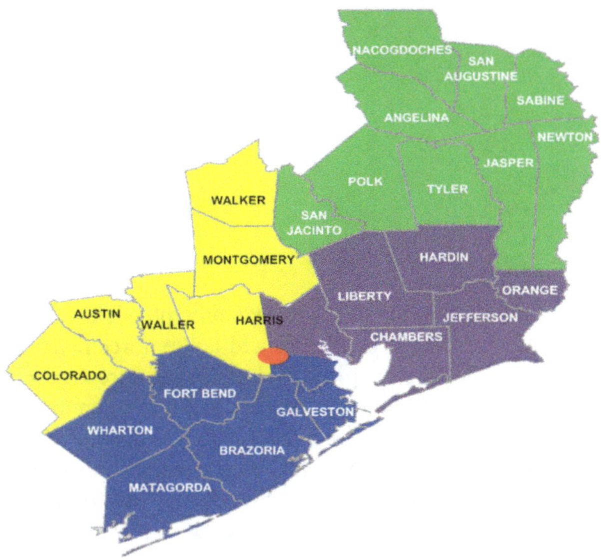

Figure 1: SETRAC regional coverage. Note: Yellow = West Corridor; Blue = South Corridor; Red = Downtown Corridor; Purple = East Corridor; and Green = North Corridor.

From March 2020 to July 2021 the CMOC was activated and staffed by SETRAC for COVID-19, Tropical Storm Beta, Hurricane Laura, civil protests in Houston, and Winter Storm Uri; each crisis presented unique sets of compounding challenges for the health and medical infrastructure of Southeast Texas. The role of SETRAC in the COVID-19 response was twofold: (a) monitoring healthcare capacity and building surge capability; and (b) distributing personal protective equipment (PPE) to healthcare agencies. Subsequently, SETRAC created a public PowerBI report which monitored COVID-19 hospitalization and bed capacity data. Daily each healthcare organization would complete a series of 137 questions eliciting data which would be corroborated by SETRAC and visualized on the public PowerBI COVID reports. Moreover, SETRAC was responsible for distributing PPEs throughout the SETRAC 25-county region to all healthcare organizations (to include dentists) through a "push pack" model. A push pack is a predesignated allotment of PPE based on the facility type, size, and inventory that SETRAC had received from the State of Texas. Facilities or agencies then were given a three-day window to come and pick up the PPE (Fig. 2 shows PPE distributed as of 22 May 2021). Coincident to tracking healthcare capacity via data collection and visualization, as well as PPE distribution, COVID-19 concurrent disasters such as Hurricane Laura, and Winter Storm Uri required SETRAC to expand additional support to healthcare infrastructure. Conceptualizing the role of HCC as fostering healthcare system resilience, we adopt the concept of linking [30] under which social capital gets strengthened by vertical ties that allow access to information, knowledge, resources, and influence. This research focuses on two questions specifically: (a) *What are the capabilities of SETRAC as an HCC that strengthened the overall regional healthcare system resiliency during COVID-19 concurrent crises?* and (b) *What are the SETRAC's capabilities in need of strengthening during COVID-19 concurrent crises in order to make the regional healthcare system more resilient?*

Facility Type	Total N95 Mask	Coveralls	Face Shields	Total Gowns	Surgical Mask	Gloves	Shoe Covers	Non-Medical Gloves	Goggles
Clinic/Physician Office	3,406,566	676	141,302	398,736	8,189,600	7,099,184	1,612	2	5,632
Dialysis	46,660	385	766	6,302	70,050	20,720	0	0	0
First Responders	4,448,598	98,895	106,978	226,095	8,089,900	4,691,490	4,500	0	50,173
Home Care Services	3,328,460	2,425	57,493	268,051	7,052,810	3,718,770	175	0	920
Hospital	5,636,796	93,072	293,397	252,804	13,529,655	5,128,692	363,400	0	1,720
Jurisdictions	1,589,926	77,410	244,144	782,731	3,530,730	2,450,200	6,700	197,500	52,235
LTAC	353,630	1,745	16,578	29,286	610,550	314,400	8,601	9	0
LTC	4,654,003	24,584	71,808	481,887	10,416,101	6,644,409	14,852	0	570
Other	66,680	650	3,546	8,451	109,400	139,650	150	10,000	570
Pharmacy	95,230	0	3,968	18,405	213,600	61,300	0	0	0
Total	**23,626,549**	**299,842**	**939,980**	**2,472,748**	**51,812,396**	**30,268,815**	**399,990**	**207,511**	**111,820**

Figure 2: SETRAC PPE distribution chart. Note: Clinic/physician office includes dentists; First responders includes EMS, fire and law enforcement; Home care services includes home health, hospice, and personal care agencies; LTC includes assisted living, skilled nursing, intermediate care, community care retirement, and inpatient hospice; LTAC includes specialty hospitals, behavioral hospitals, long term acute rehabilitation; Hospitals include freestanding ERs and urgent care facilities; Jurisdictions include city, county and tribal.

3 METHODOLOGY AND ANALYSIS

A survey design was implemented to examine data from Hurricane Laura, and Winter Storm Uri in Southeast Texas. Surveys constitute the most used research design in healthcare epidemiology research because of the low-cost, convenience, and willingness of respondents to share information without interviewers' present [31]. In tandem, an analysis of research methodology in health services found that qualitative methods are widely used and accepted in healthcare research [32]. Most recent COVID-19 study in Italy [33] adopted a survey design in healthcare as well. The methodology for collection comes as standard SETRAC. AAR institutional practice in that post-disaster or activation of the CMOC, all members of the healthcare coalition are sent an online link to answer questions about (a) individual facility strengths or weaknesses; (b) regional strengths or weaknesses; (c) the number of participating staff; (d) areas for improvement; (e) identification whether the facility utilized the CMOC when issues arose; and (f) name, Facility name, contact information, and Corridor. Even though overall, there were 16 open-ended questions on the Winter Storm Survey, and 14 on the Hurricane Laura Survey, our analysis focused on extraction of data from two items relevant to answering our research questions, namely (a) *Identify areas of SETRAC strength in support to Hurricane Laura/Winter Storm Uri*; and (b) *Identify areas for improvement based on SETRAC support to Hurricane Laura/Winter Storm Uri.*

Participation in AAR surveys is optional to all facilities; however, healthcare organizations that do participate are included in the SETRAC regional AAR. Notably, participation in a community based full-scale exercise and after-action conference is one of the Conditions of Participation a hospital must fulfil to maintain Center for Medicare and Medicaid Services (CMS) compliance [34]. The process for completing the AAR is as follows: A disaster occurs or the CMOC is activated, once the CMOC is demobilized the AAR questionnaire is sent out to all health organizations and stakeholders with a completion deadline. Subsequently, results inform the After-Action Conference which is a 4 hour meeting for all respondents to clarify or discuss strengths, weaknesses, and areas for improvements. Individual corridor meetings follow to discuss any issues in greater detail as they pertain to that geographic region. After the After-Action Conference and Corridor Meetings the AAR is completed and posted to the SETRAC website for all members to download. This document compliments the organization specific AAR that each organization has the choice of completing.

This research evaluated the 403 survey responses from the Hurricane Laura and 399 survey responses from the Winter Storm Uri AAR questionnaires and separated the organizations based on Corridor (North, South, East, West, Downtown). In total SETRAC has 1,561 total active members receiving the instrument. The Hurricane Laura survey was completed between 10 September 2020 and 15 October 2020 (with two late submissions). The Winter Weather survey was completed between 22 February 2021 and 29 March 2021 (with three late submissions). Overall, the response rate for the Hurricane Laura survey was calculated at 25.6% and the response rate for the Winter Storm Uri survey at 25.8%; as typical return rates in healthcare survey studies is 20–30% [31], our data fall within accepted range and can be viewed as representative of the Coalition and reliable. Figs 3 and 4 offer a breakdown of healthcare organizations' surveys collected for Hurricane Laura and Winter Storm Uri by agency type, respectively for each. Of the 403 responses from the Hurricane Laura survey 26 were in the North Corridor, 137 in the South Corridor, 66 in the East Corridor, 136 in the West, and 38 in the Downtown Corridor. Of the 399 Winter Storm Uri responses 26 responses were in the North Corridor, 126 were in the South Corridor, 50 in the East Corridor, 148 were in the West Corridor, and 49 in the Downtown Corridor.

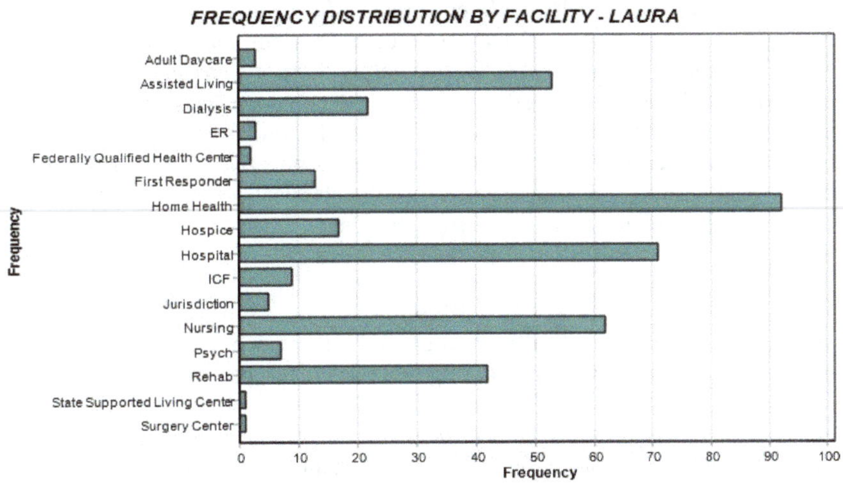

Figure 3: Frequencies of Hurricane Laura survey responses by agency type. *(Source: QDA Miner.)*

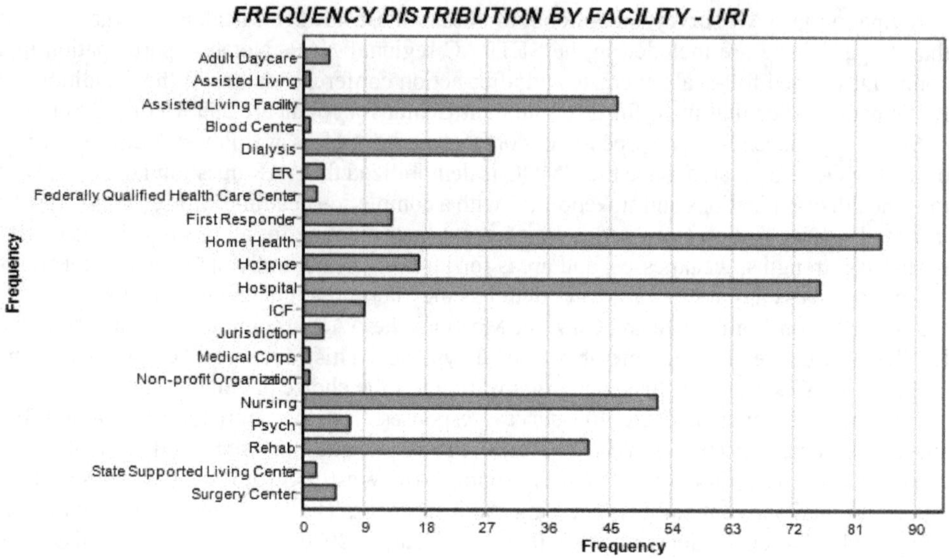

Figure 4: Frequencies of Winter Storm Uri survey responses by agency type. *(Source: QDA Miner.)*

Surveys, originally collected in Excel, were inputted into QDA Miner [35] – qualitative data analysis software – to enable classical constant comparison analysis [36], [37]. QDA Miner has been used for qualitative analyses both in disaster studies and healthcare [38]–[40] and it allows for categorization and codification of data into thematic units which in our study occurred a posteriori (identification and categorization of codes followed data-driven analysis and included assessing frequency, salience, distribution, and commonality among others). Further, the software offers functionality to split codes, rename codes, merge codes, subcategorize codes, all of which are effective means for ensuring consistency and integrity when managing a robust corpus of qualitative data. Data for each COVID-19 concurrent crisis were first categorized and coded independently form one another. This was aimed at eliminating potential transference effects; because the two events were triggered by two different hazards and two different levels of regional familiarity, we wanted to capture potential nuances. Once both cases were coded independently, we conducted cross-case evaluation for similarities and differences. Thus, the classical content analysis process was iterative and coordinated among research team members to ensure saturation and qualitative inter-rater reliability.

4 RESULTS

This study gauged participating healthcare organizations' perceptions about SETRAC's role in supporting organizational resiliency during two COVID-19 concurrent crises. In response to research question 1: *What are the capabilities of SETRAC as an HCC that strengthened the overall regional healthcare system resiliency during COVID-19 concurrent crises*, data analysis for survey responses from Hurricane Laura yielded eight final codes: (a) Communication; (b) Relationship links; (c) Accessibility; (d) Excellence; (e) Organization; (f) Reliability and knowledge; (g) Engagement; and (g) Resource management. Fig. 5 illustrates frequencies of codes across cases under analysis.

Search in: [IDENTIFY_2]

Codes: ○ All ◉ Selected: [Communication;Relationship Links;Accessibility;Excellence;Organization;Reliability and Knowledge;Engagement and Training;Resource Management]

Tree | Table

Category	Code	Description	Count	% Codes	Cases	% Cases	Nb Words	% Words
Strenghts	Communication		281	51.7%	281	69.7%	2290	64.3%
Strenghts	Relationship Links		14	2.6%	14	3.5%	148	4.2%
Strenghts	Accessibility		40	7.4%	40	9.9%	396	11.1%
Strenghts	Excellence		37	6.8%	37	9.2%	379	10.6%
Strenghts	Organization		35	6.4%	35	8.7%	254	7.1%
Strenghts	Reliability and Knowledge		39	7.2%	38	9.4%	424	11.9%
Strenghts	Engagement and Training		44	8.1%	43	10.7%	607	17.0%
Strenghts	Resource Management		54	9.9%	54	13.4%	623	17.5%

Figure 5: Analysis results of SETRAC strengths during Hurricane Laura. *(Source: QDA Miner.)*

Because communication and resource codes had two highest frequencies, specific breakdown of their types represented is provided by Fig. 6 for communication types and Fig. 7 for resource types.

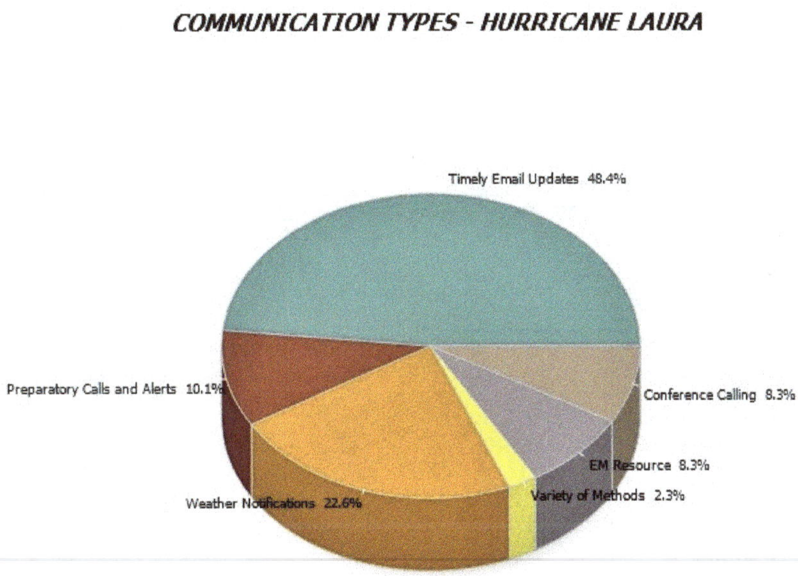

Figure 6: Types of SETRAC communications during Hurricane Laura. *(Source: QDA Miner.)*

Question 1 responses from Winter Storm Uri yielded eight final codes: (a) Communication; (b) Resource management; (c) Availability; (d) Training; (e) Reliability and knowledge; (f) Service orientation; (g) Relationship links; and (h) Organization. Fig. 8 illustrates frequencies of codes across cases under analysis.

Similarly, as communication and resource management were predominant, we offer more granular breakdown represented by Fig. 9 for communication types and Fig. 10 for resource types.

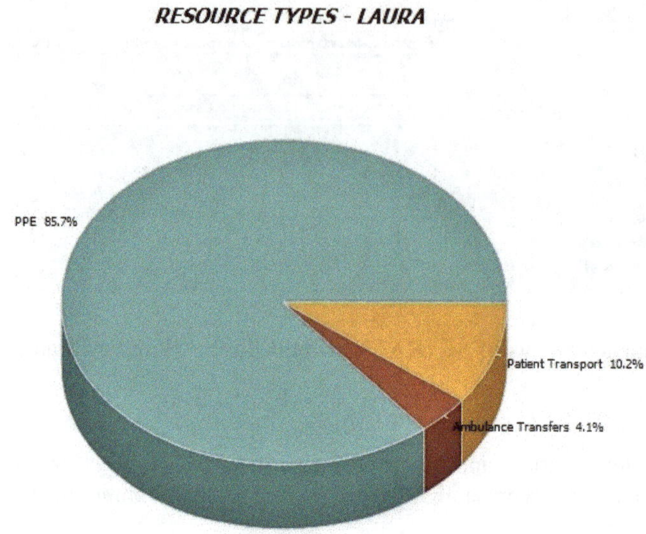

Figure 7: Types of resources provided by SETRAC during Hurricane Laura. *(Source: QDA Miner.)*

Coding Frequency

Search in: [IDENTIFY_3]

Codes: ○ All ◉ Selected: [Communications;Resource Management;Availability;Training;Reliability and Knowledge;Service Orientation;Relationship Links;Organization]

Tree | Table

Category	Code	Description	Count	% Codes	Cases	% Cases	Nb Words	% Words
Strenghts	Communications		268	51.0%	268	67.2%	2070	68.4%
Strenghts	Resource Management		72	13.7%	72	18.0%	793	26.2%
Strenghts	Availability		63	12.0%	63	15.8%	689	22.8%
Strenghts	Training		11	2.1%	11	2.8%	126	4.2%
Strenghts	Reliability and Knowledge		39	7.4%	37	9.3%	560	18.5%
Strenghts	Service Orientation		50	9.5%	50	12.5%	732	24.2%
Strenghts	Relationship Links		19	3.6%	19	4.8%	222	7.3%
Strenghts	Organization		4	0.8%	4	1.0%	39	1.3%

Figure 8: Analysis results of SETRAC strengths during Winter Storm Uri. *(Source: QDA Miner.)*

The distribution of strengths based on agency type is represented by Fig. 11 for Hurricane Laura, and by Fig. 12 for Winter Storm Uri, respectively.

In response to research question 2: *What are the SETRAC's capabilities in need of strengthening during COVID-19 concurrent crises in order to make the regional healthcare system more resilient?* Our analysis revealed six codes for Hurricane Laura as represented by Fig. 13: (a) Communications; (b) Identifying, securing, prepositioning resources; (c) Inclusivity and integration; (d) More training; (e) Evacuation preparedness; and (f) Technical improvements. For Winter Storm Uri we discerned the following five thematic categories: (a) Special resources; (b) Communications; (c) Patient transfers; (d) Inclement weather preparedness training; and (e) Inclusion summarized in Fig. 14.

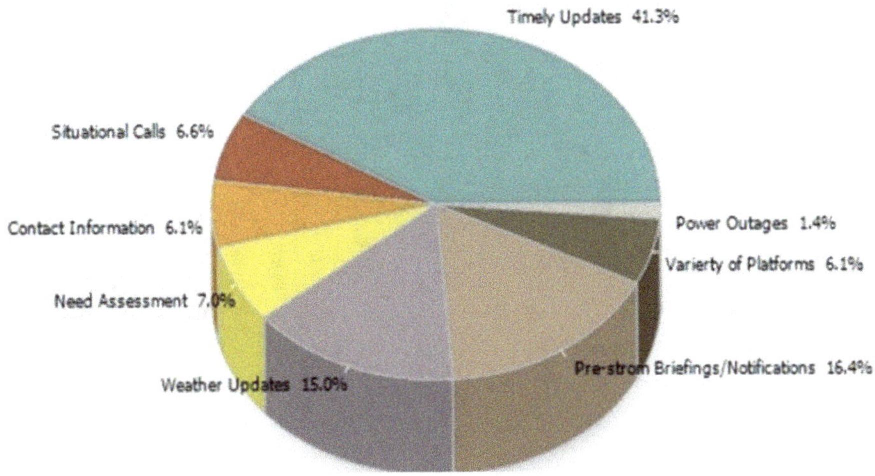

Figure 9: Types of SETRAC communications during Winter Storm Uri. *(Source: QDA Miner.)*

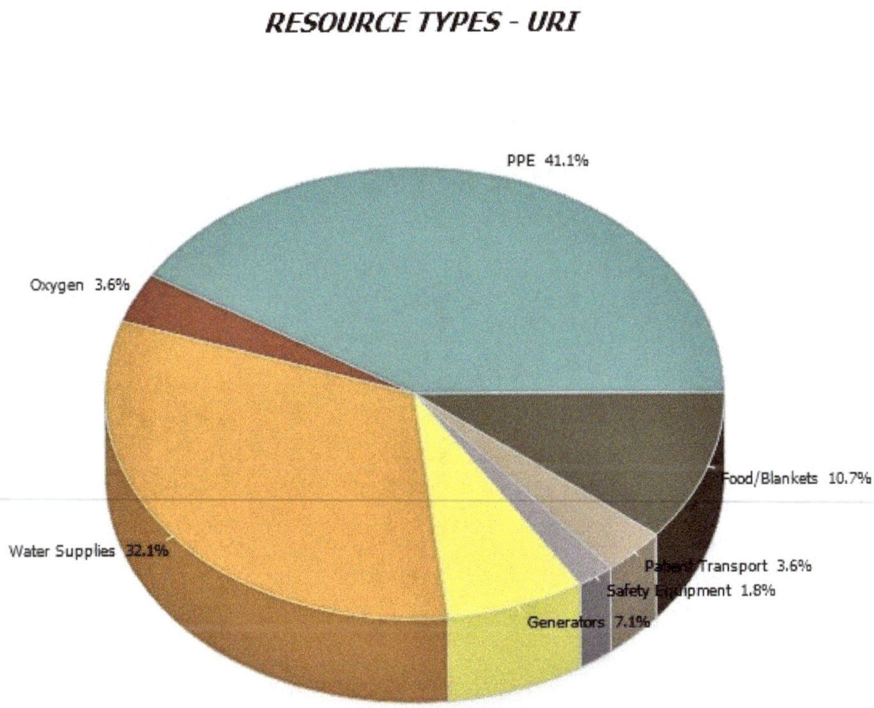

Figure 10: Types of resources provided by SETRAC during Winter Storm Uri. *(Source: QDA Miner.)*

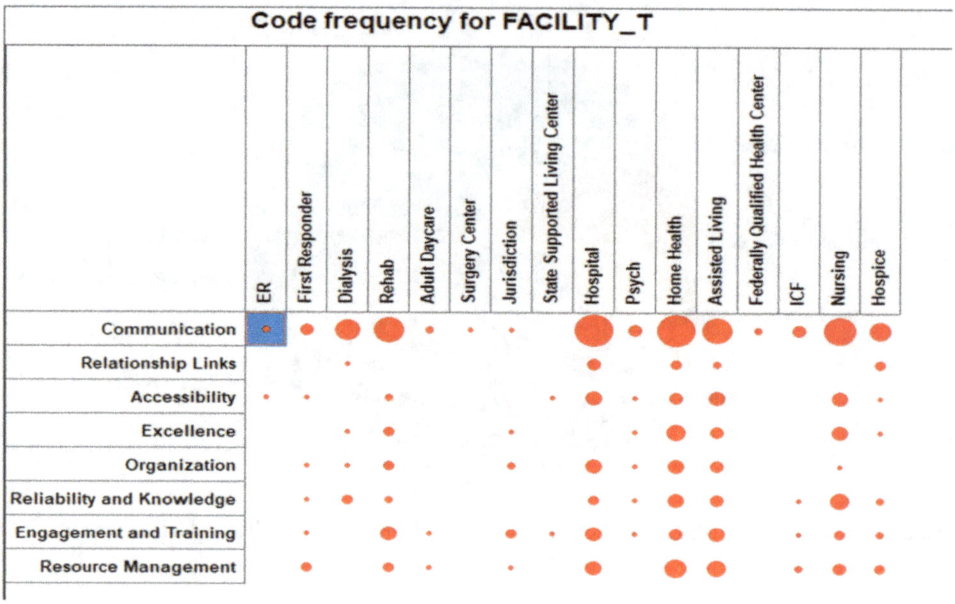

Figure 11: SETRAC strengths during Hurricane Laura recognized by agency type. *(Source: QDA Miner.)*

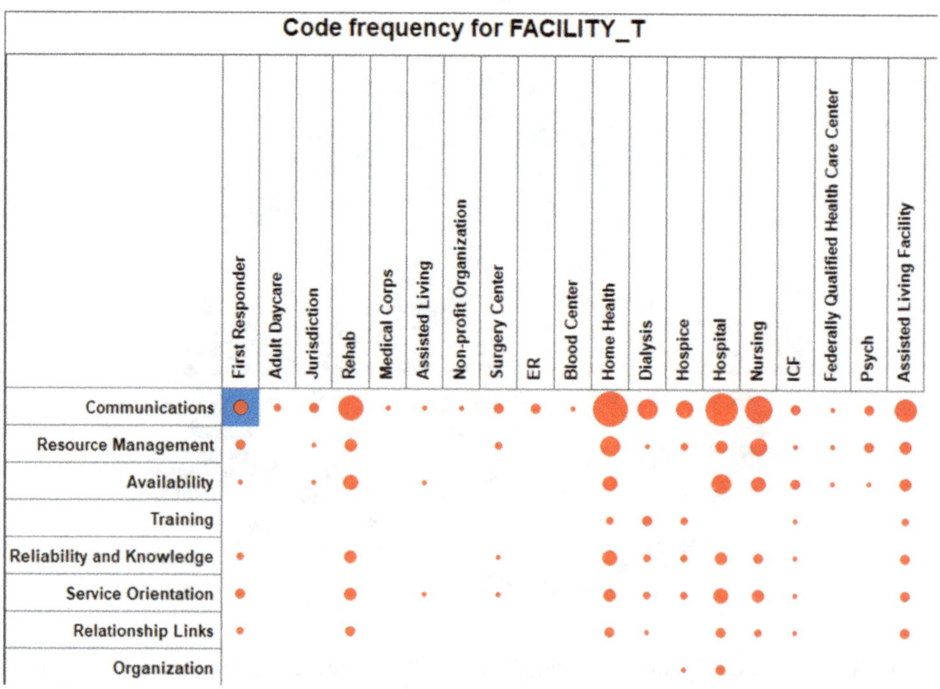

Figure 12: SETRAC strengths during Winter Storm Uri recognized by agency type. *(Source: QDA Miner.)*

Coding Frequency

Search in: [IDENTIFY_3]

Codes: ○ All ◉ Selected: [Technical Improvements;Inclusivity and Integration;More Training;Evacuation Preparendess;Communications;Identifying/Securing/Prepositioning Resources]

Tree | Table

Category	Code	Description	Count	% Codes	Cases	% Cases	Nb Words	% Words
Needs Work	Technical Improvements		10	4.8%	10	2.5%	87	2.8%
Needs Work	Inclusivity and Integration		27	13.0%	25	6.2%	288	9.2%
Needs Work	More Training		17	8.2%	17	4.2%	188	6.0%
Needs Work	Evacuation Preparendess		12	5.8%	12	3.0%	212	6.7%
Needs Work	Communications		105	50.5%	98	24.3%	1630	51.9%
Needs Work	Identifying/Securing/Prepositioning Resources		37	17.8%	37	9.2%	479	15.2%

Figure 13: Analysis results for SETRAC areas in need of improvement during Hurricane Laura. *(Source: QDA Miner.)*

Coding Frequency

Search in: [IDENTIFY_4]

Codes: ○ All ◉ Selected: [Special Resources;Communcations;Patient transfers;Inclement Weather Preparedness/Training;Inclusion]

Tree | Table

Category	Code	Description	Count	% Codes	Cases	% Cases	Nb Words	% Words
Challenge	Special Resources		71	34.0%	64	16.0%	875	32.7%
Challenge	Communcations		75	35.9%	74	18.5%	1013	37.9%
Challenge	Patient transfers		8	3.8%	8	2.0%	200	7.5%
Challenge	Inclement Weather Preparedness/Training		28	13.4%	26	6.5%	344	12.9%
Challenge	Inclusion		27	12.9%	27	6.8%	355	13.3%

Figure 14: Analysis results for SETRAC areas in need of improvement during Winter Storm Uri. *(Source: QDA Miner.)*

Finally, Figs 15 and 16 illustrate distribution of categories based on agency type for Hurricane Laura and Winter Storm Uri, respectively.

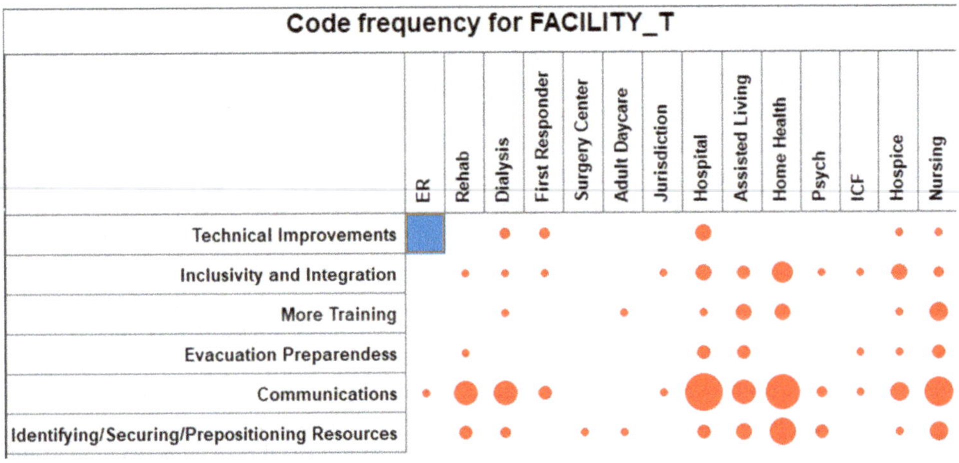

Figure 15: SETRAC's areas for improvement during Hurricane Laura by agency type. *(Source: QDA Miner.)*

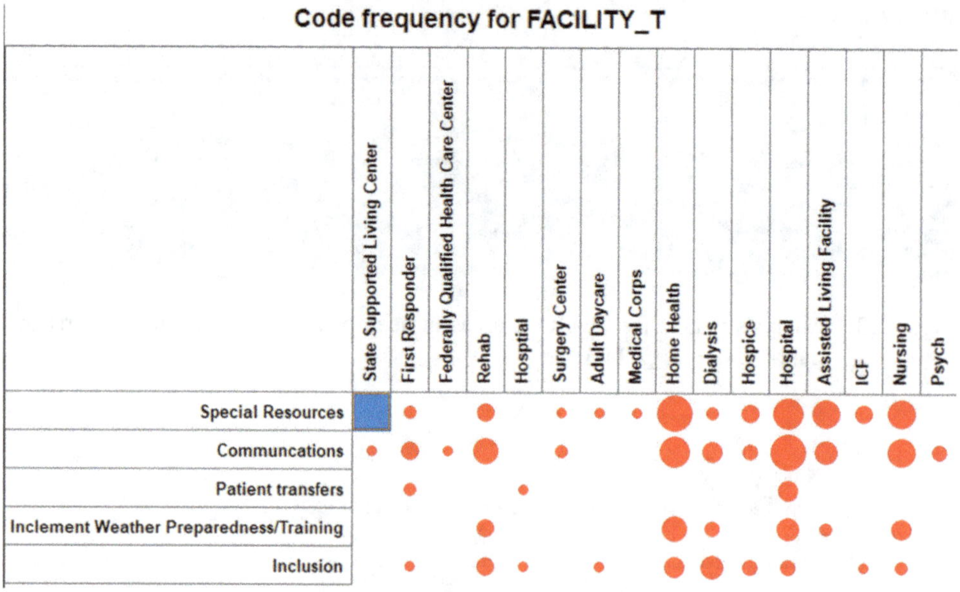

Figure 16: SETRAC's areas for improvement during Winter Storm Uri by agency type. *(Source: QDA Miner.)*

5 DISCUSSION AND IMPLICATIONS

The results of our study point to SECTRAC HCC as a formidable force for information sharing through *Communications* during COVID-19 concurrent crises. During Hurricane Laura, nearly 70% of respondents discussed the impact of communications on their ability to make disaster decisions. Among communication types, nearly half of the participating agencies (48.4%) stressed the value of constant emails with timely updates about the regional healthcare system situation prior to and during the storm; moreover, nearly quarter of the participants (22.6%) specifically stressed the enormous value of timely and credible weather reports (compiled for healthcare and based on National Weather Service (NWS) and other meteorological outlets). Continuous and reliable weather updates were critical because the path of the Hurricane Laura was constantly changing, and evacuation decisions were predicated based on the most realistic and probable landfall location; because of COVID-19, evacuation decisions were the issue of last resort for most healthcare agencies in the region. Thus, gauging the situational email updates that addressed among others COVID-19 area status as well as continuous weather tracking allowed for balanced approach to necessity of evacuation and measures/considerations during potential evacuation. In many surveys, participants stressed that they did not possess the capacity, nor the expertise to assess the situation on their own and they were already short-handed preparing themselves and patients for continuity of operations internally. Public media outlets were cited as highly unreliable for decisions in healthcare system as they lacked required strategic regional healthcare focus and specificity. Moreover, SETRAC's continually updated emergency resource management platform, was instrumental for situational awareness about regional resources while multiple communication modes (emails, calls, website, conference, webinars, or narrated PowerPoints) were also listed as invaluable. Markedly, more than 10% of those surveyed stressed the value of SETRAC's preparatory calls and alerts which provided a strategic picture for expectations, operations, collaborative capacities, and potential challenges during the storm. Finally, regular conference calls held by SETRAC allowed for continuous engagement and keeping stock of the situation as it unfolded.

For Winter Storm Uri, nearly 70% of participating agencies selected SETRAC communications as the biggest asset to them during this COVID-19 concurrent crisis as well. Beyond email updates and weather notifications though, communication types represented slightly different anatomy; particularly, pre-event notifications and alerts were given more weight as critical during Winter Storm Uri. The proactive focus by SETRAC in this crisis is particularly noteworthy because of hazard novelty. Moreover, SETRAC particularly engaged in individual situational calls and needs' assessments to gauge status of specific facilities – those were found critical by many because of the unprecedented nature of the crisis as well. In tandem, contact information and power outage information supplied by the coalition were listed as valuable. Communication differences, albeit subtle point to SETRAC's ability as a governance model to adapt in chaotic systems (something that is very difficult for traditional and centralized bureaucratic structures).

Notably, SETRAC was instrumental in resource identification, management, and fair and equitable distribution. Indeed, *Resource Management* was the second most common category of disaster capacity created by SETRAC for healthcare system in the region. For Hurricane Laura, more than 85% of the healthcare entities confirmed dire need, request, and subsequent receipt of PPE by the coalition; PPE supplies were indeed critical during Hurricane Laura as the hurricane cooccurred at the tail end of the first phase of the pandemic. Moreover, when evacuations were necessary, SETRAC facilitated ambulance transfers and provided patient transport as indicated by those surveyed. *Resource Management* in Winter Storm Uri, second only to *Communications*, was discussed by nearly one-fifth of respondents (18%). Interestingly, during Winter Storm Uri less than half of respondents than in Hurricane Laura expressed dire need, requested, and receipt of PPE supplies; this is understandable as in mid-February of 2021 Texas was two months into mass vaccination process and supply needs were slowly being met (albeit that number was still at more than 40%). However, SETRAC was cited as absolutely critical to providing needed water supplies to many healthcare system entities that lost power and had catastrophic pipe breakage. Moreover, specialized resources such as food items, generators, blankets, safety equipment were also provided. Admittedly, needs and concerns during Winter Storm Uri were vastly different from those encountered during the hurricane.

Beyond communication and resource management, during Hurricane Laura SETRAC was credited with *Reliability and Knowledge* – this theme related to trust in all communications provided by the coalition and value of disaster management expertise at the local, state, and federal level that they possess and were able to share. *Engagement and Training* – was the category mentioned by more than 10% of those surveyed; inarguably, SETRAC's regional engagement in disaster planning and preparedness to include joint trainings, conferences, and scenarios fostered capabilities such as presence of emergency plans, or general disaster awareness many agencies leveraged during the hurricane. *Accessibility* related to SETRAC's access by phone, email, or conference call on moment's notice – this was confirmed by nearly 10% of agencies as mission-critical. In nearly every case described under this theme, respondents expressed the value of getting a hold of coalition's representative in the very middle of the crisis. Both *Excellence* and *Organization* themes expressed gratitude for leadership and high level of organization. Finally, a small cluster emerged among analysed responses that listed very specific *Relationship Links* created by SETRAC for participating agencies that were found valuable. It is quite significant that both home health and assisted living facilities were two categories of responders by whom all those themes were listed as critical even though hospitals, hospices, nursing homes, and rehabilitation centres endorsed the majority of SETRAC-created resiliency capabilities as well, thus underscoring the coalition's relevance as a resilience builder across a wide range of healthcare system entities.

Beyond *Communication* and *Resource Management*, during Winter Storm Uri several themes such as *Reliability and Knowledge*, *Training*, *Organization*, and *Relationship Links* speak to similar resilience capabilities as those discussed during Hurricane Laura. Two dimensions, however, reveal a slightly different assessment of SETRAC's value in Winter Storm Uri as compared with the hurricane. Notably, *Service Orientation* was the category clearly discernible in Winter Storm Uri data. Survey participants expressed gratitude for help, empathy, courtesy, understanding, compassion, and support by SETRAC. Even though very nuanced, this difference is important as it points to SETRAC's ability to adapt situationally – in Winter Storm Uri, more individualized approach was needed to reassure, and support healthcare partners as the event was out of the ordinary realm of cyclical hurricane. Aligned with that interpretation is the theme *Availability*; while in Hurricane Laura SETRAC was cited as *Accessible* at a moment's notice, in Winter Storm Uri the coalition actively engaged with entities needed support as they were *Available* to take on novel tasks (such as water distribution) as needed. As in Hurricane Laura, Home Health, Assisted Living Facilities, or Hospitals attached importance to most of the resilience capabilities created by SETRAC while other agencies such as emergency rooms (ERs) benefitted the most from specific capability only (i.e., communications).

Coincidentally, *Communications* and *Resource Management* categories were also those categories that emerged as needing ongoing strengthening. Interestingly, during Hurricane Laura responders pointed to the need for text communication as an alternative to phone, conference, or email formats and many stressed technical and scheduling problems with conference calls as well as the need for better dissemination of contact information. It appears as those were resolved by SETRAC by February as most comments about Communication after Winter Storm Uri discussed issues related to lack of information and warnings related for example to city's issuance of such (e.g., water boil ordinance). For resource management, most of strengthening in both crises related to either pre-positioning of resources or better, more robust access to specialized resources such as transportation or generators. After Hurricane Laura, agencies stressed the need for more evacuation planning under pandemic conditions and after Winter Storm Uri more planning, training, and preparedness for adverse/inclement (non-hurricane) weather events. It is noteworthy that most comments within these analysed about SETRAC's improvements referred to building on existing capabilities. Perhaps the most notable in our assessment was the need expressed both in Hurricane Laura and Winter Storm Uri (13% or more in each event) for better agency inclusion. New, small, or non-traditional agencies such as law enforcement or dialysis were particularly interested in being more included and represented in SETRAC's outreach network, training, and relationship links. They also expressed the need for mass communication such as emails or weather updates to include specific content better tailored to their unique needs.

The implications of our study are twofold. First, based on findings related to an HCC operating in the largest medical centre in the world, we confirm that during concurrent disaster events in the United States, healthcare coalitions do fulfil a resilience-building mission. In fact, communication and resource management stewarded by SETRAC align with desired core missions [41] of healthcare coalitions. Moreover, relationship links, organizational structure, domain disaster expertise, or training provided by SETRAC are consistent with the concept of linking for social capital in disasters [30], [42]. The second implication is that SETRAC offers a promising adaptable governance model for the successful integration of decentralized, varied mission, varied jurisdiction, and varied service-type entities in healthcare system as represented by the sample used in this study. This is particularly important due to historical marginalization and exclusion of healthcare system agencies like long-term care facilities [43] or home care facilities [44] in disaster management in the United States.

REFERENCES

[1] Fakhruddin, B., Blanchard, K. & Ragupathy, D., Are we there yet? The transition from response to recovery. *Progress in Disaster Science*, **7**, pp. 1–5, 2020. DOI: 10.1016/j.pdisas.2020.100102.

[2] Trump, B.D. & Linkov, I., Risk and resilience in the time of the COVID-19 crisis. *Environment Systems and Decisions*, **40**, pp. 171–173, 2020. DOI: 101007/s10669-020-09781-0.

[3] Potutan, G. & Arakida, M., Evolving disaster response practices during COVID 19 pandemic. *International Journal of Environmental Research and Public Health*, **18**, pp. 1–11, 2021. DOI: 10.3390/ijerph18-63137.

[4] Robinson, L. et al., Cascading crises: Society in the age of COVID-19. *American Behavioral Scientist*, pp. 1–15, 2021. DOI: 10.1177/00027642211003156.

[5] Kouadio, I.K., Aljunid, S., Kamigaki, T., Hammad, K. & Oshitani, H., Infectious diseases following natural disasters: Prevention and control measures. *Expert Review of Anti-infective Therapy*, **10**, pp. 95–104, 2012. DOI: 10/1586/eri.11.155.

[6] Simonovic, S.P., Kundzewicz, Z.W. & Wright, N., Floods and the COVID-19 pandemic: A new double hazard problem. *WIREs Water*, Wiley, pp. 1–18, 2020. DOI: 10.1002/wat2.1509.

[7] Quigley, M.C., Attanayake, J., King, A. & Prideaux, F., A multi-hazards earth science perspective on the COVID-19 pandemic: The potential for concurrent and cascading crises. *Environmental Systems and Decisions*, **40**, pp. 199–215, 2020. DOI: 10.1007/s10669-020-09772-1.

[8] Finzi, Y., Ganz, N., Limon, Y. & Langer, S., The next big earthquake may inflict a multi-hazard crisis: Insights from COVID-19 – Extreme weather and resilience in peripheral cities in Israel. *International Journal of Disaster Risk Reduction*, pp. 1–9, 2021.

[9] Čivljak, R., Markotić, A. & Capak, K., Earthquake in the time of COVID-19: The story from Croatia (CROVID20). *Journal of Global Health*, **10**(1), 2020. DOI: 10.7189/jogh.10.010349.

[10] Atalić, J., Uroš, M., Šavor Novak, M., Demšić, M. & Nastev, M., The M_w5.4 Zagreb (Croatia) earthquake on March 22, 2020: Impacts and response. *Bulletin of Earthquake Engineering*, pp. 1–29, 2021. DOI: 10/1007/s10518-021-01117-w.

[11] Stepinac, M.P., Lourenço, B., Atalić, J., Kišiček, T., Uroš, M., Baniček, M. & Šavor Novak, M., Damage classification of residential buildings in the historical downtown after the ML5.5 earthquake in Zagreb, Croatia in 2020. *International Journal of Disaster Risk Reduction*, **56**, pp. 1–20, 2021. DOI: 10.1016/j.ijdrr.2021.102140.

[12] Mujeri, M.K. & Mujeri, N., Unexpected challenges: COVID-19 and Cyclone Amphan. *Structural Transformation of Bangladesh Economy*, pp. 249–273, 2021.

[13] Srivastava, S., Protecting the most vulnerable amidst COVID-19 and Cyclone Amphan, 27 May, 2020.

[14] United Nations Women, As Bangladesh battles COVID-19 and the aftermath of Suer Cyclone Amphan, Women's organizations lead their communities through recovery, June 15, 2020.

[15] Ishiwatari, M., Koike, T., Hiroki, K., Toda, T. & Katsube, T., Managing disasters amid COVID-19 pandemic: Approaches for response to flood disasters. *Progress in Disaster Science*, **6**, pp. 1–4, 2020. DOI: 10/1016/j.pdisas.2020.100096.
[16] Shultz, J.M., Fugate, C. & Galea, S., Cascading risks of COVID-19 resurgence during an active 2020 Atlantic hurricane season. *Journal of American Medical Association*, **324**(10), pp. 935–936, 2020. DOI: 10.1001/jama.2020.15398.
[17] Meyers, T., Multiple disasters strain response systems, slow recovery, and deepen inequity. Direct Relief, 7 Oct. 2020.
[18] Tracey, M.R., Plemmons, A. & Belasen, A.R., Throwing caution to the wind: COVID-19 and hurricanes, pp. 1–8, 2021.
[19] National Weather Service, Hurricane Laura.
[20] Miller, R.W., Hughes, T., Jervis, R. & Rice, D., Hurricane Laura live coverage: Trump to visit damaged areas; At least 6 dead after 'extensive' damage to Louisiana. *USA Today*, 1 Sep., 2020.
[21] Hiller, J. & Hampton, L., Pandemic stresses massive Hurricane Laura evacuations in Texas, Louisiana. *Reuters*, 26 Aug., 2020.
[22] McCullough, J. & Garnham, J.P., As Hurricane Laura heads for Southeast Texas, officials scramble to evacuate and shelter people during a pandemic. *The Texas Tribune*, 26 Aug., 2020.
[23] Doss-Gollin, J., Franham, D.J., Lall, U. & Modi, V., How unprecedented was the February 2021 Texas cold snap. *Environmental Research Letters*, **16**, pp. 1–7, 2021. DOI: 10.1088/1748-9326/ac0278.
[24] Busby, J.W. et al., Cascading risks: Understanding 2021 winter blackout in Texas. *Energy Research and Social Science*, **77**, pp. 1–10. DOI: 10.1016/j.erss.2021.102106.
[25] Courtney, B., Toner, E., Waldhorn, A., Franco, C., Rambhia, K., Norwood, A., Inglesby, T.V. & O'Toole, T., Healthcare coalitions: The new foundation for national healthcare preparedness and response for catastrophic health emergencies. *Biosecurity and Bioterrorism*, **7**(2), pp. 153–163, 2009. DOI: 10.1089/bsp.2009.0020.
[26] US Department of Health and Human Services, Public health emergency: Public health and emergency medical support for nation prepared. https://www.phe.gov/Preparedness/planning/hpp/reports/pmi-guidance-2019-2023/Pages/Background.aspx. Accessed on: 10 Jun. 2021.
[27] National Association of County and City Health Officials, Statement of policy, healthcare coalition structure, pp. 1–6, 2021.
[28] Hathaway, W. & Meyer, D.S., Competition and cooperation in social movement coalitions: Lobbying for peace in the 1980s. *Berkeley Journal of Sociology*, **38**, pp. 157–183, 1993.
[29] Murphy, G., Coalitions and the development of the global environmental movement: A double-edged sword. *Mobilization*, **10**, pp. 235–250, 2005. DOI: 10.17813/maiq.10.2.8u3626408607643t.
[30] Aldrich, D.P. & Meyer, M.A., Social capital and community resilience. *American Behavioral Scientist*, **59**, pp. 254–269, 2015. DOI: 10.1177/0002764214550299.
[31] Safdar, N., Abbo, L.M., Knobloch, M.J. & Seo, S.K., Research methods in healthcare epidemiology: Survey and qualitative research. *Infection Control Hospital Epidemiology*, **37**(11), pp. 1272–1277, 2016.
[32] Mays, N. & Pope, C., Assessing quality in qualitative research. *BMJ*, **320**, pp. 50–52, 2000.

[33] Felice, C., Luca Di Tanna, G., Zanus, G. & Grossi, U., Impact of COVID-19 outbreak on healthcare workers in Italy; Results from a national e-survey. *Journal of Community Health*, **45**, pp. 675–683, 2020. DOI: 10.1007/s10900-020-00845-5.

[34] Centers for Medicare and Medicaid Services (CMS), Emergency preparedness requirements by provider type. www.cms.gov/Medicare/Provider-Enrollment-and-Certification/SurveyCertEmergPrep/Downloads/EP-Rule-Table-Provider-Type.pdf. Accessed on: 5 Jun. 2021.

[35] Provalis Research, QDA Miner, Computer software, Version 5.0.5, Montreal, 2017.

[36] Berelson, B., *Content Analysis in Communication Research*, The Free Press: New York, 1952.

[37] Krippendorff, K., *Content Analysis: An Introduction to its Methodology*, SAGE: London, 1980.

[38] Denham, M.A. & Khemka, A.K., Homeland Security and emergency management in institutions of higher education (IHE): Texas case study. *International Journal of Safety and Security Engineering,* **7**(3), pp. 1–15, 2017. DOI: 10.2495/SAFE-V7-N3-337-351.

[39] Clavel, N. & Pomey, M.-P., Enhancing patient involvement in quality improvement: How complaint managers see their roles and limitations. *Patient Experience Journal*, 7(3), pp. 112–118, 2020. DOI: 10.35680/2372-0247.1460.

[40] Pomey, M.-P. et al, The patient advisor, an organizational resource as a lever for an enhanced oncology patient experience (PAROLE-onco): A longitudinal multiple case study protocol. *BMC Heath Services Research*, pp. 1–12, 2021. DOI: 10.1186/s12913-020-06009-4.

[41] Hafling, D., Role of regional health care coalitions in managing and coordinating disaster response, 2013, NCBI, White Paper.

[42] Warren, M.R, Thompson, P.J. & Saegert, S., The role of social capital in combating poverty. *Social Capital and Poor Communities*, eds S. Saegert, J.P. Thompson & M.R. Warren, Russell Sage: New York, pp. 1–28, 2001.

[43] Pierce, J.R. Jr, Morley, S.K., West, T.A., Pentecost, P., Upton, L.A. & Banks, L., Improving long-term care facility disaster preparedness and response: A literature review. *Disaster Medicine and Public Health Preparedness*, **11**(1), pp. 140–149, 2016. DOI: 10/1017/dmp.2016.59.

[44] Trautsamwieser, A., Gronalt, M. & Hirsch, P., Securing home health care in times of natural disasters, *OR Spectrum*, **33**, pp. 787–813, 2011. DOI: 10.1007/s00291-011-0253-4.

INSTITUTIONS OF HIGHER EDUCATION IN COVID-19 RISK COMMUNICATIONS: TEXAS STUDY

ANDREA BAUER & MAGDALENA ANNA DENHAM
Department of Security Studies, Sam Houston State University, USA

ABSTRACT

This case study research aimed to explore risk communication processes and strategies adopted by Institutions of Higher Education (IHEs) in the state of Texas during COVID-19. Because higher learning and interactions with students during the initial stages of the pandemic resided predominantly in virtual space, the study design considered website content analysis as appropriate in addressing risk communications chosen by universities. Markedly, website content analysis was conducted among 43 IHEs accredited by the Southern Association of Colleges and Schools Commission on Colleges and part of a Texas University System. This sampling replicates previous IHE research in the state. The conceptual model entitled "Conceptual Model for Evaluating Emergency Risk Communication (EERC)" developed in public health was used to ground focused risk communication dimensions to (a) review whether or to what degree messaging from IHEs aligned with the components used by experts; (b) record the types of patterns that are found in messaging as well as characteristics that foster a Disaster Resilient University (DRU); and (c) provide insights on areas that may need more focus to ensure greater efficacy in the future. The ERC model yielded the following dimensions used for analysis: (a) accurate/reliable; (b) open/transparent; (c) clear; (d) tailored messages; (e) consistent/ timely; (f) sufficient; and (g) actionable. In addition to ERC, content regarding technological solutions was assessed from websites of chosen universities to gauge how universities have responded to tools needed during the pandemic to make informed judgments for the campus and to share information openly for the community. Our findings point to various components that IHEs in Texas targeted when communicating COVID-19 related information. For example, less than 60% of IHE in our sample defined terms, reviewed general COVID-19 information, and identified at risk groups on their respective campuses and only 6% provided an option for translating information. Meanwhile, IHEs fared generally better at providing timely updates regarding campus operations (i.e., 80%). Implications for our finding are discussed within the lens of Disaster Resilient University (DRU).

Keywords: institutions of higher education, Disaster Resilient University, risk communications, college students and COVID-19, public health, digital health technology, evaluating emergency risk communication.

1 INTRODUCTION

Risk communication defined as the "exchange of real time information, advice and opinions between experts and people facing threats to their health, economic, or social well-being" [1, para. 1] supports prevention and mitigation pre-disaster, preparedness among the population, sharing of information during a disaster, and recovery [2]. Researchers reviewing historical cases of pandemics and crisis communication [3] point out the intricacies of mitigating, preparing, and responding to a public health emergency. These complexities result from maintaining security, supporting the advancement of the economy, and the continuation of routine activities while managing a public health crisis [3, p. 6]. The communication adopted during the Spanish Flu of 1918 stands as a lesson for risk communicators. At the start of the 1918 pandemic, the US government disseminated information on the pandemic in war-news style concealing the truth from the public. Stability and "prosperity" were prioritized over transparency, as fear was perceived as a higher threat. Ultimately, the lack of risk communication from the U.S government and medical field hid the gravity of the virus from the publics placing them at higher risk. Trust in leadership was

WIT Transactions on The Built Environment, Vol 207, © 2021 WIT Press
www.witpress.com, ISSN 1743-3509 (on-line)
doi:10.2495/DMAN210041

lost due to the concealment of people ill with the virus and the underreporting of deaths. Conversely, in cities, where detailed information was disseminated and mass masking recommended, action replaced fear. Albeit the sole responsibility of the Environmental Protection Agency (EPA) in the 1970s, risk communication in the United States evolved to comprise open communication with the public in the 1980s [4] and to foster urgency in the public where needed and decrease panic where risk was low. Earlier period [5] lacked empathy and understanding for public risk perception, causing outward resentment for public response. Risk communication adopted today emerged from the need to form an alliance between government and corporate entities with the public [4] to communicate – based on the hazard – a level of gravity or confidence needed, and fostering appropriate responses based on the information collected. From this, the field grew as the media, public interests, inclusion of the public, and importance of communicating risks to influence the public became important factors of risk communication and persuasive messaging [5]. This phase took into consideration traits of the audience as well as public perception and in the 1990s risk communications transitioned from a purely persuasive communication focus to focus on "social context" [5, p. 90]. After 9/11 and subsequent anthrax attacks, the need for risk communication was heightened [6]; practices recommended included (a) involving stakeholders; (b) listening to the audience; (c) being transparent and open; (d) utilizing credible sources; (e) providing information to the media; (f) communicating clearly; (g) communicating with compassion; and (h) planning ahead of an incident [6]. Risk communications for public health crises have been especially important because they might address "high rates of infection, significant morbidity, lack of therapeutic measures, and rapid increase in cases" [7, para. 1]. Strategies for risk communication in public health emergencies need to account for individual circumstances, uncertainty, and the lack of trust based on a divided political and public health environment [8]. The strategies to better communicate COVID-19 messaging included communicating with honesty, alleviating fear by sharing "that uncertainty is intrinsic to the practice of environmental and public health science" [8, para. 10], consistency, and coordinating with experts. The World Health Organization (WHO) adds trust, transparency, early communication, listening, and planning to the rules of effective risk communication [9]. Glik provides similar strategies that are viewed as necessary in communicating risks such as trust, credibility, transparency, and accountability [10, para. 6] while others [11] present practices with a focus on risk perception and its impact on public response while acknowledging the complexity of communicating risk. They suggest recommendations to enhance operations which require knowing the audience, understanding that delivery impacts efficacy, and adapting during emergencies. The study [11] determined that (a) trust in leadership leads to risk acceptance; (b) tailoring messages to address public emotions can positively impact risk perception; and (c) transparency and sufficient communication can impact public response. Researchers in Baltimore [12, para. 1] found that communicating with transparency, "timeliness, completeness, and clarity of information…" were important drivers of maintaining trust with the community. Early COVID-19 research on risk communication in Wuhan China [13] provides evidence on the impact of timely information on managing risks and containing the crisis. Indeed, lack of timely information on COVID-19 to the public in Wuhan, China, due to political and stability concerns, had deleterious consequences. Risk communications are inherently linked to risk perception defined as "people's subjective judgments about the likelihood of negative occurrences such as injury, illness, disease, and death" [14, para. 1]. Risk perception affects the effectiveness of risk communications [7] through factors such as trust, origin, control, nature, scope, awareness, imagination, dread, age affected, uncertainty, familiarity, specificity, personal impact, and fun factor [15]. These factors play a role in an individual's perception of a risk, potentially impacting the health of the recipient and the community. Ultimately, trust in risk communicators as well as in the risk evaluation processes were found to decrease fear and increase accurate risk perception. During the COVID-19 pandemic, communities have relied on information from various outlets and experts. Reportedly, "infodemic" – an over-abundance of information, of which some is accurate, and some is not, has increased during the pandemic and made absorbing reliable information a challenge subsequently posing risk to public health [16, para. 7]. The danger of an over-abundance of information and unreliable communicators leads to the dissemination of false information which in turn has the potential to cause harm or death at worst and increase anxiety among the population at best [17].

Finally, of importance are risk communication tools that can be leveraged to disseminate risk communication. Digital health technologies empower communities to manage "health and wellness" [18]. Risk communication aims to maximize "public risk awareness", providing up to date data that impacts the community, strengthens transparency, and supports informed decision making [19, para. 1]. The COVID-19 pandemic has displayed the various applications of digital health in support to public health emergencies. For example, WHO recommended the use of monitoring tools in public health emergencies to increase situational awareness about the ongoing crisis [20]. In tandem, researchers [21] agree that technologies "play an integral, increasing, and evolving role in supporting public health responses..." (para. 1). For instance, CDC recommended that reopening universities possess systems/processes set up for students, faculty, and staff to report different COVID-19 details such as positive results, exposure to COVID-19, and symptoms [22]. Digital technology in public health emergencies can play a significant role and in COVID-19 in has been used across surveillance, screening, telemedicine, and mobile applications for various services [23]. Clearly, the scale of the pandemic has emphasized digital health technology as tremendous asset in managing public health emergencies. This research aims to explore risk communication processes and strategies adopted by Institutions of Higher Education (IHEs) during COVID-19 crisis and evaluate their effectiveness in helping to form accurate risk perceptions by the campus community; moreover, the extent and nature of adoption of various platforms and techniques for risk communication dissemination with the campus will be appraised as well. This study supports the mission of emergency management to lessen the vulnerabilities of communities through its ability to respond to natural or manmade disasters [24].

2 INSTITUTIONS OF HIGHER EDUCATION

As the coronavirus emerged and spread globally, communities and institutions have been challenged with an atypical disaster. Unlike cyclical or recurring hazards, the COVID-19 pandemic has defied traditional emergency management phases and required ongoing implementation and sharing of recommendations through risk communication due to its continuing spread. Institutions of higher education (IHEs), recognized as critical infrastructure under Homeland Security Presidential Directive 7 (HSPD-7), have had an obligation for ensuring safety through practice of risk communication on their campuses [25]. Uncovering the components of effective messaging found across universities as well as identifying potential gaps supports the concept of a Disaster Resilient University (DRU), which is "an institution capable of adapting to the aftermath of a disaster" [26]. A DRU, encouraged under Presidential Directive-5, requires universities to prepare for and develop systems that follow the National Incident Management System (NIMS). As preparing for all kinds of disasters is challenging and may lead to institutions preparing for the more likely disaster, there is a gap of preparation and planning for atypical disasters [27]. During COVID-

19, many IHES have had to face the challenges of the virus while maintaining the continuity of operations both in person and online. A coronavirus tracking survey [28] disseminated to universities and colleges in Fall 2020 revealed more than 1,000 universities reporting more than 300,000 coronavirus cases on their campuses and 50 deaths. In Texas specifically, more than 80 schools have reported more than 31,000 cases [28]. According to the Centers for Disease Control and Prevention (CDC), more than 15% of coronavirus cases have been from individuals aged 18–29, which includes the age group of the majority of college and university students [29]. The prevalence of coronavirus cases on campuses and the potential threat of infection among college students and staff, needed to be managed through "detection and containment of clusters of infection and the interruption of community transmission to mitigate the impact on human health" [30, para. 3]. Measures in managing outbreaks such as surveillance, case identification, preventing community spread, and communication with the public have thus been vital to IHEs in controlling COVID-19. Moreover, because the ages of individuals working, teaching, and learning on campuses range from "younger to older adults", disseminating an accurate level of risk to the community supports the application of appropriate protocols and a safe environment [22]. As the "normal" college experience has included in person classes, housing on and off campus, large gathering school events, and social events, managing COVID-19 on campuses has required communication with the community. The techniques used to reach students on campus can influence the safety and sanitary practices used and the overall safety of the campus. Universities adapting to virtual learning at the beginning of the crisis are now offering a mix of online and in person classes, a hybrid method, with students returning to class and adapting to the new college environment.

As many students have become accustomed to life on campus before the pandemic, risk communication permits information to be grasped and practices to be adopted in order to meet the safety requirements of life on and off campus during COVID-19; online information-sharing necessitates reliable COVID-19 information as well as information on campus operations. The priority of "preserving the health of students, faculty, and staff", has required universities to take on the role as emergency managers [29]. In addition, resilience from institutions of higher education is reflected through the generation of "new capabilities" as well as the "shift to new points of stability" [31, p. 7]. Evaluating the risk communications adopted by universities can help strengthen communications in the midst of the pandemic as well as prepare institutions for future disasters that may require long term risk communication. A "DRU has to be created to ensure an overall culture of preparedness is developed and a campus is properly prepared for all dangers that are unique to it" [27, p. 5]. Indeed, research form the start of the pandemic [32] revealed that students' safety adoptions were necessary, as were students' inclusion in messaging and planning, and leveraging of various communication channels. Recognizing obstacles to risk communication is also essential, in order to raise awareness on the factors that may influence the audience's acceptance and acknowledgment of the information.

Emerging literature on risk communication and college campuses has addressed various subjects from students understanding of the public health emergency to the platforms that students more often use to receive information. Data collected from a study of 483 college student activities and knowledge during the H1N1 influenza [33] administered at a university in the southwestern US determined that more than half of students were actively taking health and safety precautions; however, more than half of students were not knowledgeable of the symptoms of H1N1. The majority could not differentiate H1N1 flu signs and symptoms from the seasonal flu. The findings also determined that less than half of students would take proper isolation measures in case of symptoms or illness. Survey research on risk communications on college campuses during COVID-19, targeting 1,000 university students in a Midwestern city and their knowledge of the virus [34] found that majority indicated using "the internet and social media" for news on the coronavirus, less than a quarter (i.e., 18%) accurately identified the symptoms of the coronavirus, more than 50% incorrectly identified at risk groups, yet more than 50% identified the appropriate hand washing time to prevent the spread of the coronavirus. This pattern infers a potential vulnerability of safety, which could be addressed by communicating risks to college students and information relating to public health. With respect to research on risk perception, researchers in China [35] assessed the concept of *origin* or the decrease in risk perception college students had of their own health versus others. Alarmingly, less than 7% "of college students thought they were more likely to be infected by COVID-19" and more than 80% of college students worried about a family member becoming infected by COVID-19 [35, para. 13). Origin amongst college students during COVID-19 pointed to an increase in risk perception of the community exposure to COVID-19, but a decrease in risk perception of students' own risk of virus exposure. Effective and targeted risk communications by IHEs are thus critical considering more than 60% of information sought by college students in a 5,000-sample survey [36] originated from higher education sources. Others [37] suggested such sources be diversified to connect effectively with young adults (e.g., text message, internet, print media). Because most of research on public health risk communications at IHEs consisted of surveying student, our study aims to contribute to the literature by examining ways in which IHEs themselves disseminated risk communication to student community during COVID-19. As campuses adapt to the "new normal" during the ongoing pandemic, collecting data and disseminating information from and to the public will continue to increase awareness and uphold transparency. As research and updates on COVID-19 and latest developments emerge, risk communication will remain vital in ensuring that individuals have accurate information and practices that are recommended to mitigate risks. Due to restrictions being softened and institutions opening again for operations, fortifying safety and security means delivering information in a manner that permits populations to not only change behaviour but also to continuously practice safety guidelines as the pandemic remains active and extends into latter part of 2021.

3 METHODOLOGY

This case study research adopts multimodal content analysis (MMCA) that includes written language, illustrations, links, videos, and applications [38], [39] to explore risk communication processes and strategies adopted by selected IHEs in the state of Texas during COVID-19. Because higher learning and interactions with students during the initial stages of the pandemic resided predominantly in virtual space, websites of selected universities were considered valuable repository to assess risk communication approaches in sharing COVID-19 information to the users, determining its efficacy, and identifying areas that could be enhanced in continuing operations. Our sample follows filtering criteria applied in a previous research study [26] and comprises 43 IHEs in Texas that are (a) accredited by the Southern Association of Colleges and Schools (SACS) Commission on Colleges and are part of a Texas University System. Table 1 lists IHEs whose website content was explored in this research. While there is no universal tool to measure the effectiveness of risk communication, the Conceptual Model for Evaluating Emergency Risk Communication (EERC) was used to review to what degree messaging from IHEs align with the components used by experts [40]. Fig. 1 represents components of the EERC as well as criteria for each used in website MMCA. The EERC identifies the following components (a) accurate/reliable; (b) open/ transparent; (c) clear; (d) tailored messages; (e) consistent/timely; (f) sufficiency; and

Table 1: Texas IHEs represented in the study.

University of Houston System	University of North Texas System
University of Houston	University of North Texas
University of Houston – Clear Lake	University of North Texas Health Science Center
University of Houston – Downtown	University of North Texas at Dallas
University of Houston – Victoria	
University of Texas System	**Texas A&M University System**
University of Texas at Arlington	Texas A&M University
The University of Texas Southwestern Medical Center	Prairie View A&M University
The University of Texas at Austin	Texas A&M University – Commerce
The University of Texas Medical Branch at Galveston	Tarleton State University
The University of Texas at Dallas	West Texas A&M University
The University of Texas Health Science Center at Houston	Texas A&M University Kingsville
The University of Texas at El Paso	Texas A&M University – Corpus Christi
The University of Texas Health Science Center at San Antonio	Texas A&M International University
The University of Texas Permian Basin	Texas A&M University – Texarkana
The University of Texas MD Anderson Cancer Center	Texas A&M University – Central Texas
The University of Texas Rio Grande Valley	Texas A&M University – San Antonio
The University of Texas Health Science Center at Tyler	
The University of Texas at San Antonio	
The University of Texas at Tyler	
Texas State University System	**Texas Tech University System**
Lamar University	Angelo State University
Sam Houston State University	Texas Tech University Health Sciences Center
Sul Ross University	Texas Tech University
Texas State University	Texas Tech University Health Sciences Center at El Paso
Lamar State of Institute of Technology	
Lamar State College Orange	
Lamar State College Port Arthur	

(g) actionable [40]. Fig. 2 expands the EERC to include dissemination through multiple channels using technology. The *Accurate and Credible* criterion defined as information being up to date and scientifically accurate and noted as a vital component to effective risk communication, uncovers the activities that would support this characteristic when reviewing the websites of institutions of IHEs (e.g., coordination of information from credible resources). The *Transparent and Open* criterion designates candid information provided to

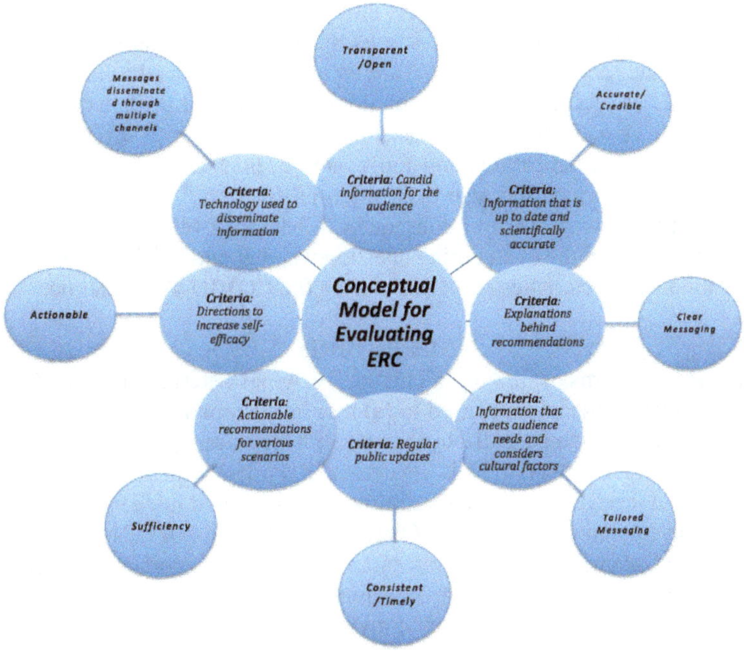

Figure 1: EERC model. *(Source: Adapted from Seeger et al. [40].)*

key audiences in an immediate and accessible way [40]. Information regarding the number of cases from the community as well as components that are still evolving in relation to the pandemic would be categorized under "transparent/open". The *Timely and Consistent* criterion ensures that messages are consistent to maintain the public's trust and decrease scepticism. The conceptual model uses consistency to portray the importance of information being aligned with other sources. While information does not have to be a replica from one source to another, providing coherent context maximizes the public's decision to take protective actions [40]. Another aspect of timeliness and consistency in risk communication is updating of information on a regular basis. In collecting information from websites of selected IHEs, information regarding a record of regular updates as well as the date from the latest updates were examined. The *Actionable* criterion consists of translating information into recommendations to populations that can lessen the risks and mitigate the threat. Actionable messaging can be as simple as referring an individual to a medical professional, a credible source, and offering protective measures or practices that can be used to mitigate the threat. The *Tailored Messaging* component involves sharing information that meets the needs of the audience and considers cultural factors. Tailored messaging can increase the perceptions of an audience's ability to utilize actionable recommendations in order to decrease risks. Information from selected IHEs that demonstrated the sharing of information specifically for the college community was grouped under "tailored messaging". The *Sufficient* dimension referred to the amount of information an individual needs to address a given risk; it is a precise guide that includes information and actionable recommendations that the audience can use to decide how to act. In gathering information from websites of IHEs, information that exhibited characteristics of being robust (e.g., steps and actions to take in various scenarios during COVID-19) was recorded *Sufficient* category. The *Clear* criterion referred to an individual's ability to process information; for example, using plain language to explain recommendations and activities or explanation of terms and scientific jargon were evaluated in this category. Finally, the *Dissemination of COVID-19-related Information via Multiple Channels* as well as through technological tools used to collect community data, was added to reflect elements of a disaster resilient university (DRU). Notably, during COVID-19 virtual communications and tools have been vital in raising awareness while maintaining functions, which has also been essential for IHEs. Taken together, those eight criteria provided framework for our subsequent MMCA of selected IHEs' websites. Our research sought to answer the overarching question whether and to what degree selected IHEs in Texas exhibited characteristics of a DRU during the COVID-19 pandemic.

4 FINDINGS

Table 2 represents patterns of findings along the 7 original criteria of the EERC while Figs 2 and 3 represent technological tools used to (a) monitor symptoms and (b) manage COVID-19, respectively. Our findings indicate that under Open/Transparent category, operations were generally labeled under "levels", "phases", "codes", "plans", "color codes", and "approaches", and consisted of sharing information related to campus guidelines and approaches being implemented during COVID-19 for Spring 2021. Notably, less than half (i.e., 46%) of IHEs examined provided information on campus operations to users. However, majority of IHEs provided users with information regarding the plans the campus was utilizing, sharing information regarding individuals permitted back on campus, learning styles, safety recommendations, and more. Transparency on the dissemination of COVID-19 cases from the community were also recorded and categorized under this section. Accordingly, 88% of the institutions of IHEs in the sample provided the number of COVID-19 community cases. Those were reported based on (a) active totals; (b) total of students; staff, and faculty cases; (c) recovered cases; and (d) cases of isolation/quarantine. IHEs that were a health program or hospital themselves, were difficult to categorize because their cases included the numbers of cases from the county, city, and or hospital. Nonetheless, the majority of IHEs did provide the number of COVID-19 cases. Findings grouped under the *Accurate/Credible* criterion and based on the use of credible sources in sharing COVID-19 information revealed that every IHE provided the sources used in developing COVID-19 messaging as well as recommended guidelines based on credible sources. The sources named from IHEs were as follows: Centers for Disease Control and Prevention (CDC), World Health Organization (WHO), Texas Department of Health and Human Services, US Department of State, and local, state, and federal authorities. The findings further indicate that 95% IHE used the CDC as a resource, 48% IHE used the Texas Department of Health Services (TDHS) as a resource, 25% IHE used the WHO as a resource, and 32% IHE used other credible resources. The findings indicate that many IHEs used *multiple credible sources* in sharing COVID-19 information and raising awareness. Analysis of website content that provided the basis for recommendations as well as clarification of meanings behind terms and recommendations and included in the *Clear* category revealed that more than half of IHEs (62%) provided general COVID-19 information exploration on their websites. COVID-19 messaging included information on the virus that causes the disease, ways of transmission, symptoms, and in rare cases, the developing variants. Slightly less than half of IHEs however (48%) supplied definitions for terms such as isolation, quarantine, or close contact, etc. Findings related to *Tailored Messaging* revealed that cultural factors were rarely included in their risk communications. Notably, a minimal number of IHEs (6%) provided options for immediate COVID-19 information translation. Much of the information collected and

Table 2: Patterns of findings based on EERC criteria.

Open/Transparent	Accurate/Credible	Clear Messaging	Tailored Messaging	Consistent/Timely	Sufficiency	Actionable
About 43% of IHE provided information on campus operations *55% of IHE did not provide campus operations* About 82% of IHE provided COVID-19 community cases updates *18% of IHE did not provide community case updates*	About 89% of IHE used the CDC as a credible resource *11% of IHE did not use the CDC as a source* About 45% of IHE used the Texas Health and Human Services as a credible resource *55% of IHE did not use the Texas Health and Human Services* About 23% of IHE used the WHO as a credible resource *77% of IHE did not use the WHO as a credible source* 30% provided other credible resources not listed above	About 58% of IHE provided an overview of COVID-19 *42% of IHE did not provide an overview of COVID-19* About 45% of IHE defined terms *55% of IHE did not define terms*	About 6% of IHE provided an immediate option for translation *94% of IHE did not provide an immediate option for translation*	About 81% of IHE provided recent updates within the past two months About 19% provided updates before the last two months	100% of schools provided reporting guidelines	100% of IHE provided safety recommendations and protocols

inserted in this section referred community members to other information provided on websites of campus operations, buildings opened and closed, information on signage on campus, COVID-19 testing on campus and guidelines, process of case reporting, guidelines for positive, exposed, and symptomatic cases, and return to work guidelines.

Markedly, under both *Clear* and *Tailored Messages* categories, the EERC suggests components have a duplicate impact to outcomes. When searching for data that aligned with the "clear" less than 60% of IHE defined terms, reviewed general COVID-19 information, and identified at risk groups. Although, IHE provided information on COVID-19, the information listed above altogether was not posted on each website. Thus, while some websites provided the definitions for "close contact", "quarantine", "isolation", and an overview of COVID-19, others provided other useful information. Alarmingly, less than half were found to have provided information regarding at risk groups. Findings related *to Timeliness and Consistency* of updates pointed to the majority of the IHEs disseminating information to the campus regarding COVID-19 since the beginning of 2020; by the time of our analysis in March 2021, 81% of IHEs provided updates within the two previous months, yet only 19% of IHEs published their last updates in the months before March 2021. With respect to *Sufficiency,* 100% of IHEs provided reporting guidelines for their community for positive, exposed, or symptomatic cases and all IHEs displayed detailed information for their campuses describing enabling actions to be taken when it comes to reporting, safety recommendations, and other operations on campus during COVID-19. Finally, results related to *Actionable* messaging (safety recommendations provided in order to lessen the impact or risk of a hazard, recommendations during COVID-19) revealed 100% of IHEs displayed safety recommendations and protocols for the community and recommendations such as wearing a face covering, practicing good hand hygiene, following the guidelines for hand washing, social distancing, daily self-monitoring and screening, practicing respiratory etiquette, and staying home if sick.

As digital health is tied to risk communication, Fig. 2 provides an overview of the ways that technology and digital health have been used in managing risks by selected IHEs in Texas. More than half (68%) of IHEs recommended self-monitoring and self-screening; however, the use of apps, online symptom checklists, health surveys, self-screening questions, and self-monitoring guides were provided and detailed only on 12% of websites. Only 16% of IHE leveraged technology to implement daily health screenings, and only 4% of IHE leveraged technology for daily in person health screenings.

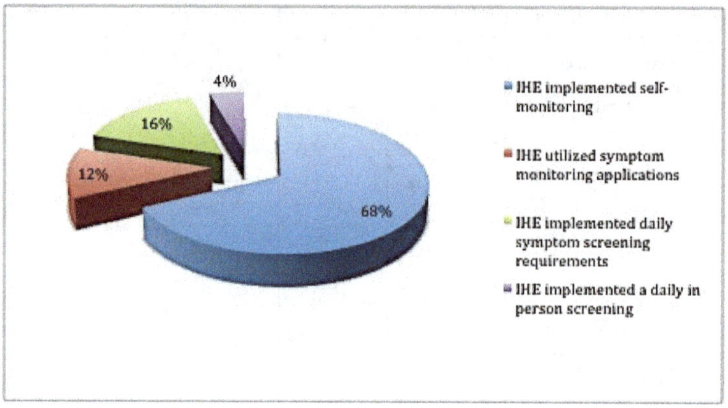

Figure 2: IHEs' use of technology to monitor COVID-19 symptoms.

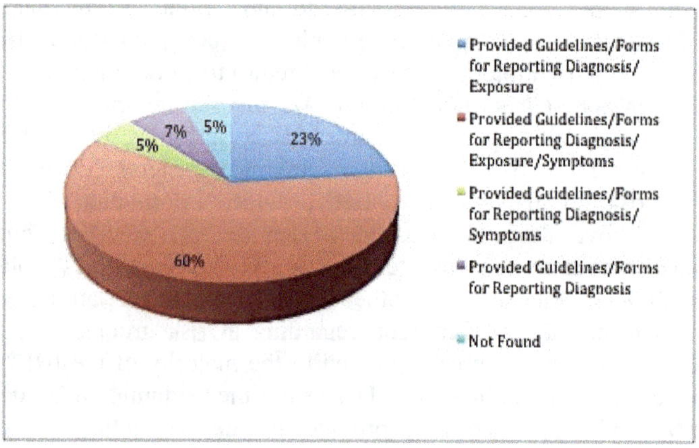

Figure 3: Technological tools to manage COVID-19.

Fig. 3 provides data collected from selected IHE used to collect COVID-19 community case information as well as enhance transparent risk communication. The findings indicate that all IHEs in our sample have developed a system to collect data on positive, exposed, or symptomatic community cases. The findings indicate that nearly a quarter (23%) of IHE supplied guidelines/forms for reporting COVID-19 diagnosis and exposure, more than half (60%) provided guidelines/forms for reporting COVID-19 diagnosis, exposure, and symptoms, merely 5% of IHE provided guidelines/forms for reporting COVID-19 diagnosis and symptoms, and only 7% provided guidelines for reporting COVID-19 diagnosis; 5% were not found.

5 DISCUSSION AND IMPLICATIONS

The findings indicate that there are various components that IHEs targeted when communicating COVID-19 related information to their campus communities. Based on EERC criteria, IHEs biggest strengths resided in *Sufficient* and *Actionable* categories. The findings indicate that institutions disseminated robust information on their websites on safety recommendations, COVID-19 testing information, travel guidelines, reporting guidelines, face covering guidelines, social distancing guidelines, quarantine and isolation guidelines, vaccine information, campus guidelines, and returning to campus guidelines. Indeed, sufficient risk communications impact the outcome of reducing uncertainty as well as support self-efficacy [40]. To that end, risk communications from IHEs in Texas might have had a positive impact on reducing uncertainty among the campus community. In tandem, IHEs addressed both expert recommendations such as daily screening or self-monitoring, wearing a face covering, social distancing, staying home if sick, practicing good hand hygiene, respiratory etiquette, cleaning guidelines, and limiting gatherings as well as their own recommendations such as reporting positive, exposure, or symptomatic cases, returning to campus guidelines, and completing COVID-19 training. The findings also indicate that institutions of higher education adopted recommendations of establishing a report schedule [21]. This was reflected through the notification from websites of updating COVID-19 case reports daily or weekly. Actionable messaging leads to increased self-efficacy, supporting the community's belief in executing preventative actions. Actionable messaging has long-term impact on behaviour change, which is also associated with "reduced levels of mortality and morbidity…" [40, p. 199]. Clearly, both federal (CDC) and state (TDHS) – and to a lesser degree international – guidelines were sourced by the majority of IHEs; thus, IHEs can be credited with becoming a clearinghouse of accurate and credible risk communications during the pandemic which is quite important given research on college students points to 60% of this population [36] accessing its information from higher education sources. While *Consistent/Timely* messaging was characterized by timely and regular updates provided by 80% of IHEs during initial onset of the pandemic with earliest update marked in January 2020, risk communication messaging waned in Spring of 2021 with only 19% of the IHEs making updates monthly; this might be troublesome due to the ongoing need for vaccination information and updates among college student population at that time. Indeed, college student COVID-19 vaccine hesitancy was reported by researchers [41] in March of 2021 to be at 47.5%. Clearly, the pandemic has been moving beyond the crisis response stage, but because its nature is protracted, consistency and timeliness of ongoing risk communications remain paramount particularly among student populations that have been identified as a sub-group with high infectiousness concurrent with high vaccination hesitancy. Admittedly, IHEs in our study used technology and digital health to collect data such as positive, exposed, or symptomatic cases of COVID-19, to foster self-awareness as well as support community safety by monitoring symptoms. However, it is noteworthy that the overwhelming stress on self-monitoring by students ((nearly three fourths of IHEs) was not paralleled by availability of technological solutions to monitor symptoms (i.e., displayed only by 12% of IHEs). That divergence can be problematic from the self-efficacy standpoint. Finally, results on clarity as well as tailoring of risk communications point to IHEs needing to strengthen how they define terms, review general COVID-19 information, and most importantly identify at risk groups. Issues of identification of at-risk groups in pandemics by college students has been previously reported in literature whereby more than half college students incorrectly identified at risk groups and less than 25% accurately pointed out the signs and symptoms of COVID-19 [34]; our study confirms IHEs efforts in risk communications identifying at-risk groups remain ineffective with potential vulnerability in regard to knowledge of COVID-19 and precautionary data. Similarly, the element of tailored messaging was found to be a weakness of selected IHEs in our research. Only 6% of IHEs in our sample were culturally/linguistically responsive in their risk communications, yet universities exist in social contexts of communities in which they operate to include the microsystem of students' lives and their associated households. Thus, uncovered lack of cultural consideration could impact in turn the surrounding community understanding of the virus overall.

This research is not without limitations; first, not included in analysis were email communications IHEs might have used to communicate with students, faculty, and staff. Moreover, while this study gauges to what extent IHEs leveraged elected criteria of effective risk communications through design of its website platforms, it does not address students' perceptions of such. Even though technologies and digital health solution are displayed, their levels of use could not be ascertained. We propose that ongoing COVID-19 and IHE research focus on college communities' use of specific digital health tools to determine the degree and effectiveness of use and effectiveness. As Institutions of Higher Education (IHE) strive at DRU classification, exploring other channels of communication such as social media and text (SMS) in communicating risks to college students could further expand research on risk communication dissemination and consumption.

ACKNOWLEDGEMENT

The authors would like to thank Mr. ShuQi Zhang for sponsoring graduate student research.

REFERENCES

[1] World Health Organization (WHO), General information on risk communication. 2015. https://www.who.int/risk-communication/background/en/. Accessed on: 1 Jul. 2021.

[2] Bradley, D.T., McFarland, M. & Clark, M., The effectiveness of disaster risk communication: A systematic review of intervention studies. *PLoS Currents*, **6**, 2014. DOI: 10.1371/currents.dis.349062e0db1048bb9fc3a3fa67d8a4f8.

[3] Rubin, H., OECD/IFP project on "Future Global Shocks". Organisation for Economic Co-operation and Development, 2011. https://www.oecd.org/fr/gov/risques/Pandemics.pdf.

[4] Covello, V. & Sandman, P.M., *Risk Communication: Evolution and Revolution*, John Hopkins University Press, 2004. https://www.psandman.com/articles/covello.htm.

[5] Leiss, W., Three phases in the evolution of risk communication practice. *The Annals of the American Academy*, 1966. DOI: 10.1177/0002716296545001009.

[6] Covello, V.T., *Best Practices in Public Health Risk and Crisis Communication*, Taylor and Francis Health Sciences, 2003. DOI: 10.1080/713851971.

[7] Abrams, E.M. & Greenhawt, M., Risk Communication during COVID-19. *The Journal of Allergy and Clinical Immunology*, 2020. DOI: 10.1016/j.jaip.2020.04.

[8] Paulik, L., Keenan, R.E. & Durda, J.L., The case for effective risk communication: Lessons from a global pandemic. *Integrated Environmental Assessment and Management*, **16**(5), pp. 552–554, 2020. DOI: 1010.1002/ieam.4312.

[9] World Health Organization, Risk communication strategy for public health emergencies in the WHO south-east Asia region: 2019–2023, 2019. https://apps.who.int/iris/handle/10665/326853. Accessed on: 1 Jul. 2021.

[10] Glik, D.C., Risk communication for public health emergencies. *Annual Review of Public Health*, **28**, pp. 33–54, 2007. DOI: 10.1146/annurev.publhealth.28.021406.144123.

[11] Janoske, M., Liu, B. & Sheppard, B., Understanding risk communication best practices: A guide for emergency managers and communicators. Report to Human Factors/Behavioral Sciences Division, Science and Technology Directorate, US Department of Homeland Security, START: College Park, MD, 2012.

[12] Holroyd, T., Oloko, O., Salmon, D., Omer, S. & Limaye, R., Communicating recommendations in public health emergencies: The role of public health authorities. *Health Security*, **18**(1), 2020. DOI: 10/1089/hs.2019.0073.

[13] Zhang, L., Li, H. & Chen, K., Effective risk communication for public health emergency: Reflection on the COVID-19 (2019-nCoV) outbreak in Wuhan, China. *Healthcare*, **8**(64), pp. 1–14, 2020. DOI: 10.3390/healthcare8010064.

[14] Paek, H. & Hove, T., *Risk Perceptions and Risk Characteristics, Communication. Online Risk Perceptions and Risk Characteristics*, Oxford Research Encyclopaedia of Communication.

[15] The psychology of risk perception, *Harvard Mental Health Letter*, **27**(12), p. 6, 2011. https://www.health.harvard.edu/newsletter_article/the-psychology-of-risk-perception. Accessed on: 1 Jul. 2021.

[16] Porat, T., Nyrup, R., Calvo, R.A., Paudyal, P. & Ford, E., Public health and risk communication during COVID-19-enhancing psychological needs to promote sustainable behavior change., *Frontiers in Public Health*. https://www.ncbi.nlm.nih.gov/pmc/articles/PMC7652763/. Accessed on: 1 Jul. 2021.

[17] Barua, Z., Barua, S., Aktar, S., Kabir, N. & Li, M., Effects of misinformation on COVID-19 individual responses and recommendations for resilience of disastrous consequences of misinformation. *Progress in Disaster Science*, **8**, pp. 1–9, 2020. DOI: 10.1016/j.pdisas.2020.100119.

[18] Alwashmi, M.F., The use of digital health in the detection and management of COVID-19. *International Journal of Environmental Research and Public Health*, **17**(8), 2020. DOI: 10.3390/ijerph17082906.

[19] Dransch, D., Rotzoll, H. & Poser, K., The contribution of maps to the challenges of risk communication to the public. *International Journal of Digital Earth*, **3**(3), pp. 292–311, 2020. DOI: 10.1080/17538941003774668.

[20] World Health Organization, Communicating risk in public health emergencies. 2018. https://www.who.int/publications/i/item/9789241550208. Accessed on: 11 Jul. 2021.

[21] Hamilton, J.J. & Hopkins, R.S., Using technologies for data collection and management, Centers for Disease Control and Prevention, 2021.

[22] Centers for Disease Control and Prevention (CDC), Considerations for institutions of higher education, 2021. https://www.cdc.gov/coronavirus/2019-ncov/community/colleges-universities/considerations.html. Accessed on: 1 Jul. 2021.

[23] Observations from the ongoing coronavirus (COVID-19) disease outbreak, 2020. https://www.ehtel.eu/covid-19-and-digital-health/public-health-emergencies-and-digital-health.html. Accessed on: 1 Jul. 2021.

[24] Blanchard, B., Canton, L., Cwiak, C., Gess, K., McEntire, D., Newsome, L., Selves, M., Sorchik, E., Stenson, K., Turner III, J., Waugh Jr., W. & West, D., *Principles of Emergency Management Supplement,* 2007.

[25] US Department of Homeland Security, US national infrastructure protection plan education facilities subsector, Homeland Security Digital Library, 2011. https://www.hsdl.org/?view&did=763303. Accessed on: 1 Jul. 2021.

[26] Denham, M.A. & Khemka, A.K., Homeland security and emergency management in institutions of higher education (IHE): Texas case study. *International Journal of Safety and Security Engineering*, pp. 1–15, 2017.

[27] Kapuçu, N. & Khosa, S., *Disaster Resiliency and Culture of Preparedness for University and College Campuses*, SAGE. DOI: 10.1177/0095399712471626.

[28] *The New York Times*, Tracking the coronavirus at U.S. colleges and universities, 2021. https://www.nytimes.com/interactive/2020/us/covid-college-cases-tracker.html. Accessed on: 1 Jul. 2021.

[29] Wrighton, M.S. & Lawrence, S.J., Reopening colleges and universities during the COVID-19 pandemic. *Annals of Internal Medicine*, 2020.

[30] Budd, J., Miller, B.S., Manning, E.M., Lampos, V., Zhuang, M., Edelstein, M. & McKendry, R.A., Digital technologies in the public-health response to COVID-19. *Nature News*, 2020. https://www.nature.com/articles/s41591-020-1011-4. Accessed on: 1 Jul. 2021.

[31] Kapucu, N. & Khosa, S., *Disaster Resiliency and Culture of Preparedness for University and College Campuses,* SAGE, 2013. DOI: 10.1177/0095399712471626.

[32] American College Health Association, Considerations for reopening institutions of higher education for the spring semester 2021. https://www.acha.org/documents/resources/guidelines/ACHA_Considerations_for_Reopening_IHEs_for_Spring_2021.pdf.

[33] Mas, F.S., Jacobson, H.E. & Olivarez, A., Communicating H1N1 risk to college students: A regional cross-sectional survey study. *Journal of Homeland Security and Emergency Management*, 2012. https://www.researchgate.net/publication/271382910_Communicating_H1N1_Risk_to_College_Students_A_Regional_Cross-Sectional_Survey_Study.

[34] Chesser, A., Drassen, A. & Woods, N.K., Assessment of COVID-19 knowledge among university students: Implications for future risk communication strategies. *Health Education and Behavior*. DOI: 10.1177/1090198120931420.

[35] Ding, Y., Du, X., Li, Q., Zhang, M., Zhang, Q., Tan, X. & Liu, Q., Risk perception of coronavirus disease 2019 (COVID-19) and its related factors among college students in China during quarantine. *PLOS One*, 2020. https://journals.plos.org/plosone/article?id=10.1371%2Fjournal.pone.0237626. Accessed on: 1 Jul. 2021.

[36] Callahan, M., Faced with a daily barrage of news, college students find it hard to tell what's real and what's "fake news", 2018. https://news.northeastern.edu/2018/10/16/faced-with-a-daily-barrage-of-news-college-students-find-it-hard-to-tell-whats-real-and-whats-fake-news/. Accessed on: 1 Jul. 2021.

[37] Stanley, M., Communicating risk to student populations. DRU Workshop 2011, Presentations: Disaster Resistant University Workshop: Building Partnerships in Mitigation. Paper 32. https://scholarworks.uno.edu/dru2011/32. Accessed on: 1 Jul. 2021.

[38] Serafini, F. & Reid, S.F., Multimodal content analysis: Expanding analytical approaches to content analysis. *Visual Communications*, pp. 1–27, 2019.

[39] Onwuegbuzie, A.J. & Denham, M.A., *Qualitative Data Analysis Techniques*, Oxford University Press: New York, 2014.

[40] Seeger, M.W., Pechta, L.E., Price, S.M., Lubell, K.M., Rose, D.A., Sapru, S. & Smith, B.J., A conceptual model for evaluating emergency risk communication in public health. *Health Security*, 2018. DOI: 10.1089/hs.2018.0020.

[41] Sharma, M., Davis, R.E. & Wilkerson, A.H., COVID-19 vaccine acceptance among college students: A theory-based analysis. *International Journal of Environmental Research and Public Health*, **18**(9), p. 4617, 2021. DOI: 10.3390/ijerph18094617.

"SEDUCTIVE ATTRACTION" OF "APOCALYPTIC DESIRE": PANIC, TRAUMA NEWS, AND PATHOGEN ARMEGGEDONS

NATALIE D. BAKER
Sam Houston State University, USA

ABSTRACT

This paper uses the cases of Ebola and SARS-CoV-2 in the United States, and examines discourses within specific infectious disease outbreaks as apocalyptic crises in mediated popular culture (i.e. mass news media and social media). In 2014, the US was dealt its first cases of Ebola, a horrific virus not typically occurring outside of West Africa. However, an outbreak never materialized. In 2020, Ebola paled in comparison to the slash and burn global spectacle of SARS-CoV-2. Many approaches in US mass media took a certain doom orientation in different ways as the pandemic unfolded over time. With both Ebola and SARS-CoV-2, there was a politicization of science in tandem with apocalyptic spectacle within media discourses. It is argued, the SARS-CoV-2 pandemic filled a void of apocalyptic longing Ebola left unfulfilled. News coverage of both Ebola and SARS-CoV-2 uses manipulation of the idea public panic is common in approaching disaster situations. The concept of the "three acts of trauma news" is to explore these issues, especially with respect to the social construct of panic. Such spectacles are used in trauma journalism to heighten excitement and fear and are a fundamental element of security theater. Panic discourses underscore a problem of the human propensity for fantasy that undermines responses to disaster.

Keywords: Ebola, SARS-CoV-2, coronavirus, emotions, three acts of trauma news, crisis communication, panic, spectacle, mediated popular culture.

1 INTRODUCTION

"In his 1994 bestselling book *The Hot Zone*, Preston had focused on the most lurid and visually shocking symptoms of Ebola, such as the way that in the last stage of illness patients sometimes 'bled out', leaking blood and haemorrhagic fluids from their eyes, noses and intestines. Even though such symptoms occur in only about half of Ebola cases they helped fix the idea that Ebola was, as Preston put it, 'a perfect parasite…[that] transforms virtually every part of the body into a digested slime of virus particles'. Through the imaginative use of flypapers marked with biosafety hazard warnings, Preston also reinforced the impression of Ebola as a potential biowarfare agent, one that could emerge from the jungles of Africa, or the laboratory of a deranged terrorist, to threaten global health security at any time" [1].

Medical fictions such as the *The Hot Zone* can be hyperbolic in descriptions of infectious disease. However, these stories have immense power. The blending of science with spectacle creates visions of hellish visions of possible futures. From the potential product of a laboratory to the eating of bats by humans, Ebola, and COVID-19, along with other infectious agents are often cast as existential threats to the United States, despite the fact they are not [2]–[4]. This approach has colored national security discourses since the anthrax attacks that occurred almost immediately after 9/11. Biosecurity and its bedfellow, counterterrorism, capitalized on and repackaged the remnants of societal dread of nuclear winter.

Without threats of communism and atomic annihilation, there was little left to reinforce American hegemony built on sociotechnical advancements in an obsession with containment in the post-WWII context [5]. While tragic and demonstrative of some of the worst behaviors of man, the terrors of 9/11 and anthrax in the fall of 2001 were perfect excuses to reinvigorate large-scale societal fears lost as the Cold War fizzled out. A host of threats under these schemes after 2001 manifested to perpetuate burgeoning industries of homeland and biosecurity. In 2014, Ebola came to the United States as part of this cast of seemingly endless existential threats. Ebola is a deadly hemorrhagic virus that spreads via human-to-human transmission through bodily secretions. Even though outbreaks of the disease are almost exclusively confined to Central and West Africa, its arrival to the US *appeared* to have shocked the nation into panic.

In 2014, various mouthpieces of American society such as media, experts, politicians and members of the public manufactured a reality where stateside Ebola became just as bad, if not potentially worse, than the large outbreak that was happening in West Africa [6]. Misinformation abounded. As an example, the WHO stated, "Americans were told that the United States' national security would be threatened by both an Ebola-triggered state collapse in Africa and a large number of panicked people (some of whom also carry the virus) flooding into the United States from Central and South America" [7]. Such statements provoke an imaginary where hordes of the diseased invade the country from areas not even touched by the virus, spilling wells of blood from every orifice, infecting all within their path. Ebola as a "global killer" rhetorics had been advanced in American mass media. This was prior to the virus crossing the Atlantic as press focused attention on the rapid spread of the disease in Africa [8]. Once US patient zero presented with the virus in Dallas on 20 September 2014, the mass outbreaks in West Africa did not seem to matter as much within press circles. Rather, major news networks did not disappoint for heightened drama. Headlines such as, "Ebola, the ISIS of biological agents" [9] or "Ebola victim thought dead comes back to life minutes before cremation" [10] underscored a quick turn to spectacle through a legitimate source of information for people [11]. The virus was portrayed as impending doom by certain factions of US news media, health providers and politicians [12].

Despite media spectacle, there were a total of 10 infected with Ebola [13], most of whom were brought to the States for treatment within a population of 320 million, according to the US Census. Thus, there was virtually no risk of an outbreak. Americans, at least according to mass discourses, were terrified by the prospect of Ebola [14]. Whether this is true or not, is much more complicated [15]–[18]. While people can be legitimately worried about the threat of emerging pathogens, the notion there is mass panic in the face of potential or actual pandemics is not well-established [19]. Books and movies such as *The Hot Zone*, *Outbreak*, and *Contagion* have come to influence views of how infectious disease works in reality, as much, if not more than scientific information, a panic is portrayed as a common feature in these media. The Ebola situation in 2014 foreshadowed a similar, but different approach to SARS-CoV-2 (COVID-19). This new virus presented as potential pandemic in the United States in early 2020. As the prospect for a massive outbreak unfolded, there were similar theatrical approaches in media towards the virus, as there were with Ebola. Whether or not this adversely affected the response to the pandemic in the United States is not the focus of this paper. Rather, it is the spectacle, and the panic discourses evoked in similar ways as Ebola that are central here. There are obvious differences, but in both cases, there was a distinctly American brand of sociocultural catastrophe fulminated by media largely divorced from reality on the ground.

This paper uses Ebola and SARS-CoV-2 in the US to understand how pandemics, or their prospect, reinforces aspects of the security state. It is argued, panic discourses of SARS-CoV-2 filled a void of apocalyptic longing Ebola, unrealized as a pandemic, left unfulfilled. This is common with approaching disaster situations, however why this is the case is less clear beyond the obvious answer of selling stories. The paper uses deconstruction to examine panic discourses as framed through the "three acts of trauma news" as discussed by Ochberg [20].

2 LITERATURE REVIEW

Before SARS-CoV-2, the framing of a possible outbreak of the Ebola virus in key political, public, and mass media discourses foreshadowed the social reaction to COVID-19, although the political climate was different. Science, when it comes to issues elevated to the level of controversial, can become politicized (e.g. climate change or vaccinations) [21], [22]. This makes it challenging to meaningfully deliberate on alternative viewpoints and approaches, as politics can work to dichotomize science – as was seen with both Ebola and COVID-19. In the case of Ebola, the virus became either certain doom or a non-event, where absolutist versions of science and facts, or nationalistic rhetoric [18] were used to justify a variety of responses deemed appropriate under the umbrella of particular political logics. When a disease is "foreign" it exacerbates this issue, and made worse when a need for global health security is built on the fact that an "envisioned outbreak must remain on the near-term horizon" [23, p. 300]. One of the mechanisms by which this is transmitted is mediated popular culture, or in the case here, mass news media.

The news reflects many aspects of a society and its cultures, but it also shapes them. Luhmann [24] argues mass media in general, is a social system created through recursive, self-referential modes of communication rather than objective truths. As media is a public sphere [25]–[27], it is a "communicative space in which ideas, information, discourses and opinions find public representation and elaboration, and thereby help constitute civil societies" [28, p. 20]. News functions as an outlet where "cultural struggles over meaning are carried out" [28, p. 59], and it enables the creation of multiple realities and truths [24]. Media is one way how modern societies reflexively construct illusions of reality. Information constituted through the production and mass consumption of news guised as authoritative knowledge can actually constitute multiple realities, rather than a set of objective truths by which audiences make rational decisions [29]. Audiences, in many ways, believe in the authority of news and make decisions about the world based on knowledge obtained from this medium. As Schäfer [30] suggests, this does not just include civilians, but also government officials. Thus, the influence of media on lived reality can be very problematic given the extent of its trustworthiness.

Problems arise when news acts as conduits of authoritative knowledge. This is because there are often major discrepancies between expert knowledge and how related information are portrayed in such mediums [31], [33]. One issue is reports tend towards over simplification to sensationalist narratives because they sell stories [30]. Publics are then exposed to content that can distort the meaning of relevant and important information [32]. Knowledge creation through news is also informed by preferences and values that work to evoke emotion [33]. Fear is one such emotion mass media uses to frame the concepts like threats and security. This is a major way in which "fear culture" is both created and reconstituted.

A socio-cultural disposition whereby people use anxiety to respond in a world positioned as threatening is known as fear culture. Scholars like Furedi [32] and Altheide [33, p. 26] argue, "fear does not just happen; it is constructed and manipulated by those who seek to benefit". Similarly, Guzelian [37] claims rapid communication of risk through the Internet, for instance, shapes societal fear more than actuality of threat. Thus, the media, as a major societal system, has a role in crafting and re-creating fear [36]. For potential threats, doubt helps shape how people understand risk and informs resultant actions [37]. Uncertainty becomes troubling when "what appears to be certified and objective knowledge is transformed into decisions in a highly selective and often seemingly irrational manner" [38,

p. 25]. Uncertainty and fear are crucial components in the construction of modern risk societies.

Situating fear within the larger context of risk society enacted in the United States is germane to pandemics. Viruses like Ebola and SARS-CoV-2 both threatened the nation and produced the pretence of panic as an "emergency trap" [39] among government actors, medical professionals, and the public. This is a feature of risk societies, which is perhaps best known through the works of Beck [40], [41] and Giddens [42], [43]. Risk society is an omnipresence of low probability, high consequence risks that heighten a collective conscious of anxiety. The risk society is constituted around dual notions of risk and safety [44]. An ominous appearance of pandemics, which existed well before Ebola and COVID-19, is situated within the larger bioterrorism/counterterrorism approach within American culture defined as risk society. Masco [3] explains:

> "Biosecurity policies now anticipate foreign-born infections and mount national media campaigns amplifying the worst-case outcomes for a global contagion, and thereby mobilizing experts and publics in advance of the disease. This anticipatory commitment to disease outbreaks is enabled by advances in the real-time surveillance of infectious disease and scenario-driven calculations of potential risks" (p. 182).

Biosecurity as an extension of counterterror, engages in "scripting futures as well as of responding to contingent fears: it is a world-making project, but one structured at every level by apocalyptic institutions and commitments" [3, p. 146]. Masco also suggests that institutional attempts to manage future terrors make these very terrors seem as if they are real and always on the horizon. Thus, the very institutions intended to "create" security and safety on the societal level are built entirely on the prospect of future crises. The cases of Ebola and SARS-CoV-2 reiterate this point.

Here, Derrida's deconstruction [45] is a way to explore discourses of pandemic in US media. Hansen [46, p. 372] adds, "Deconstruction shows how dichotomies are crucial to the construction of meaning by first identifying the hierarchical relationship between the two concepts of the dichotomy, secondly, reversing the hierarchy, and thirdly, by this procedure, undo the pairing". This "anti-method" method permits an examination of contestation in meanings embedded within master-dichotomies that run through discourses in the public sphere.

The perspective of deconstruction holds any duality must be unpacked so that oppositions (e.g. good versus evil) can provide insight into what the polemics of interest hide. In this case, safety and its associates (e.g. security and rationality) are believed to be superior to opposite states (e.g. danger and panic), as they are dependent on the exclusion and marginalization of their subordinate counters [47]. Danger and unpredictability, as fundamental to threat, crisis, and panic is unacceptable and should be eliminated, much like evil should be vanquished in the advancement of good, for example within a state built on a premise of security in the face of good versus evil [48]. What happens, then, if the polemic is collapsed? If it is asked, what happens if safety, rationality and even panic exist on a continuum, as opposed to a polemic as opposites from each other? This is an antenarrative of safety and all of its synonyms.

The focus is on tearing apart polemics, specifically those induced by the prospect of panic, as this was a focus within discourse surrounding Ebola and SARS-CoV-2. The battle over what the viruses meant to the overarching safety of the nation, and by extension the World, represented in a clash of different versions of reality. Ultimately, what such contestations of meaning suggest about modern American society beyond the obvious in the application of a deconstruction of pandemic master-narratives into antenarratives [49]. The prospect of public panic as an expected, normal response to threat is of interest here.

3 METHODS

A key consideration of this project was how mediated popular culture positioned both Ebola and SARS-CoV-2 and resultant projections within mediated culture. The research was an immersing into news reports and articles, as well as associated commentaries during the SARS-CoV-2 pandemic as well as Ebola situation in 2014. The analysis built on a prior study of Ebola [18] upon which data and findings for were expanded on for COVID-19. Initial considerations were to explore how news media and other relevant actors position pandemics as polemics (e.g. threat or not). As Brummet [50, p. 4] argues, "understanding how social issues may be disguised in discourse is an important goal. If we need to disguise social issues, we should do so with awareness, and we should especially be aware of how those issues may be disguised in the messages we encounter".

Data was extracted from a variety of sources. A major portion of the research consisted of online news (e.g. CNN, FOX News, Al-Jazeera America, among others). Multiple search terms were used, (e.g. Ebola in the US, Ebola panic, SARS-CoV-2 in US, coronavirus panic) in websites to find data (with reader commentary). Data for the Ebola study were constrained from September 2014 to January 2015 as this was the height of the scare. Data for COVID-19 were more expansive for obvious reasons, as the temporal period of the pandemic was much longer. Theoretical sampling was used to guide data collection and analysis [51]. As data analysis occurs simultaneously with collection, subsequent steps were based on emergent analysis.

Analysis occurred until theoretical saturation was achieved, or when no new codes emerged and major themes that informed findings were solidified after iterative rounds of theoretical memoing [51]. There were well over 400 articles alone analyzed for this project. Codes and themes for the two viruses differed slightly in terms of aspects such as how threat and panic was positioned. This stage elicited major themes that represented the predominant ways of constructing pandemics in the public sphere. These were called "threat, crisis, and panic". Because there was not much in between within these representations (i.e. Ebola and COVID-19 were either cast as threatening or not, or there was panic amongst the public, or not) the postmodern critique [52] was appropriate given what came out of the initial analysis were categories of distinct dichotomies.

4 THE DECONSTRUCTIVE PROCESS

The deconstructive process is not black and white (no pun intended). However, as a deconstruction focuses on texts, and this paper was a reading of mediated texts focused on pandemic threat, relevant data in this case, were mediated forms of information [53]–[55]. Thus, this required the analysis to ascertain central themes of textual outputs. The main theme of texts centered around the presentation of both Ebola and COVID-19 as legitimate threats to the United States and its overall quality and pre-existent way of life, by and large. Of course, the extent to which each virus was characterized as threatening was different, however both were positioned as potentially existential. This theme is crucial here.

The next step in a deconstructive form to identify binary oppositions within texts. As one could imagine, there were an abundance; too many to recount here. Prior analysis of Ebola data indicated relevant dichotomies also discussed with COVID-19 (e.g. foreign disease dangerous/homeland safe, broken down further, foreign equals danger, home equals safe). While these polemics are fascinating, the focus here is on panic discourse, and what its binary opposites indicate about American society. If there were concrete steps to deconstruction, the last would be to provide a critique of a body of texts based on what it attempts to assert as taken for granted and the meaning that is being asserted. The purpose is to create new meaning in the destruction of the original polemic. Panic discourse for the two viruses was examined through a lens of the acts of trauma news [20]. The reason for its use was given in the following questions:

> "Why do we, collectively, overdose on some elements of trauma and fail to digest others? Is there a better way to communicate facts and feelings after a horrifying event?" [20, p. 1]

The examination of panic discourses for an imagined pandemic, and a very real one that vastly affected everyday life, is undertaken to get at those questions still left unaddressed. An illustration of a deconstruction along the lines of the three acts of trauma news is undertaken below.

4.1 Act One: If it bleeds it leads

The first act is demonstrated in the following quote:

> "Competition for audience attention is part of media business, and we the audience attend to violence. We always have, always will, and MUST. Act One of any factual story about crime, cruelty, violence or trauma will create arousal, interest and paradoxical emotion. The only time a trauma story is pure horror is when it is about us" [20, p. 3].

Press over Ebola and COVID-19 by and large, mostly fall within the realm of Act One. This point is important. With respect to panic and Ebola, the role of mediated culture was to shame the public for its apparent panic over the virus, and then argue the fact the public were panicking. The polemic here was thus "panic/composure (or antonyms of panic)". The public were simultaneously panicked, but also non-panicked in the eyes of the media. Therefore, both states of public behavior happened at once, creating two versions of reality. A panicked public and a not-panicked public. One quote from an expert as interviewed in CBS News demonstrates [56]:

> "Public health professor Andrew Noymer studies infectious diseases at University of California, Irvine. He says people do not need to be afraid of Ebola in the U.S. 'I would say the panic is harder to contain than the spread of the disease itself', he said. 'People are focusing on that it's a scary disease from far away, and they see scary images of people caring for sick people with protective gear, and it looks frightening'".

To highlight panic, news reports would discuss actions undertaken to prevent disease exposure, such as stories of unnecessary school closures, people frightened about contracting the virus, and government actions (e.g. quarantines or calls for border closures). Diametrically opposed, was a denial of panic, the argument being it was not a reality, that more Americans, for example, were afraid of "witches" than the virus [57]. One case was the derision CNN and Joe Biden received from critics in their reference to Ebola as the "ISIS of disease" [58]. The quote below demonstrates:

> "Is Ebola the ISIS of biological agents? Is Ebola the Boko Haram of AIDS? Is Ebola the al-Shabaab of dengue fever? Some say Ebola is the Milosevic of West Nile virus. Others say Ebola is the Ku Klux Klan of paper cuts. It's obvious that Ebola is the MH370 of MH17. But at some point, the question must be asked whether Ebola isn't also the Narendra Modi of sleeping sickness. And I don't mean to offend anyone's sensitivities, but there's more and more reason to believe that Ebola is the Sani Abacha of having some trouble peeing".

References to panic saturated discourses surrounding the public reaction to the threat of Ebola – where not in between – or more nuanced discussions of human behavior in the context of this situation were explored. Panic or not, as the story, was in itself, a media creation. And this dichotomy works to create a version of reality for people, when the reality of lived life is probably nothing like what is portrayed in media.

A similar situation occurred with the SARS-CoV-2 virus, however in different ways given how vastly different this was from Ebola. Panic discourse was delayed as there was much uncertainty about how the virus would affect the US. When it became clear SARS-CoV-2 would indeed make a significant impact on the nation, there were two ways panic discussions in media came about. The first was toilet paper panic buying that occurred in the early stages of initial lockdown in the US. What happened though, was people purchased emergency supplies in advance of an event they perceived as threatening – a reasonable response to an impending disaster and one in line with legitimate preparedness actions – not panic. Instead, articles available on the Internet and reported in news reflected the following sentiments:

> "Don't panic. It's what our mothers and neighbors are telling us, newspaper headlines are announcing, and public health officials are urging. It's excellent advice in any situation, and is especially apt as we learn to adjust to the collective changes we're making to slow the spread of COVID-19. However, if you look around at some of the ways many are responding, including rushing to local supermarkets and big box stores to stock up on food, cleaning products, and paper products, we see behavior that looks an awful lot like panic" [59].

These sentiments saturated reportage of the SARS-CoV-2 virus, even though political orientation of an outlet did reflect how seriously the virus was taken to some extent. It must be underscored, purchasing supplies in advance of a foreseeable disaster is not panic.

Visual imagery did not help with either Ebola or COVID-19, where the spectacle of viral apocalypse was made more possible through a focus on biohazard suits, patients on ventilators (for COVID-19), body bags, microscopic images of the virus, all heightened the trauma of both situations, and in fact spectacularized both to their detriment. This is all part of the leading story of trauma news – Act One. The macabre focus on crafting chaos and destruction as experiential reality. This is the fundamental purpose of Act One.

4.2 Act Two: The stages of trauma and recovery

> "The later scenes of Act Two have a sadder, wiser tone. A person reflects upon loss but is thankful for friends. Spiritual themes emerge. The survivor considers the meaning of life. People note their attachments and their desire to help others" [20, p. 3].

There were no examples of Act Two for Ebola. It disappeared once it was evident there was no outbreak or apocalypse in the States. With COVID-19, there was a slight hint at Act Two. These are the "lessons learned" pieces, the reflecting on those lost, and the more in-depth examinations of how the pandemic has changed the course of life. Panic disappears in Act Two.

While Ebola was short-lived, COVID-19 was not. Thus, Act One was running simultaneous to Act Two and ate up much more of the coverage than the latter. There is an attempt by media to keep going the doom and gloom of the first act to supplant the moral lessons and melancholy of the second. Namely, because what bleeds does indeed lead. Panic becomes constantly replaced with the numbers of the dead. Not so much attention to those who recover. There is no need to focus on public unrest when horror is so easy to come by.

4.3 Act Three: Abandon hope all ye who enter here

> "In Act Three there is no meaning, no moral, no transcendent truth" [20, p. 4].

The lesson of Act Three with respect to both Ebola and COVID-19 in terms of media had dissipated. Ochberg [20, p. 4] again elaborates

> "I believe there are Act Threes, but they are rare. Act Three is an idea rather than a set of scenes that follow Acts One and Two. It is the idea in the mind of a witness (a direct witness or one who 'witnessed' the recorded word) that a destructive act was so extreme, so complete, that no light escaped. Nothing good came of it, or came out of it. It made no sense and left no helpful legacy".

There was a sense of nihilism reflected in media stories more outside of the regular 24/7 news cycle for both Ebola and COVID-19 [60]. Again, the panic element is decentralized and/or non-existent. Nihilism in COVID-19 emerges as a direct contrast to panic. Visceral fear is filled with action, whereas the nihilism of something purported to be an apocalyptic experience of death (which it was not) was relegated to inaction – suffering at home, in lock-down, isolated from everyone, and poor mental health outcomes. Not a soul wants to read about Act Three. It is a futile existence, a shadow of apocalypse of inaction – perhaps just as much created by the media as its diametric – panic was as well.

5 CONCLUSIONS

> "We have been systematically educated in a way that makes violence enjoyable" [61].

The modern US state, in many ways, has a bit of a flair towards the oppressive. And similarly, also needs crises and horror to construct its identity to continue to oppress [62], [63]. Thus, crises are both generated by, and generative of, a state, which is also an imaginative set of structures [62], [64] and spectacles. We see this also with infectious diseases such as HIV/AIDS [63], [65]–[67]. Safety expected of the developed world presents a crisis of meaning where the dominant culture is "seduced" [46] in part, by an adrenaline of fear and existential ends. Panic discourse is crucial here.

Media treatment of both viruses provided a glimpse into an American brand of sociocultural catastrophe. The SARS-CoV-2 pandemic would come to fill the void of apocalyptic longing, or a morbid human desire for the end of all things [68] Ebola left this unfulfilled. News coverage of both Ebola and COVID-19 uses manipulation of the idea public panic is common in approaching disaster situations. This is highly unsupported in empirical research. It is argued here, panic discourses are used to heighten excitement and fear in support of the production of security theater [69]. This is a collection of institutions and practices that enact security for audiences – and it is only a performance – rather than meaningful creation of societal safety (e.g. the TSA or sanitation practices during COVID-19 [70]). This played out in dramatic fashion as the US response to the coronavirus deteriorated and the pandemic became politicized to its detriment. There was an explosion of polemics within the positioning of the virus as threat or not, or similarly crises or not, which lends to discussions of people enacting panic or not. Panic discourses, in the prospect and/or experience of pandemic, were used to heighten excitement and fear and completely constrained as a first act of trauma news. Panic discourses underscore a problem of the human propensity for fantasy that undermines just responses to disaster.

The paper does not take a stance on whether or not Ebola or COVID-19 threats were "real or not" as it fits within the postmodernist tradition. Here there is an inherent "skepticism about the human capacity to grasp or define what is ultimately and totally real" [47, p. 13]. Deconstruction helps reveal the social consequences of polemics in a meaningful way to show what is suppressed and marginalized in master narratives. News of both Ebola and COVID-19 both used manipulation of the idea public panic is common in approaching disaster situations as existential threats. Crisis can both foreshadow intense disruption or indicate its arrival. As kin, threat and crisis are built on urgency that safety is no longer guaranteed. Panic, as the presumed default human behavior is assumed a natural response to threat and/or crisis. All are incestuous descriptions of chaos; their implied opposites are safety, predictability, peacefulness, and security, or more simply put, order. Order is the supposed desired state.

This antenarrative [49] to the assumed contiguousness of social order as defined by safety, is that threat and danger are fundamental characteristics of the natural state of human social interactions and their resultant worlds. An antenarrative produced within a deconstruction, is a story about representations "told without the proper plot sequence and mediated coherence, as 'they are too unconstructed and fragmented to be captured by retrospective sensemaking'" [49, p. 2]. Thus, the normality of threat, crisis, and subsequent panic are the antenarrative within this reversal; here safety is no superior to chaos.

Given this argument, deconstruction does not suggest it is one end of a polemic that is an accurate representation of reality, or that an antenarrative is the one truth, or that there even exists such thing as a one truth in the social world. Rather, there is insight here into "otherwise fragmented and multi-layered experiences of desire" [49, p. 2] that deconstruction reveals about associates of safety.

Public panic about both Ebola and COVID-19 was a myth imagined in media discourses. People tend to act reasonably in disaster situations, as is the same in normal life. Mostly, the concept of crisis lives within a collective imagination of harms to the future, but reinforces a need for security in the face of harm. Or as Ochberg [20, p. 4] profoundly states:

> "When we, the audience, tire of the formulaic, repetitive treatment of trauma in the news, the formula will change. But we may not tire. We may be destined by biology to feed endlessly on other peoples' horror, distorting our perception and understanding of reality".

The labor of both sanitizing and pushing back death is fruitless for certain – it comes anyway for, "Normal life is the grim reaper who's almost certainly going to get you sooner or later" [71]. It comes as a passing in one's sleep after a life toiling away at meaningless work. Or, for the unlucky, a sluggish and unrepentant torture. However, it is pretty much assured we will not die panicking in some version of an apocalypse as much as we scream about it in the unremitting supply of mediated popular culture.

ACKNOWLEDGEMENTS

I would like to thank Dr. Spiro Samonas for his prior work on the Ebola project. I would also like to thank Dr Frank Ochberg for being an unlikely friend.

REFERENCES

[1] Honigsbaum, M., Between securitisation and neglect: Managing Ebola at the borders of global health. *Medical History*, **61**(2), pp. 270–294, 2007. DOI: 10.1017/mdh.2017.6.

[2] Bostrom, N., Existential risk prevention as global priority, *Global Policy*, **4**(1), pp. 15–31, 2013.

[3] Masco, J., *The Theater of Operations: National Security Affect from the Cold War to the War on Terror*, Duke University Press: Durham, 2014.

[4] Millet, P. & Snyder-Beattie, A., Human agency global catastrophic biorisks, *Health Security*, **15**(4), 2017. DOI: 10.1089/hs.2017.0044.

[5] Ambrose, S.E. & Brinkley, D.G., *Rise to Globalism: American Foreign Policy Since 1938*, Penguin Books, 1997.

[6] Abeysinghe, S., Ebola at the borders: Newspaper representations and the politics of border control. *Third World Quarterly,* **37**(3), pp. 452–467, 2016. DOI: 10.1080/01436597.2015.1111753.

[7] Council on Foreign Relations, The downside of securitizing the Ebola virus. Expert Brief, Yanzhong Huang. https://www.cfr.org/expert-brief/downside-securitizing-ebola-virus. Accessed on: 23 Oct. 2015.

[8] Rothstein, M.A., The moral challenge of Ebola. *American Journal of Public Health,* **105**(1), pp. 6–8, 2015. DOI: 10.2105/AJPH.2014.302413.

[9] Hooten, C., "The ISIS of biological agents": CNN is asking stupid Ebola questions. *The Independent*, 2014. https://www.independent.co.uk/news/world/americas/isis-biological-agents-cnn-asking-stupid-ebola-questions-9779584.html. Accessed on: 7 Oct. 2014.

[10] Dovey, D., Ebola victim thought dead comes back to life minutes before cremation. *Newsweek*, 2014. https://www.newsweek.com/ebola-zombie-comes-back-life-minutes-cremation-275149. Accessed on: 15 Oct. 2014.

[11] Altheide, D.L., The news media, the problem frame, and the production of fear. *The Sociological Quarterly*, pp. 647–668, 1997.

[12] Ashkenas, J. et al., How many Ebola patients have been treated outside of Africa? *New York Times*, 2015.

[13] Centers for Disease Control, Cases of Ebola diagnosed in the United States. http://www.cdc.gov/vhf/ebola/outbreaks/2014-west-africa/united-states-imported-case.html. Accessed on: 18 Oct. 2014.

[14] NBC News, Why are Americans so scared of Ebola? https://www.nbcnews.com/storyline/ebola-virus-outbreak/why-are-americans-so-scared-ebola-n188806. Accessed on: 18 Oct. 2014.

[15] McInnes, C. & Lee, K., Health, security, and foreign policy. *Review of International Studies*, **32**(1), pp. 5–23, 2006.

[16] Elbe, S., Haggling over viruses: The downside risks of securitizing infectious disease. *Health Policy and Planning*, **25**, pp. 476–485, 2010.

[17] Wagner-Egger, P., Bangerter, A., Gilles, I., Green, E., Rigaud, D., Krings, F., Staerklé, C. & Clémence, A., Lay perceptions of collectives at the outbreak of the H1N1 epidemic: Heroes, villains and victims. *Public Understanding of Science*, **20**(4), pp. 461–476, 2011. DOI: 10.1177/0963662510393605.

[18] Baker, N., Samonas, S. & Artello, K., (Not) welcome to the US: Hyper-Ebola and the crisis of misinformation. *Conference Proceedings of the International Association for Information Systems for Crisis Response and Management (ISCRAM)*, 2015.

[19] Baker, N.D., Zombie experts and anarchy imaginaries: Fantasies of "crises to be" in climate change futures. *Journal of Strategic Security*, **13**(4), pp. 141–155, 2020. DOI: 10.5038/1944-0472.13.4.1867.

[20] Ochberg, F., *Three Acts of Trauma News, PTSD Resources for Survivors and Caregivers*, 1999. https://www.giftfromwithin.org/pdf/threeact.pdf.

[21] Gauchat, G., Politicization of science in the public sphere: A study of public trust in the United States, 1974 to 2010. *American Sociological Review*, **77**(2), pp. 167–187, 2012. DOI: 10.1177/0003122412438225.

[22] Schmid-Petri, H., Politicization of science: How climate change skeptics use experts and scientific evidence in their online communication. *Climatic Change*, **145**, pp. 523–537, 2017. DOI: 10.1007/s10584-017-2112-z.

[23] Lakoff, A., Global health security and the pathogenic imaginary. *Dreamscapes of Modernity: Sociotechnical Imaginaries and the Fabrication of Power*, eds S. Jasanoff & S.H. Kim, University of Chicago Press: Chicago, pp. 301–320, 2015.

[24] Luhmann, N., *Social Systems*, Stanford University Press: Stanford, 1995.

[25] Habermas, J., Lennox, S. & Lennox, F., The public sphere: An encyclopedia article, *New German Critique*, **3**, pp. 49–55, 1974.

[26] Habermas, J., *The Structural Transformation of the Public Sphere: An Inquiry into a Category of Bourgeois Society*, MIT Press: Cambridge, MA, 1991.

[27] Habermas, J., *Between Facts and Norms*, MIT Press: Cambridge, MA, 1998.

[28] Cottle, S., *Mediatized Conflict*, McGraw-Hill, 2006.

[29] Weingart, P., The lure of the mass media and its repercussions on science. *The Sciences' Media Connection: Public Communication and its Repercussions,* eds S. Rodder, M. Franzen & P. Weingart, Springer Science and Business Media: Berlin, pp. 17–34, 2012.

[30] Schäfer, M.S., Sources, characteristics and effects of mass media communication on science: A review of the literature, current trends and areas for future research. *Sociology Compass*, **5**(6), pp. 399–412, 2011.

[31] Scheufele, D.A., Communicating science in social settings. *Proceedings of the National Academy of Sciences*, **110**(Supplement 3), pp. 14040–14047, 2013.

[32] Furedi, F., The only thing we have to fear is the "culture of fear" itself: How human thought and action are being stifled by a regime of uncertainty. *Spiked*, 4 Apr. 2007. http://www.spiked-online.com/newsite/article/3053#.ViPW7tbb-hI. Accessed on: 18 Oct. 2015.

[33] Altheide, D.L., *Creating Fear: News, and the Construction of Crisis*, Aldine De Gruyter: New York, 2002.

[34] Donohue, G.A., Tichenor, P.J. & Olien, C.N., Mass media functions, knowledge and social control. *Journalism Quarterly*, **50**(4), pp. 652–659, 1973. DOI: 10.1177/107769907305000404.

[35] Dunwoody, S., United States: Focus on the audience. *Journalism, Science and Society: Science Communication Between News and Public Relations*, eds M.W. Bauer & M. Bucchi, Routledge: New York, pp. 241–246, 2008.

[36] Vainio, A. & Paloniemi, R., Does belief matter in climate change action? *Public Understanding of Science*, **22**(4), pp. 382–393, 2011. DOI: 10.1177/0963662511410268.
[37] Guzelian, C.P., Liability and fear. *Ohio State Law Journal*, **65**(4), Stanford Public Law Working Paper Number, pp. 713–851, 2004.
[38] Slovic, P., *The Perception of Risk*, Routledge: London, 2000.
[39] Hanrieder, T. & Kreuder-Sonnen, C., WHO decides on the exception? Securitization and emergency governance in global health. *Security Dialogue,* **45**(4), pp. 331–348, 2014.
[40] Beck, U., *Risk Society: Towards a New Modernity*, SAGE: Thousand Oaks.
[41] Beck, U., *World Risk Society*, Polity Press: Cambridge, UK, 1999.
[42] Giddens, A., *The Constitution of Society: Outline of a Theory of Structuration*, University of California Press: Berkeley, CA, 1984.
[43] Giddens, A., *Modernity and Self-Identity: Self, Society in the Late Modern Age*, Polity Press: Cambridge, UK, 1991.
[44] Ekberg, M., The parameters of the risk society: A review and exploration. *Current Sociology*, **55**, pp. 343–366, 2007.
[45] Derrida, J., *Positions*, The University of Chicago Press: Chicago, 1981.
[46] Hansen, L., A case for seduction? Evaluating the poststructuralist conceptualization of security. *Cooperation and Conflict*, **32**(4), pp. 369–397, 1997.
[47] Farmer, D. & Derrida, J., Deconstruction and public administration. *American Behavioral Scientist*, **41**(1), pp. 12–27, 1997. DOI: 10.1177/0002764297041001003.
[48] Baker, N.D. & Jones, N., A snake who eats the devil's tail: The recursivity of good and evil in the security state. *Media, War and Conflict*, **13**(4), pp. 468–486, 2020. DOI: 10.1177/1750635219846021.
[49] Boje, D.M., *Narrative Methods for Organizational and Communication Research*, SAGE: London, 2001.
[50] Brummet, B., *Uncovering Hidden Rhetorics: Social Issues in Disguise*, SAGE: Los Angeles, 2008.
[51] Glaser, B.G. &Strauss, A., *The Discovery of Grounded Theory: Strategies for Qualitative Research*, Transaction Publishers: New Brunswick, 2012.
[52] Derrida, J., The time is out of joint. *Deconstruction is/in America: A New Sense of the Political*, ed. A. Haverkamp, New York University Press: New York, 1995.
[53] Royle, N., *Christopher Norris, Deconstruction: Theory and Practice*, Methuen: London and New York, pp. 170–180, 1982.
[54] Culler, J., *On Deconstruction: Theory and Criticism after Structuralism*, Cornell University Press: Ithaca, 1982.
[55] Sim, S., *The Routledge Companion to Postmodernism*, 3rd ed., Routledge: London, 2011.
[56] Evans, C., Ebola panic spreading much faster than disease in U.S. CBS Evening News, 2014. https://www.cbsnews.com/news/ebola-panic-in-us-spreading-much-faster-than-disease/.
[57] Noack, R., What the world thinks of the American Ebola panic. *The Washington Post*, 2014. https://www.washingtonpost.com/news/worldviews/wp/2014/10/22/what-the-world-thinks-of-the-american-ebola-panic/. Accessed on: 18 Oct. 2014.
[58] Killough, A., Joe Biden explains how Ebola is like ISIS, *CNN*, 2014. https://www.cnn.com/2014/10/03/politics/biden-ebola-isis/index.html. Accessed on: 22 Oct. 2014.

[59] Sharp.com, Conquer your panic about Covid, 2020. https://www.sharp.com/health-news/conquer-your-panic-about-covid-19.cfm. Accessed on: 8 Jun. 2021.

[60] Kurmelovs, R., Lonely repetition and growing nihilism how the pandemic impacts your mental health, *The Guardian*, 2020, https://www.theguardian.com/australia-news/2020/dec/13/lonely-repetition-and-growing-nihilism-how-the-pandemic-impacts-youth-mental-health. Accessed on: 8 Jun. 2021.

[61] Rosenberg, M., How to express anger compassionately. *The Bible in Transmission,* ed. W. Wink, 1999. https://www2.goshen.edu/~joannab/women/wink99.pdf.

[62] Bialasiewicz, L., Campbell, D., Elden, S., Graham, S., Jeffrey, A. & Williams, A.J., Performing security: The imaginative geographies of current US strategy. *Political Geography,* **26**, pp. 405–422, 2007.

[63] Campbell, D., *Writing Security: United States Foreign Policy and Politics of Identity,* Manchester University Press: Manchester, 1992.

[64] Castoriadis, C., *The Imaginary Institution of Society*, MIT Press: Cambridge, MA, 1987.

[65] Campbell, D., Time is broken: The return of the past in response to September 11. *Theory and Event*, **5**(4), 2001. DOI: 10.1353/tae.2001.0032.

[66] Elbe, S., Should HIV/AIDS be securitized? The ethical dilemmas of linking HIV/AIDS and security. *International Studies Quarterly*, **50**(1), pp. 119–144, 2006.

[67] McInnes, C., Crisis! What crisis? Global health and the 2014–15 West African Ebola outbreak. *Third World Quarterly*, 2016.

[68] Dellamora, R., *Postmodern Apocalypse: Theory and Cultural Practices at the End*, University of Pennsylvania Press: Philadelphia, 1985.

[69] Schneier, B., *Beyond Fear: Thinking Sensibly about Security in an Uncertain World*, Copernicus Books: New York, 2003.

[70] Thompson, D., Hygiene theater is a huge waste of time. *The Atlantic*, 2020. https://www.theatlantic.com/ideas/archive/2020/07/scourge-hygiene-theater/614599/. Accessed on: 12 Mar. 2021.

[71] Rotella, C., This is not the zombie apocalypse. *The Boston Globe*, 2014. https://www.bostonglobe.com/opinion/2014/10/24/the-ebola-panic-and-zombies/r8wo8gqEjeh5VFcRpKOvLL/story.html. Accessed on: 11 Sep. 2017.

SECTION 2
SOCIO-ECONOMIC ISSUES

ROLE OF AWARENESS TO PREVENT PERSONAL DISASTERS: REDUCING THE RISKS OF FALLING FOR PHISHING BY STRENGTHENING USER AWARENESS

BETTINA ISER & ROMAN BRANDTWEINER
Vienna University of Economics and Business, Austria

ABSTRACT

Phishing still represents a main security threat in the digital world. Attackers primarily by means of email try to gain access to user's sensitive personal credentials. By using these credentials, attackers cause disasters on the individual level by inflicting, for example, severe economic losses or reputational damage due to identity theft. Awareness is one important factor to increase resilience. This paper based on recent literature first gives a general overview on social engineering as mean for phishing and then evaluates how awareness as preventive measure is considered effective in the selected literature. Information on phishing is one measure to raise awareness, others are trainings and phishing test campaigns to evaluate risk exposure and increase awareness. With regards on information sharing a case study, focusing on portals and homepages of selected Austrian financial institutions was conducted. In this case, study we emphasis on content and eventual differences in the presentation of information concerning prevention and mitigation.

Keywords: disaster prevention and mitigation, social engineering, phishing, risk awareness, security management.

1 INTRODUCTION

Phishing is a social engineering attack used by malicious threat agents in order to gain confidential logon information from users [1] either in companies as administrators or end users as bank users to commit identity theft and misuse the data for further attacks on company data or to commit fraud. Awareness of users is considered one measure to reduce the risk from disclosing sensible data to attackers [2]. In today's private and business life email is an important mean of digital communication and therefore an interesting attack vector for phishers directly targeting the end users.

Phishing mails are regularly sent to random target populations and leverage on victims being found within. The classic approach how phishers try to lure their victims to click on a malicious link is raising concerns/fear on user-side preventing them from being cautious, e.g. the text of the message refers to their account being blocked [3].

Starting from a general introduction on social engineering and phishing in general the paper will address awareness information on selected Austrian banking websites, search for available information regarding a trend about the phishing threat and based on literature evaluate the contribution of awareness to phishing prevention.

As visible on the phishing alerts on the selected internet portals and by Austrian cyber security organizations like Cert.at (Cyber Emergency Response Team Austria) phishing is a persistent threat to users in their digital communication.

2 RESEARCH QUESTION AND METHODOLOGY

Phishing represents a threat to users and companies concerning financial and reputational losses. Customers that in the course of a successful phishing campaign lost money not only consider the financial loss but may also loose the trust in their (financial) service providers [4].

WIT Transactions on The Built Environment, Vol 207, © 2021 WIT Press
www.witpress.com, ISSN 1743-3509 (on-line)
doi:10.2495/DMAN210061

2.1 Research question

Following the persistent threat of experiencing personal disasters triggered by phishing this paper will address the following research question: How can user awareness as effective measure of risk mitigation lead to avoidance or at least to the reduction of losses from phishing hazards?

2.2 Methodology

The research was executed as desk research, therefore as methodology a systematic literature review [5] has been applied accompanied by statistics and information shared by law enforcement authorities with a focus on the Austrian market. Main literature sources will be scientific articles published not before 2010 and accessible via Web of Science and Google Scholar. Additionally, information from cases especially the awareness information available on the homepages of selected banks and other financial institutions in Austria will be evaluated and compared.

3 PHISHING AS A FORM OF SOCIAL ENGINEERING

Hong [1] states that the best security concepts applied on technical level are limited if "the person behind the keyboard falls for a phish". Phishing is considered the most widely spread of social engineering and addresses emails that either are sent in the forms of spam to a large group of recipients or could also be sent to target recipient groups [6].

Social engineering following Hadnagy [6] is defined as action that motivates somebody to perform an action that is to be considered as resulting in a disadvantage. The social engineers try to create an atmosphere of trust and relationship that motivates the victim to click, e.g. on a malicious link. Alternatively the social engineer is using curiosity, threat or authority to get the victim follow the intended link and to insert personal credential in, for example, the faked website [6].

The attacker usually claims to be someone else; such mails could claim to come from a bank, social media platforms or various other platforms requiring user credentials to logon. Sending those mails can be done at very low cost from the attackers [7]. Clicking on the phishing link directs the victim to a manipulated website where usually sensitive credentials as password and user-id are to be inserted. Having those collected, the attacker – often without being recognized by the victim – has access to the different services and can continue the fraudulent activities.

Sometimes the attackers create fear with the victim claiming to have recognized malicious activity on one of the payment accounts and asking for a logon in order to proceed with clean-up of the comprised accounts [3].

As Fig. 1 shows apart from fear there are other motivators that can attract users to follow malicious links in phishing mails [3]. Once having clicked on a link the victim either is asked to insert personal credentials or (unconsciously) downloads malware that then is used to gain access to user credentials and further misuse them. These motivators can repeatedly be found in research literature [8].

Hong [1] describes the anatomy of a phishing attack following a three phases approach. The first phase is sending out phishing mails to the targeted users. Here motivation factors as described in Fig. 1 are used to get the victims fall into it. Those phishing mails in many cases direct the users to phishing sites, faked websites, claiming to be another target website users there are requested to insert their credentials assuming they are on their target websites (e.g.

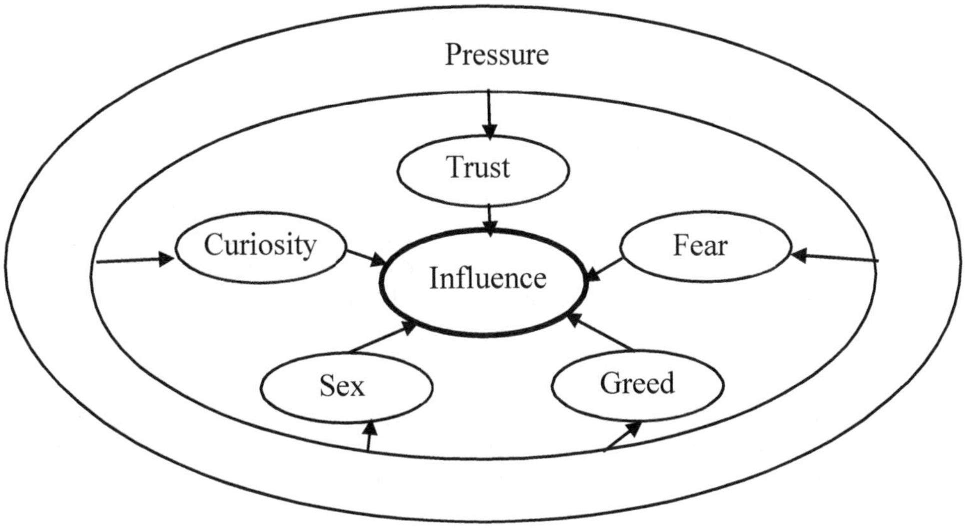

Figure 1: Motivators in social engineering [3].

their banking webpage) while instead are on a phishing site [4]. Meanwhile there exist services offered to institutions monitoring faked sites and supporting takedowns to mitigate the risk, what leads to attackers also leveraging on more enhanced and sophisticated technology. Last step of the attack is following this definition the monetarization of the stolen information, either by the attackers themselves or by the information being sold on illegal marketplaces to other criminals [1].

4 CURRENT SITUATION IN AUSTRIA

The Austrian CERT (Cyber Emergency Response Team) in its yearly report 2019 [9] identifies in the statistics of malicious web trends phishing and phishing sites as continuous threat.

Online services like Watchlist Internet [10] are additionally monitoring the phishing situation in Austria and providing updated information on their website.

In Austria, the Cyber Security Center of the Austrian Ministry of the Interior is permanently monitoring the situation of overall cybersecurity in Austria. In the respective monthly reports, the so-called *OpKoord (Operative Koordinierungsstruktur) Lagebericht* [11], a continuous trend of phishing is visible.

Throughout October, a series of phishing/SPAM waves is reported with faked postal or package deliveries – via mail as via messenger services like WhatsApp. Aim of those campaigns is to get logon credentials like payment information, e.g. credit or other payment data.

In the report of November a constant trend of phishing mails, e.g. on bank institutions or post are reported. In particular, as of end of October also a phishing campaign against users of *Finanz Online* was launched. *Finanz Online* is the official portal of the Austrian Ministry of Finance where Austrian citizens can submit their tax declarations. This phishing campaign was remarking a tax reimbursement of 1.850 Euro and linking to a phishing site where *Finanz Online* credentials were collected.

In December, the reports state an increasing number of phishing shops and phishing mails being sent around which following the report is regularly seen around Christmas time.

International research shows the continuous worldwide trend of phishing targeting end users and customers equally as employees in companies [12]. In addition, other research indicates that phishing will remain an important threat vector [13].

The global consortium "Anti Phishing Workgroup" (APWG) in a joint cooperation with users and companies providing security solutions or being in scope of phishing attacks monitor the global evolution of phishing. In its report on the third quarter of 2020, it considers a continuously high number of phishing sites. Additionally, it states that phishing sites increasingly (up to 2/3) are using SSL (secure socket layer) encryption (visible as "https://" at the beginning of URLs) to gain trust of the users [14].

5 RISK MITIGATION

Beißel [3] defines the focus of awareness to apply a set of tools and methods in order to increase the consciousness of users about malicious activities and to facilitate a desired behaviour in case of, for example a phishing attack. He defines three aspects on security-relevant behaviour that could be the desired or the not desired one: behaviour in line with the policies/rules, risk behaviour and behaviour in problematic situations.

Awareness measures could be distributed on various levels – could be print media, electronic media or also audio/video measures.

5.1 Approaches and measures to create awareness among users

Arachchilage and Love [13] in their study evaluated the self-efficacy of users to thwart from falling into phishing and see conceptual knowledge (know why) and procedural knowledge (know how) as positively influencing the threat avoidance and threat behaviour of users. Their study was considering younger users aged between 18 und 25. An age group that is often reported to be at higher risk to fall into phishing, eventually also due to minor experience [1].

With regards to awareness trainings, targeting training to users not yet sufficiently aware of the threat and avoiding "overtraining" is another aspect to be considered. The latter could result in non-fraudulent mails being misinterpreted as fraudulent [15].

Arachchilage and Love [13] identify a lack of user knowledge as a main reason for a successful attack. They propose a mix of information including even games to improve users' ability to thwart phishing attacks.

Research demonstrates that awareness among users is positively affecting the resilience and on detection capabilities of users. With increasing connectivity and use of smartphones there is identified a strong need to train people especially about phishing [16].

Ebot [8] introduces an additional aspect concerning phishing by identifying different stages that users might be in when clicking on phishing mails. This could be a potential explanation why, for example, in some evaluated literature awareness leads to less clicks and in others the perceived awareness leads users to be less cautious and click on malicious mails. Understanding the different stages and their characteristics would lead to a fine-tuned training and awareness approach allowing designing awareness based on the users current risk profile. In the study by Ebot [8], users are classified in three different stages when clicking on an email. Stage 1 refers to the naïve users, not experienced with IT and little knowledge. At this stage, the priority in trainings should be given to raising understanding of phishing and all and providing a solid background. Stage 2 includes users with already basic understanding on phishing that went through some training already. They would click on mails that do not fall into standard phishing mails and have indicators that are not familiar to the users or are more complex to identify. For those users, simulated phishing mails and more detailed training on typical phishing indicators should be the priority. The last stage then covers users with a deep understanding that went through several programs and test phishing campaigns. In addition, such users still tap into phishing mails. For those users information should be kept reminded, also avoiding creating overconfidence leading to less awareness [8].

5.2 Raising user awareness: The case of Austrian financial services providers

Awareness can be considered as the first step towards a sustainable protection from phishing fraud since it helps to make the users aware of the risk and realize exposure to this threat [17]. Financial institutions and their customers are regular targets of phishing campaigns as the information shared on the homepages of selected Austrian financial institutions demonstrate.

Having a deeper look at the homepages of selected Austrian financial institutions: Erste Bank, BAWAG as established financial institutions and N26 as fintech gets visible that phishing is an important threat to banking users. All institutions provide information on phishing, most recent samples and tips how to prevent from falling into phishing [18], [19]. The review of the warnings shared on the different webpages shows that phishing campaigns are repeatedly started targeting customers of various institutions. The phishing campaigns are regularly adopted to actual topics as the case of COVID-19 or the implementation of the *Payment Services Directive 2* (PSD2) shows.

Butler and Butler [17] underline the importance of websites as a source of security awareness effort for financial institutions. The quality of phishing related information plays therefore an important role and can significantly contribute to help users identify phishing attacks and avoid from falling into them. However, there is also reference to online users not knowing where to find the related information or users actively seeking for such information. There is also reference to users not reading related information even when receiving clear recommendation on respective information sections [17].

Erste Bank und Sparkasse on their internet representation has a dedicated portal "*Sicherheits-Center*" to provide security information to the end users. The portal provides most recent warnings of ongoing phishing campaigns, tricks to increase security and contact information to get support in case of an incident or to report an attempted fraud.

The important security tips include an overview on the 10 most important rules for digital security. The three top rules are [20]:

- to open the internet banking only via verified webpages (visible in the URL);
- to protect logon information against access by third parties;
- to use dedicated special authorization method for login and signing.

The portal further includes details on phishing and how to recognize phishing mails. An example given are the checks of the sender mail-address as also checking the destination address. The information is supported by samples to increase recognition. The information is then followed also by guidelines how to behave in case fraudulent mails received and the request to share phishing mails with the fraud-service center.

In addition, BAWAG PSK has on its website not only a collection of most recent phishing campaigns but also recommendations how to keep online and mobile banking safe. Apart from warning of fraudulent mails there are also warnings on phishing SMS.

The portal warns on 16 September [21] from phishing mails in the context of PSD2, the updated payment directive introducing various additional Security features as one-time password in addition to logon credentials. Fraudsters adopted their campaigns to be considered more trustworthy when claiming to be the respective institution and to provide information or requests for an update of logon data to its users. In addition, BAWAG asks its customers to report phishing mails to a dedicated mailbox.

N26, a Fintech offering its services in Austria, provides on its internet representation an overview on its security implementations as well as short information on phishing. The information displayed is kept shorter than on the homepages of the other institutions. It offers a security guide to be downloaded with a variety of information concerning security. The 10 important rules to safely use digital services are not focused on banking but include more general rules too, e.g. trusted sellers or careful use of unprotected Wi-Fi networks. In addition, the Fintech is asking its customers to report phishing mails or phishing sites to them [22].

Ebot [8] summarizes that in general the simple providing of information has limits with regard to user awareness. Users receive such information via books, emails or other types of instructions. Due to missing reciprocity and missing feedback, there is a lack of transforming such information into awareness. Therefore, recommendations go towards involving learning by doing and anti-phishing games.

5.3 Raising user awareness through regular test campaigns

Apart from information being provided there are different other possibilities to increase the awareness of users.

A few years ago Google has released an online test on phishing recognition accessible for any user [23]. It provides some samples of mails with embedded links or documents where the user needs to decide whether it is a legitimate mail or a phishing mail. For any answer there is then a walkthrough explaining the information embedded allowing to properly distinguishing valid from fake. This is accessible for anybody.

Within companies, simulated phishing attacks can be considered as an additional method to raise the awareness of users towards phishing attacks. As a second channel to raise awareness more related likely to a company perimeter are regular test phishing campaigns to detect weaknesses in employee awareness and provide targeted mitigation training.

Some authors, for example Jampen et al. [12], underline the importance of user awareness additional to countermeasures implemented on technical layers (e.g. spam filters). Key elements of institutional anti-phishing trainings are ranging from information material provided to the users, dedicated trainings or phishing simulations. In case of a phishing simulation, there are different steps feasible after a user fell into it. There could be immediately an information shown on phishing, and how to prevent it or the clicks could be monitored and users provided with dedicated trainings, which are then tracked. Which setting a company decides to apply depends also on risk–benefit evaluations.

Even more benefit could be gained if phishing awareness is embedded in a game format. Students are learning better from game based learning formats [16]. In case of phishing, it could be an option not to have the game purely online but in a combination of non-online and online. Non-online part could be used to identify potentially suspicious URLs what in the study revealed that certain combinations like URLs containing numbers, IP addresses or containing abbreviations are likely to be erroneously considered trustworthy by users [16], after a game walkthrough the recognition rate was more significantly higher. Additionally, crucial considered make users aware about how their data is being disclosed publicly. In this approach study participants not only had to respond to phishing but also to design by themselves phishing mails based on social information provided to them.

Purkait et al. [4] evaluated what kind of factors lead to users clicking on phishing mails and found that "the Internet user's level of phishing awareness, safe Internet habits, past experiences of being phished, years of performing financial activity over the Internet, short-term memory and attention vigilance are all positively related to his/her ability of identifying phishing sites. On the other hand, his/her age and frequency of online transactions had an inverse relationship with the same" [4, p. 22] Other evaluated criteria as technical experience, gender or family size in their study didn't have a significant impact on user's capability to identify a phishing website [4].

5.4 User awareness alone is not the solution

While various literature refers to user awareness being an important pillar of successful prevention against phishing [1], [16] there are remarks also on limitations with regards to awareness. Technical measures are equally important to reduce the risk from phishing. Such measures are transparent to the users and applied on infrastructure level [7] as with antispam solutions, web content filtering solutions or scanning solutions capable to identify malicious content in mail and quarantining the malicious content. Additionally, cooperation with security providers offering services to take down identified phishing sites [1], [12].

Ebot remarks that training might also create overconfidence with the users, feeling familiar with phishing threat and being too sure on recognizing phishing mails though overseeing less familiar indications in phishing mails [8].

The human element will remain in digital communication and there will remain a residual small percentage of users that despite technical measures and awareness programs will click on phishing mails and provide their credentials to fraudsters. Measurable, consistent and continuous awareness programs will contribute to minimizing this percentage [12].

6 LIMITATIONS

Within this paper, the whole industry of phishing and spamming, that from the literature chosen, can be considered as a significant malicious business domain [17] couldn't be further investigated in.

About the awareness measures taken by selected Austrian banks, we could consider only the information presented and shared on their homepages. It would be worth a further investigation on additional measures being imposed by the institutions such as ad hoc messages integrated with the mobile/online banking, eventual training possibilities for customers or active acknowledgement of customers about phishing updates being shared. Such measures would strongly increase the value and contribution to user awareness and represent a difference [17].

Even though law enforcement was not evaluated, it could represent an interesting additional aspect to understand possibilities and limitations of current law enforcement agencies to fight against the phishing industry considering national competences versus globally acting fraudsters and phishing sites distributed over the world.

Focus of this paper was on awareness of users as the parts of the protection chain and therefore excluded measures applied on technical level to prevent from phishing frauds. Such solutions – which were not discusses in thus paper – could be anti-spam filters or evolutions on browser side [1].

7 CONCLUSIONS

Considering the literature review and the data available from current phishing campaigns it can be summarized that phishing is a tremendous hazard, and will remain one. Even though technical solutions are increasingly available to mitigate the risk of phishing, also fraudsters are adopting their approaches to become more efficient. Focusing purely on technology and not consider the user awareness will not be sufficient [13], [17].

Phishing allows attackers to target the user as the potentially weakest link in the chain [3] and by gaining legitimate access; credentials bypass various technical protection measures. In other scenarios, users are lured to access websites where they download malicious software putting at risk personal and company data. Further Security education and awareness measures allow reaching out to the end users, and motivating them to process emails more systematically and review certain criteria to determine potential phishing mails. Organizations nevertheless should not fully rely on the user awareness since there is indication on partially ineffectiveness of trainings provided [8].

Awareness has been generally discussed as an important contribution to reduce the exposure but also referring to risks as overconfidence in case of users feeling well aware about phishing and therefore being less cautious when receiving malicious mail that are set-up well [8]. In this context, the approach of different levels of awareness trainings being provided to users with different levels of expertise was introduced [8]. Such distinction would allow taking more into consideration the underlying reasons why users may click on a phishing mail and allow to have targeted training. Such approach could also be helpful to avoid users being bored from too general trainings not reflecting their level of knowledge on the threat.

As the evaluation on selected webpages of Austrian financial institutions has shown, there are continuous campaigns ongoing that are also visible from statistical data available with Austrian organizations aiming to increase the level of Cybersecurity (e.g. Cert.at). Reporting to such structures the information on phishing sites allows them to share the information on their portals and also support to take down phishing sites reducing the risk from falling into the same phishing mail for other customers.

In the literature, examples of commercial platforms are described allowing a broad customer base to leverage on the “crowd” as source of phishing campaigns [16]. “PhishMe” as company is cited where users can report on experienced phishing attempts, experts analyse the reports and add them to their database available again to all users.

All literature indicates that it is unlikely that phishing as threat will vanish within the upcoming years especially since attackers are also leveraging on newly upcoming technologies and therefore remain in a race with those protecting user data [1], [7].

This implies jointly with the proposed stage-approach on user-awareness [8] that awareness measures cannot be considered one-time activities but need to be done repeatedly including updates with most recent information. In this regard, when it comes to users within an organization the awareness needs to be assured on all hierarchy levels since trainings and test campaigns need to be financially covered [2].

Providing and sharing information is one measure to increase the recognition of phishing mails and raise awareness among users. Information sharing can be done in different ways to address the various target groups [13].

While pure information sharing is considered as the weakest measure in raising phishing awareness, resulting from the need for users that are willing to read the information, reflect and apply the read information in their daily life. Other measures as regular training including active feedback, phishing test campaigns with dedicated particular training/information shared to users that clicked on malicious mails or phishing training in form of games are reflected as being more efficient with regards to the contribution on awareness [7], [12].

Trainings and awareness measures therefore are subject to continuous evolution. These activities need to cover either contextual knowledge with general understanding of phishing and related threats as procedural aspects including proper reaction and reporting of malicious sites as remarked on the security websites of Austrian financial institutions. These measures are also an important contribution in reducing the risk for other users.

Even though it has been reflected that there will remain a residual subset of users clicking on phishing mails awareness measures are considered to contribute positively to increase the knowledge of users on the threat and their ability to identify phishing mails. By increasing the awareness of the users, the success rates for the attackers can be reduced [7].

Our research shows that a bundle of measures reaching from technical arrangements to awareness measures is most promising to fight against phishing threats successfully. As shown in the paper attackers use social engineering techniques and stimuli known from psychological analysis to target the users and gain access to their confidential data [3]. Despite several years of anti-phishing solutions in place and awareness trainings being done, there is still a continuous threat resulting from it visible [12] that requires ongoing initiatives to reduce the risk.

REFERENCES

[1] Hong, J., The state of phishing attacks. *Communications of the ACM*, **55**(1), pp. 74–81, 2012.

[2] Miranda, M.J.A., Enhancing cybersecurity awareness training: A comprehensive phishing exercise approach. *International Management Review*, **14**(2), pp. 5–10, 2018.

[3] Beißel, S., *Security Awareness: Grundlagen, Maßnahmen und Programme für die Informationssicherheit*, Walter de Gruyter GmbH: Berlin and Boston, 2019.

[4] Purkait, S., Kumar De, S. & Suar, D., An empirical investigation of the factors that influence internet user's ability to correctly identify a phishing website. *Information Management and Computer Security*, **22**(3), pp. 194–234, 2014.

[5] Tranfield, D., Denyer, D. & Smart, P., Towards a methodology for developing evidence-informed management knowledge by means of systematic review. *British Journal of Management*, **14**(3), pp. 207–222, 2003.

[6] Hadnagy, C., *Social Engineering Enttarnt: Sicherheitsrisiko Mensch*, mitp-Verlag: Heidelberg, 2014.

[7] Purkait, S., Phishing counter measures and their effectiveness: Literature review. *Information Management and Computer Security*, **20**(5), pp. 382–420, 2012.

[8] Ebot, A.T., Using stage theorizing to make anti-phishing recommendations more effective. *Information and Computer Security*, **26**(4), pp. 401–419, 2018.

[9] Cert.at, Bericht Internet Sicherheit Österreich 2019. https://cert.at/de/berichte/. Accessed on: 21 Dec. 2020.

[10] Watchlist Internet, Internet-betrug, fallen und fakes im blick. https://www.watchlist-internet.at/news/jahresrueckblick-2020-diese-themen-beschaeftigten-uns-heuer/. Accessed on: 21 Dec. 2020.

[11] OpKoord (Operative Koordinierungsstruktur) Lagebericht, Report cyber security, 2019. https://www.bundeskanzleramt.gv.at/dam/jcr:0a5d5734-a53f-4a60-9e64-70092b554c09/EN-Cybersicherheit_Bericht_2019.pdf. Accessed on: 30 Dec. 2020.

[12] Jampen, D., Gur, G., Sutter, T. & Teilenbach, B., Don't click: Towards an effect anti-phishing training. A comparative literature review. *Human-Centric Computing and Information Sciences*, **10**(1), pp. 1–41, 2020.

[13] Arachchilage, N.A.G. & Love, S., Security awareness of computer users: A phishing threat avoidance perspective. *Computers in Human Behaviour*, **38**, pp. 304–312, 2014.

[14] Anti Phishing Workgroup (APWG). https://apwg.org/trendsreports/. Accessed on: 2 Jan. 2021.

[15] Jansson, K. & von Solms, R., Phishing for phishing awareness. *Behaviour and Information Technology*, **32**(6), 2013.

[16] Rubia, F., Affan, Y., Liu, L. & Wang, J., How persuasive is a phishing email? A phishing game for phishing awareness. *Journal of Computer Science*, **27**(6), pp. 581–612, 2019.

[17] Butler, R. & Butler, M., Assessing the information quality of phishing-related content on financial institutions' websites. *Information and Computer Security*. **26**(5), pp. 514–532, 2018.

[18] Bawag. https://www.bawagpsk.com/BAWAGPSK/Sicherheit. Accessed on: 27 Oct. 2020.

[19] Erste Bank. https://www.sparkasse.at/sicherheitscenter-en/current-warning. Accessed on: 27 Oct. 2020.

[20] Sparkasse. https://www.sparkasse.at/sicherheitscenter-en/important-security-tips. Accessed on: 27 Oct. 2020.

[21] BAWAG. https://www.bawagpsk.com/BAWAGPSK/Sicherheit/475006/20190916.html. Accessed on: 27 Oct. 2020.

[22] N26. https://n26.com/en-at/security. Accessed on: 27 Oct. 2020.

[23] Google. https://phishingquiz.withgoogle.com. Accessed on: 30 Dec. 2020.

COASTAL SOCIO-ECOLOGICAL SYSTEM ASSESSMENT IN MUNICIPALITIES: SYSTEM DYNAMICS MODELLING APPLICATION IN LATVIA

IVETA ŠTEINBERGA, IVARS KUDREŅICKIS & RAIMONDS ERNŠTEINS
Environmental Science Department, University of Latvia, Latvia

ABSTRACT

Understanding functionality of coastal ecosystem and, specifically, coastal socio-ecological system (SES) and its governance from environmental, social and economic point of view at existing state is challenging and complex. Also, analysing future scenarios under various stressors could be solved only at site level or by using complex mathematical systems for the wider area. SES based tool is Systems Analysis Framework (SAF) methodology that includes definition of system borders, relevant system elements, flows and interactions within the system, and understanding of all these aspects. The methodology also includes identification of crucial components, as well as building of model, testing and verifying processes. Coastal ecosystems are very sensitive to various factors, such as environmental changes and, particularly, anthropogenic impacts. Even minor changes in temperature, salinity and nutrient availability can lead to substantial changes in life cycle. For example, the runoff can result in higher nutrient and/or pollutant levels in coastal waters – algae blooms can be dangerous to both humans and marine life. While coastal ecosystems should be prevented by special care, they are still very popular as a tourist destination. That causes requirement to link tourism impact to any other environmental impact and, vice versa, to estimate coastal state as promoter of tourism. This paper describes systemic approach used for the Baltic Sea coast at the local Salacgrīva municipality level in Latvia in the case of topical municipal development scenario of coastal tourism perspective. All impacts are modelled and evaluated with STELLA software.

Keywords: coastal management, Systems Approach Framework, system dynamics modelling, Latvia.

1 INTRODUCTION

Numerous studies of coastal ecosystems [1]–[4] shown substantial role of these ecosystems as socioeconomic systems, interacting with environmental stressors. These socioeconomic systems should be evaluated from spatial and temporal point of view. Integrated environmental management of coastal ecosystems could be performed by understanding the links between ecosystem services and their social effects. To ensure successful policy implementation it should include a participatory approach, integrating social and ecosystem considerations into climate adaptation measures. Moreover, it is essential to involve stakeholders in environmental management and decision-making. Adaptation to the climate change includes evaluation of the risks for coastal ecosystems through direct impacts that could result in various ways: (1) loss or changes to habitats and their associated organisms; (2) warming could force species to move to higher latitudes or elevations to maintain the same environmental temperatures; (3) sea level rise will be accompanied by saltwater intrusion into freshwater habitats, and, as a result, some could be forced to relocate or die, thus removing predators or prey that are critical in the existing food chain [5].

At the same time, the problem of integrative assessment of much more complex socio-ecological systems (SES) (more specifically, socioeconomic-ecological systems) that change under natural and anthropogenic drivers, is much less studied and modelled. This article covers complex assessment of coastal socio-ecosystem in local municipality of Salacgrīva (Latvia), done by Systems Approach Framework (SAF) methodology. This methodology was used in order to understand the system model that was built by STELLA software. SAF

WIT Transactions on The Built Environment, Vol 207, © 2021 WIT Press
www.witpress.com, ISSN 1743-3509 (on-line)
doi:10.2495/DMAN210071

common approach and our applied governance basic principle (that coastal area is governed as socio-ecological system (SES)) were used to develop system dynamic coastal territory characterizing model and to assess the most essential SES functioning and its development components. In general, 40 parameters were selected and included in the model, characterizing (both in general and also specifically) all three sustainable development basic dimensions of concrete coastal territory as mandatory: environmental quality, economic and social sphere, and governance dimension.

This research was conducted with some specific purposes: (1) to identify and describe key drivers for selected territory – Salacgrīva, Latvia; (2) to build system dynamic model and test it under real world conditions; (3) to select one locally interested and possible development scenario for local coastal municipality (in this case, tourism increase) and to quantify it; (4) to model expected changes in coastal socio-ecological system and formulate consequences. Built conceptual complex model system, currently including 15 different variables, shows progress in applied research fields connecting basic local municipal development dimensions – environmental/natural, social, economic and governance fields. Moreover, in order to be designed, the model shall be flexible and open. Eventually, after some revisions, the model should be available for other territories and different scenarios, depending on new specifics.

2 MATERIALS AND METHODS

2.1 Study site description: Salacgrīva coastal municipality

Salacgrīva coastal municipality is located in the northeastern part of Latvia, in Vidzeme region (57°45′6″ N, 24°21′26″ E). It has a border with Estonia and stretches along the Baltic Sea coast for 55 km (covering around 10% of Latvia's coastline), but extends inland only for 5–15 km. Salacgrīva town is the centre of the area, located in the mouth of Salaca River. The total area of Salacgrīva municipality is 637.6 km^2, but, according to statistical census in 2015, population for this long coastal territory is only 8,685 inhabitants, with average population density of 14.8 inhabitants/km^2. Dominating commercial activities are connected with Salacgrīva port and export of timber, woodworking industry and food production. The port is located in 100 km from the capital Riga and 13 km from the border with Estonia. Its inland territory in Salacgrīva is 14.4 ha, in the sea – 28.6 ha. There are six fully functioning berths in the port, and the main activities are cargo handling, fishing and vessel mooring.

2.2 The Systems Approach Framework (SAF) methodology

Systems analysis, as well as participatory systems approaches, applying system dynamics for urban and countryside planning, are widely known and generally used [6], [7]. In the described study, specialized methodological approach was used for system research – System Approach Framework (SAF). It has been developed and approbated in several European Union cooperation research institutes and universities, with the focus on coastal problems analysis. Particularly, SAF [8] was developed within the EU SPICOSA project as a practical tool for integrated coastal zone management, including possible and probable future scenarios, managing and forecasting eventual activity. SAF is based on coastal socio-ecological system (SES) approach. In order to improve the zone's ecological sustainability, economic efficiency, and social equity, it provides multidisciplinary advice to environmental managers and policymakers, concerning with environmental problems in the coastal zone.

Second SAF key requirement is that during the SAF application a considerable cooperation and consultation between the relevant stakeholders (end-users, policy makers, scientists, governance agencies, other relevant institutions, and nongovernmental organizations) should be included. To sum up, the general task of SAF is to ensure methodology providing framework conditions for cooperation of coastal society target groups. It would be based on the multi-thematical analysis and discussion of complex natural scientific and social scientific studies results. The aim of the methodology is to jointly develop proposals and to agree on how to improve ecological sustainability, economic efficiency and social equality of the particular territory [9]. The methodology contains four main steps: system design, formulation, appraisal, and output.

2.3 Design of system dynamics model

In the first step of SAF we identified and characterised local bottom-up governance elements. Also, top-down governance institutions are characterised. Thus we offer the overall institutional and stakeholders environment, in which SAF method is implemented. The generic lessons can be drawn out from our approach on problem analysis, adopted from general SAF method. As there is no single dominating problem, the negative environmental and socioeconomic impacts are caused due to synergy effects of a number of problems. The 1st part of SAF consists of: (1) issue identification; (2) identifying all stakeholders (stakeholder mapping); (3) holding the first stakeholder meeting; (4) institutional mapping; (5) describing the cause–effect chain; (6) listing the main ecosystem goods, services and economic drivers.

In order to study complex problems and environmental systems, it is possible to use both models of specific fields (air quality, water dynamics, etc.), descriptive statistical analysis methods, and system dynamics methodology models. Systems dynamics is a stock-flow methodology that helps to explore complexity, interconnectedness, and change over time. Because of their structural simplicity, usage convenience and open system, system dynamics models are becoming more and more popular in the spatial development research. They simulate different development scenarios and potential impacts in environmental, economic, and social fields. In the system dynamics modelling, it is assumed that behaviour of the system depends not only on its components' (attractors') behaviour, but also on their mutual interactions [10]. By using essential basic indicators or criteria describing system behaviour, based on scenarios included in system dynamics models and gained modelling results, it is possible to study more deeply and develop hypothetical governance system change before implementation process.

In general, 40 parameters were selected and included in the model, characterizing all three sustainable development basic dimensions of concrete coastal territory – both in general and, specifically, in environmental quality, economic and social sphere. Thus the second step of SAF includes: (1) system definition; (2) conceptual model; (3) data and methods; (4) problem scaling.

Traditionally, this type of modelling is a five-step process that is performed as follows: (1) defining the problem and the goals; (2) understanding the real life system, figuring out crucial components of the system; (3) building the model with connections between model variables; (4) testing and revising the model, and (5) verifying the model. STELLA model for Salacgrīva case study was selected also according to literature analysis. Various publications show successful usage of STELLA model in coastal development research and SAF methodology application [11]–[15].

2.4 Scenario description

As coastal tourism in the municipality is extended for 55 km along eastern coast of Riga Bay of the Baltic Sea, it may improve economic development. Therefore, this scenario has been selected for further territory development and investigation. For Salacgrīva case analysis it is assumed that number of visitors could increase by 30% in the future (until 2020). Model with existing data has been developed for baseline scenario. Data selection and covered periodare data-dependent, e.g. climatological data are analysed for at least 30 years' period to see trends, while changes in air pollution sources are analysed only for last 15 years. Data selection is defined by scenario. In Salacgrīva case, the coastal tourism development by 30% increase was selected as more reliable scenario. Respectively, tourism increase is linked to wide versa parameters, e.g. change of visitors, number of mobile toilets, number of equipped places (for passenger cars), traffic flow impact as increase of noise and air pollution sources, etc.

2.5 Input data

Several independent or semi-dependent parameters were selected as actors in scenario, and they are: number of biologically active days, number of strong wind annual cases, precipitation extreme cases, and coastal erosion. These parameters are changing due to natural processes, including climate change as well. Climate change and variability of meteorological factors were estimated according to measurements in national meteorological station near Ainaži town (57°52`04.45`` N and 024°21`57.48``E; 6.1 m ASL BAS-77), which is located exactly within the territory of our case study area – in Salacgrīva municipality, near the border with Estonia.

Number of biologically active days indicate biological activity and could lead to watersheds flowering and decrease of water biological quality. This temperature indicator has been selected after discussion with water quality expert. Expert judgement was based on biological physiology of blue-green algae life cycle, flowering process and biological activity of other water plants that starts after stable temperature periods (over 10°C on daily average). Number of strong wind cases (very strong wind day cases according to seventh (over 13.8 m/s) Boffort class) indicate potential impact to erosion process. Precipitation extreme cases are analysed as number of cases exceeded 75th percentile. Analysis of historical data (1961–2015) shows strong decreasing tendency ($R^2 = 0.2297$; $p = 0.05$), and it is possible that this tendency will remain. In the nearest future potential decrease could achieve even 28% or precipitation extremes could decrease in average by 19 days during the year.

Coastal erosion process in Salacgrīva case is described by two variables – volume of primary dune (classified as a descriptor of volume) and volume of seashore silt (classified as a descriptor of volume). Measurements of both variables are done during 2016. Historical data were overviewed for trend analysis. According to historical data analysis, Salacgrīva coastal line is merely affected by erosion. It is estimated that erosion process would not exceed 7 m in average until 2023, which means that it could be approximated that biogenic effect on erosion process is non-substantial.

Initially hypothetical universal system dynamics model structure was built for the whole coastal SES situation. This model included big amount of indicators, acting in different local coastal municipal development scenarios for Salacgrīva municipality: tourism extension, economic enhancement or "sleeping wagon scenario" for people working in national capital city, etc. Within expert and stakeholder discussions it was proposed that tourism development scenario should be analyzed more detailed due to several reasons: due to existing territory development plans, existing stakeholders' interest, past development and financial inputs. According to this assumption, some of variables could be neglected, e.g. birth rate, migration rate, etc. All parameters/variables included in tourism development scenario are listed in Table 1. In order to collect all needed data, external expert's advice and long-term measurement data, results from previous studies and research projects were used.

Table 1: Overview of base input data in the model.

No	Parameter	Baseline value	Unit	Period
1	Number of biologically active days with average air temperature over 10°C	114–158	Days/year	1951–2015
2	Number of strong wind cases	12–89	Days/year	1961–2015
3	Precipitation extremes as 75th percentile exceedances	38–130	Days/year	1961–2015
4	Visual change quality	1–3	Indicator (code)	2015
5	Volume of primary dune, class	1–6	Absolute value	2009–2016
6	Volume of seashore silt, class	1–6	Absolute value	2009–2016
7	Marine litter	104	#(pieces)/100 m beach length	2012–2015
8	Noise pollution according to number of vehicles	< 2500 2500–3500 > 3500	dBA/daily	1996–2015
9	River (Salaca) ecological quality indicator	Good	Indicator	2010–2015
10	Swim water quality	Good	Indicator	2010–2015
11	Air pollution sources	19 (stationary) 4399 (mobile)	#number and abs/year	2000–2014
12	Number of mobile WC	15	Number (total)	2016
13	Traffic flow	2888–4399	#number (vehicles)/daily	1996–2015
14	Equipped sites	10	Number (total)	2016
15	Number of visitors	100–1000	#number/day	2015

Complete conceptual system dynamics model is shown in Fig. 1. Complex modelling system consists of variables acting in different levels and subfields. These levels and subfields are based on SES approach, triple bottom line sustainability dimension and mandatory governance dimension (SEGS), e.g. economic development, stagnation, etc. In this case, tourism scenario was chosen and tested as the most realistic and stakeholders desired local coastal territory development scenario. Variables outside the tourism scenario were neglected (excluded) and input data preparation was performed depending on data availability and data sense. For example, climate change was estimated for long period (according to World Meteorological Organization, at least 30 years' period should be considered), while number of mobile WC is estimated only at present and as the function of attenders.

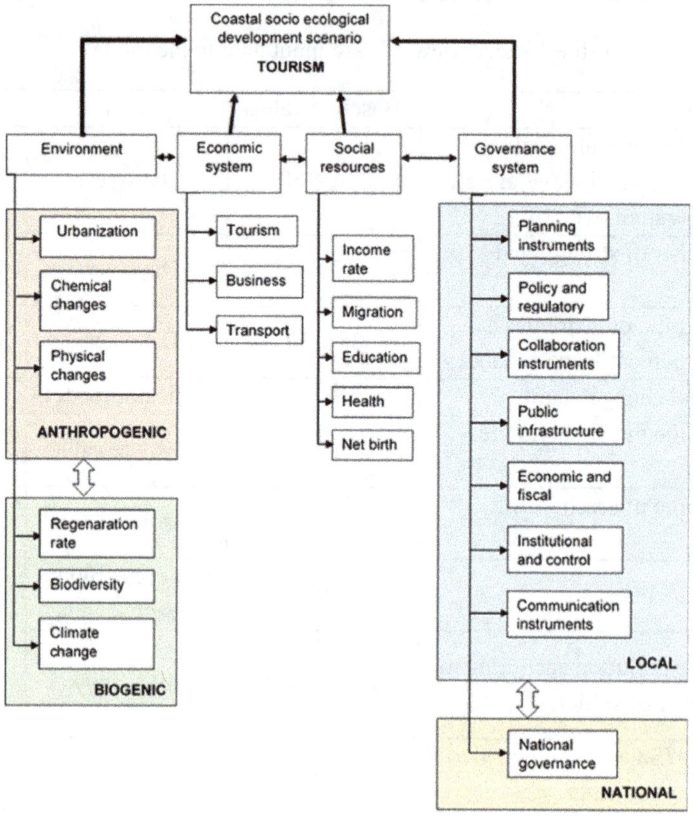

Figure 1: Conceptual system dynamic coastal SEGS model with general adaptation for Salacgrīva municipality.

2.6 The model mathematical description

From mathematical point of view, system model could be described as ARX polynomial black box model, consisting of linear structures, which have traditionally been useful for representing dynamic systems. Firstly, model system was built according to our and experts' understanding of direct, indirect and reverse links between variables set. Secondly, the model was optimized by measurement data and process of calibration that was implemented. Afterwards, scenario of tourism increase (by 30%) was developed in order to find changes in other dependent variables. Some of parameters were set as constant, because non-substantial changes are not detected during period of interest. As a basic scenario, data from Table 1 were used.

As inputs used in model are not comparable (e.g. number of tourist visitors during the days could reach about 1000 persons per day, but swim water quality is assessed as "good" based on Coli form bacteria), numerical analysis is not available. According to other authors publications [10], [16] non-comparable data could be converted to percentile form which is done in this case. Procedure used for data conversion is quite complex. For example, in case of marine litter assessment: (1) average number of marine litter collected along 100 m beach length is 104 pieces; (2) number of marine litter varies from 215 to 95, depending on year; (3) there is statistically substantial, but weak decreasing trend which varies in line with number of beach visitors; (4) according to above-written, in the interest of tourist increase, the number of marine litter in the future is calculated.

Substantiality of long-term climate changes was initially detected by trend analysis, while in case of substantial trend the extrapolation (or forecast) was done. In Salacgrīva case there is positive trend, detected in case of number of biologically active days, number of strong wind days, but decreasing trend in precipitation extremes.

Mainly, 4 driving forces were described: (1) climate change, (2) coastal changes, (3) environmental changes, and (4) structural changes that could be described as follows:

$$ClC = BAD \times 0.13 + SWD \times 0.35 - PED \times 0.28, \quad (1)$$

where ClC = climate change driving force, dimensionless; BAD = biologically active days, number; SWD = strong wind days, number; PED = precipitation extreme days, number.

$$CoC = VCQ \times 0.05 + VPD \times 0.01 + VSS \times 0.01 + ML \times 0.05, \quad (2)$$

where CoC = coastal changes driving force, dimensionless; VCQ = visual change quality, index; VPD = volume of primary dune, class; VSS = volume of seashore silt, class; ML = marine litter, number of pieces.

$$EC = NP \times 0.05 + SEQI \times 0.05 - SWQ \times 0.01 + APS \times 0.1, \quad (3)$$

where EC = environment changes driving force, dimensionless; NP = noise pollution, index; SEQI = Salaca ecological quality indicator, index; SWQ = swim water quality, index; APS = air pollution sources, number.

$$SC = NM_{WC} \times 0.25 + TF \times 0.1 + ES \times 0.15 + EBF \times 0.10 + E \times 0.05 + NV \times 0.3, \quad (4)$$

where SC = structural changes driving force, dimensionless; NMWC = number of mobile WC, number; TF = traffic flow, number of vehicles; ES = equipped sites, number; EBF = employment in biological farming, number; E = employment, number; NV = number of visitors, number.

All above-mentioned driving factors, including internal components, were mathematically described as a result of linear regression analysis. Prioritization of driving forces could look as follows (from most to less important): structural changes in ecosystem (80%), coastal processes (10%), environmental changes (5%) and climate changes (5%) that could be explained by the following expression (5):

$$IA = SC \times 0.8 + EC \times 0.05 + CoC \times 0.1 + ClC \times 0.05, \quad (5)$$

where IA = overall impact assessment, dimensionless; SC = structural changes driving force, dimensionless; EC = environment changes driving force, dimensionless; CoC = coastal changes driving force, dimensionless; ClC = climate change driving force, dimensionless.

Main system development drivers are described according to belonging activities and changes in parameters, while overall impact assessment was proposed to be calculated according to drivers impact importance: (1) climate change structural parameters are independent and cannot be affected by local human activities; (2) climate change impact as process driving force in whole system is minimal (about 5%), which is mainly connected to system temporal frame (modelling is done for next five years); (3) the main changes are expected in "structural changes" force – about 80%; (4) coastal changes are expected within 10%.

3 MODELLING RESULTS AND DISCUSSION

Results are represented as expected changes (semi-quantitative grading) in the dependent variables from baseline condition, based on model runs. Following grading and criteria were used:

- "unimportant" (+/–), expected changes are less than 1%;
- "important" (++/– –), expected changes are 1–35%;
- "very important" (+++/– – –), expected changes are over 35%.

Description of modelling results is given in Table 2.

Table 2: Overview of modelling results.

Parameter/Process	Baseline	Change from baseline (%)
Number of biologically active days	137 days	++ (13)
Number of strong wind cases	51 cases	++ (35)
Precipitation extreme cases	83 cases	– – (–28)
Volume of primary dune	2–4 m^3/m	+ (<1)
Volume of seashore	10–15 m^3/m	+ (<1)
Visitors (tourists)	1000 #/day	++ (30)
Mobile WC	15	++ (25)
Traffic flow	4399	+ (10)
Air pollution sources	4418	+ (10)
Noise pollution sources, level	High	+ (5)
Marine litter	104 pieces	+ (5)
Swim water quality	Good	– (–1)
Employment (crafts)	15	+ (5)
Employment in biological farming, forestry	10	+ (10)
Visual quality change	Good	+/– (5/–5)
Salaca ecological quality indicator	Good	– (–5)
Equipped sites	10	– (–15)

Modelling results show that even 30% increase of tourism flow could lead to some minor negative anthropogenic influence to ecosystem (e.g. increase of mobile pollution sources). On the other hand, there are expected some positive changes in socio-ecosystem – tourism scenario promotes economic income and other factors in the municipality. In some cases, influence is unclear, e.g. visual changes could be positive and negative as well. Also, participation of municipality in effective waste management is essential. Nevertheless, most changes are expected to be promoted by natural processes, e.g. climate change. Also, this result shows that tourism flow could increase by 30%, indicating that tourism still has been underdeveloped and underestimated as area development promoter. More detailed discussion on this is provided further.

Overall, system dynamic approach shows high applicability for territory development planning purposes. Until now, other authors have been mainly focusing on specific species modelling in ecosystem (according to Foresters theory in 1970), close systems or impact of urbanization factors [6], [17]. This research describes wider approach, showing ability to include and describe complex governance, natural, social, economic, and other factors in the simplest possible way [18]. Authors found only one similar research concerning effect of fishing tourism in coastal municipality in Norway [10], where changes within system are analysed only based on stock accumulation (fishing intensity factor, fish life cycle), while other factors (social, governance and natural) were neglected.

It could be assumed that spatial coverage of results is strongly dependent on driving forces and modelling level. In case of governance impact assessment, even European legislation has some impact on situation in any single municipality. However, if there are modelled changes in natural processes, depending on process sporadicity, spatial coverage could vary from one specie meta niche (some kilometres) to even global processes (wind regime changes under climate change driving force). For Salacgrīva study spatial coverage borders are selected as municipality borders.

Additionally, model construction process was challenging because of insufficiency of some of data requiring developing the methods for data identification or interpretation of existing data. Some computation between data holders and problems to identify key processes and indicators required more detailed analysis, as dominantly external experts highlighted particular sphere as more important. Applied SAF methodology clearly shows practical applicability in various development strategies, even if development of territory is planned in opposite way. However, it should be noted that stakeholder's participation is essential to obtain vision of overall system. Unfortunately, passivity of interested groups is a limiting factor to develop site specific information within the model. In order to improve the model and test how it works in real life, some case activities, similar to other cases, should be considered [19]. During SAF implementation some failure to identify stakeholders was observed, which is explained by time consuming process during whole activity period. Alternatively, different communication ways should be considered, otherwise stakeholders will lose interest.

According to the obtained results, increase in the number of tourists by 30% in the coming years will understandably worsen environmental quality and practically in all fields (waste, water quality, air quality, noise level). However, many essential positive trends for municipal social economic development should also be observed, e.g. increase in the number of jobs, etc. Social economic development will occur in the context of local development, based directly in district coastal territory resources. In order to ensure softening/prevention of negative impacts, developed model can be used directly in planning of municipal budget expenditures–revenues, assessing as necessary investments part in appropriate fields – waste management, improvement of water supply systems, development of specific infrastructure for tourists in the coastal area. This would make tourism a potentially expected financial contribution source.

Used general/theoretical methodology during the research approved it's constructively practical applicability:

1. System dynamics model identifies both the benefits of a particular development scenario and critical points. Therefore, it can serve as a basis for argumentation and target groups discussion about measures for critical impact reduction, development of new scenarios and assessment of different governance measures effectivity during the pre-introduction stage.
2. Municipality is perceived as closely interlinked system, and different level stakeholders are capable to identify their role and importance in the development process, regardless of whether it is local household, craftsmen, spatial development planner or top civil servant.
3. Developed complex system dynamic model is serving as a general spatial development basic dimensions framework model. In case of sectoral development, the general model can be comfortably "narrowed/actualized" depending on the needs of municipality and interested groups of society, excluding minimally influencing scenarios of development of those sectors, and more deeply studying essential parameters for specific sector.
4. Model identifying has also got very essential practical benefit both for the assessment of present situation and future planning. Sometimes there is an inadequacy of data in separate essential development areas – data might be too fragmented, too general or might not cover all involved areas and, therefore, development possibility assessments become limited. Therefore, society and decision makers have limited possibilities for adequate discussion and planning, which is particularly essential in the complex and controversial interests development areas (such as coastal resources and territory governance) and in their scenarios assessment (modelling).

4 CONCLUSIONS

The system dynamics conceptual model, designed in the framework of SAF methodology and further elaborated for STELLA application, presented here for complex coastal socio-ecological system, is a first attempt in Latvia to understand the main interactions between climate change, coastal changes, environmental and structural changes in coastal rural municipalities at an integrated quantitative approach. In this research, the case of coastal tourism intensive development in Salacgrīva coastal municipality was studied.

The study has highlighted challenges related to the development and implementation of the model with components in different fields. Within the SAF process, the initially prepared system description with selected 40 factors has been generalized by selected 15 key factors, which allows to describe the system more sufficiently and completely. The key factors were included into the model to describe live and open, and simultaneously complex coastal system. Detailed analysis, formulation process indicates main four agglomerated process drivers – environmental, coastal, climate and structural, where structural changes indicate the most impressive impact.

Elaborated tourism scenario shows that increase of tourists will have comparatively minor influence on coastal socio-ecosystem in case of adequate planning measures. It is expected that even climate change impact will be more evident. During the experts and stakeholders' discussions, SAF application has shown positive practical results for several reasons:

- during the model development phase crucial data and knowledge gaps were identified, leading to practical steps and solutions between any new measures and future scenarios;
- discussion within stakeholders group gives them opportunity to actively participate in municipality development and spatial planning processes;

- developed model shows kind of universality, which means that the model could be reduced or made wider according to necessity of particular socio-ecological governance system at any development point.
- Key leverage points for application measures to be developed within following overall decision-making process are visible within the model structure.

Approbated conceptual complex system dynamics model and STELLA modelling software approach show their eventual applicability and flexibility, which makes it possible to use them for other processes, territories and different local coastal/municipal development scenarios.

ACKNOWLEDGEMENTS

The authors would like to thank Salacgrīva municipality, all the experts and stakeholders, who were very cooperative and helpful for designing and testing of this case study. Funding acknowledgements are given to the EU BONUS programme BaltCoast project – A Systems Approach Framework for Coastal Research and Management in the Baltics (call 2014–2015). BONUS BaltCoast has received funding from BONUS (Art 185), funded jointly from the European Union's Seventh Programme for research, technological development and demonstration and from the State Education Development Agency national grant of the Republic of Latvia (B/2015/4).

REFERENCES

[1] Ali, M., Hadi, S. & Sulistyantara, M., Study on land cover change of Ciliwung downstream watershed with spatial dynamic approach. *Procedia – Social and Behavioural Sciences,* **227**, pp. 52–59, 2016.

[2] Ballé-Béganton, J. et al., A modelling platform for complex socio-ecosystems: An application to freshwater management in coastal zones. *Proceedings of International Congress on Environmental Modelling and Software*, pp. 264–274, 2010.

[3] Kok, J.-L., Engelen, G. & Maes, J., Reusability of model components for environmental simulation: Case studies for integrated coastal zone management. *Environmental Modelling and Software*, **68**, pp. 42–54, 2015.

[4] Moksness, E. et al., Effects of fishing tourism in a coastal municipality: A case study from Risør, Norway. *Ecology and Society*, **16**(3), p. 11, 2011.

[5] European Commission, Science for Environment Policy. Thematic issue: Coastal zones: Achieving sustainable management, Issue 46, 2014.

[6] Mavrommati, G., Bithas, K. & Panayiotidis, P., Operationalizing sustainability in urban coastal systems: A system dynamics analysis. *Water Research,* **47**, pp. 7235–7250, 2013.

[7] Rich, K.M., Rich, M. & Dizyee, K., Participatory systems approaches for urban and peri-urban agriculture planning: The role of system dynamics and spatial group model building. *Agricultural Systems,* **160**, pp. 110–123, 2018.

[8] Hopkins, T.S., Bailly, D. & Stottrup, J.G., A systems approach framework for coastal zones. *Ecology and Society*, **16**(4), pp. 25–37, 2011.

[9] Systems Approach Framework (SAF) Handbook, 2016. http://www.coastal-saf.eu/introduction/index.shtml. Accessed on: 10 Jun. 2020.

[10] Costanza, R. & Voinov, A., Modelling ecological and economic systems with STELLA: Part III. *Ecological Modelling*, **143**, pp. 1–7, 2001.

[11] Kotira, J.H. et al., A system dynamics simulation model for sustainable water resources management and agricultural development in the Volta River Basin, Ghana. *Science of the Total Environment,* **573**, pp. 444–457, 2016.

[12] Marzouk, M. & Azab, S., Analyzing sustainability in low-income housing projects using system dynamics. *Energy and Buildings*, **134**, pp. 143–153, 2017.

[13] Park, S., Sahleh, V. & Jung, S.Y., A system dynamics computer model to assess the effects of developing an alternate water source on the water supply systems management. *Procedia Engineering*, **119**, pp. 753–760, 2015.

[14] Waltersa, J.P. et al., Exploring agricultural production systems and their fundamental components with system dynamics modelling. *Ecological Modelling,* **333**, pp. 51–65, 2016.

[15] Wang, Y.F., Li, B. & Zhang, B., Probability prediction and cost benefit analysis based on system dynamics. *Process Safety and Environmental Protection*, **114**, pp. 271–278, 2018.

[16] Espinoza, A. et al., Sustainability assessment to support governmental biodiesel policy in Colombia: A system dynamics model. *Journal of Cleaner Production,* **141**, pp. 1145–1163, 2017.

[17] Kudreņickis, I. et al., System approach methodology framework adaptation for governance development scenarios: Salacgriva coastal municipality. *Proceedings of 75th Scientific Conference*, University of Latvia, Riga, Latvia, pp. 314–318, 2017. (In Latvian.)

[18] Elsawah, S. et al., An overview of the system dynamics process for integrated modelling of socio-ecological systems: Lessons on good modelling practice from five case studies. *Environmental Modelling and Software*, **93**, pp. 127–145, 2017.

[19] Guo, L., Qu, Y., Wu, C. & Wang, X., Identifying a pathway towards green growth of Chinese industrial regions based on a system dynamics approach. *Resources, Conservation and Recycling*, **128**, pp. 143–154, 2018.

SOCIAL CRISIS, PROTESTS AND EFFECTS ON PUBLIC EMERGENCY SERVICES: ECONOMETRIC ANALYSIS OF THE CHILEAN SOCIAL OUTBREAK

VALERIA SCAPINI[1], CINTHYA VERGARA[2] & JASMINA CONTRERAS[1]
[1]Centro de Investigación en Innovación, Desarrollo Económico y Políticas Sociales (CIDEP), Universidad de Valparaíso, Chile
[2]Universidad Chile, Chile

ABSTRACT

Chile is known for being one of the richest and most stable countries in Latin America, however it is also the most unequal country in the Organization for Economic Cooperation and Development (OECD) with an income gap that is 65% higher than the OECD average. This situation has generated, for more than a decade, protests to improve living conditions. Citizens, seeing no response from the authorities and faced with a 30-peso hike in the public transport rate in October 2019, went out to protest with greater intensity, generating a large-scale social unrest, producing a series of demonstrations throughout the country where there were violent clashes with the police that ended with the application of toxic gases and riot guns, as measures of repression. The purpose of this work is to determine the impact that the demonstrations had on the number of admission to emergency care in public health centres, through a difference-in-differences analysis where two municipalities are compared, one that was exposed to demonstrations and disorders daily, while the other was not. The results are statistically significant and indicate that there was a 4% increase in the number of emergency care visits, and a 7% increase in the number of care for respiratory symptoms in health centres located in the municipality that was exposed to manifestations and public disorder. Finally, the results can be useful to dialogue and discuss "the protocols of police action in Chile" and take into account the health-related effects on citizens and their impact on other public services.

Keywords: socio-economic crisis, manmade disasters, social outbreak, citizen protests, public health management.

1 INTRODUCTION

From the beginning of its history, Chile has proven to be a country with great socio-economic inequalities, reflected both in income differences and geographical segregation by social segment [1]. During 2019, it is estimated that the richest 1% capture close to 17% of tax revenues, while the richest 10% receive more than 50% of all income and, considering the undistributed profits of the companies (retained earnings within firms), 1% of the population receive around 24% of all income generated [2]. Additionally, and according to the GINI index, Chile has remained the most unequal country within the OECD countries [3].

However, the government authorities, led by President Sebastian Piñera, celebrated the country's economic growth as a success of their management with great optimism, while the unemployment rate was growing [4], [5] together with the prices of basic services, and in general the living cost [6]. In this scenario and after a 30-peso hike in the cost of public transport fares, Chilean citizens, mainly high school students, began a series of protests calling to evade the payment of public transportation tickets. On 14 October, facing a lack of response from the authorities to solve economic problems and excessive police violence, spontaneously, many cities in the country came out to protest, generating one of the largest protests, most massive and extensive geographical areas in the country's history, marking the beginning of a period of periodic demonstrations marked by strong confrontations between

www.witpress.com, ISSN 1743-3509 (on-line)
doi:10.2495/DMAN210081

the police and protesters, leading even to cases of death and mutilation, particularly of the eyes [7], [8].

Under this context, it is hypothesized that an increase in emergency care was generated in the public health services in the main area of demonstrations in the capital, known as demonstrations of the "Chilean Social Outbreak". With this, this research seeks to quantify the effect of the social outbreak on the admission of emergency care visits to the public health centres of two municipalities of the Metropolitan Region: the municipality of Santiago and the municipality of La Florida, selected because of their similarity in socioeconomic aspects, where both municipalities have a low social priority classification in the Metropolitan Region, ranking 41st and 42nd [9] and a similar population with 404,495 and 366,916 inhabitants each [10], respectively.

By applying the difference-in-difference method, the impact of the social outbreak (exogenous shock) is studied, using income data from a total of 18 public emergency care centres before and after the event, in order to determine the percentage variation. Finally, the results are expected to provide information on the effects of the demonstrations on emergency care and, thereby, improve the riot, demonstration, or protest management, police protocols and reactions to imbalances that are generated in the health system.

2 CHILEAN CONTEXT: CHILEAN SOCIAL OUTBREAK

After a period with increases in the cost of living, an increase in unemployment in addition to the historical inequality in the distribution of income that has marked the history of Chile [1], on 14 October 2019, there was a large-scale demonstration throughout the country, with cross-sectional citizen support, which started the so-called "Chilean Social Outbreak". This period, characterized by large mobilizations and protests, prompted the questioning of the Chilean socio-economic model and generated, as one of the main results, the beginning of a constitutional reform process.

However, the process was not free of costs neither for the citizens nor for the government. According to the updated data of the National Institute of Human Rights as of 19 March 2020, as a result of the police action, 3,878 injured people were registered in health centres. The origin of the injuries is broken down as follows: bale 53, pellet 193, shots 1,687, tear gas 298, blows 1,411 and unidentified causes 196 [10]. Of which 460 people have eye injuries. Where eye trauma ranges from irritation or inflammation of the conjunctiva to loss of the eyeball, trauma that affects various parts of the body. In addition, the use of tear gas, which has been questioned for its harmful effects on the respiratory system, not only causes a transitory effect, but also a long-term effect, which may cause in some people; pulmonary oedema, laryngospasm and reactive dysfunction, up to ten months after exposure to riot control agents, reaching in extreme situations to trigger the death of the person. To date, around 36 deaths linked to the protests have been declared [7], [11]–[13].

Some cases are known that have shocked society; like the young photographer Gustavo Gatica, who lost both eyes after being wounded by police pellets in a protest. Or the case of Fabiola Campillay, a young woman who was hit by a tear gas canister while waiting for the bus on her way to work, resulting in a loss of sight. Also the case of Abel Acuña, a young man with a heart history who died after suffering a cardiorespiratory arrest as a result of tear gas canisters. The inappropriate use of riot control agents as a dispersal measure to control protest actions has serious consequences for those exposed to the demonstrations, both directly and indirectly.

Since mid-October 2019, protests and greater social unrest have continued, which, in turn, have generated an increase in emergency care [14]. In the period from 18 October to 18 December 2019, throughout the country 13,046 people attended emergency services, of which 4,784 people belong to the metropolitan region, reaching 37% of the total number of care cases and the remaining 63% belonging to other regions with 8,262 people served [15]. Figs 1 and 2 show some emblematic images of what happened.

Figure 1: Chile Protest – Image of Chile woke up [16].

Figure 2: Protesters take cover as security forces spray them with a water cannon [17].

In this scenario, it is of interest to quantify the variation in care in percentage terms and to identify if there is any level of impact on public health services. With the results of this analysis, it is expected to provide input for the government's management to develop and exercise public policies, to focus programs, projects and resources in those areas that present greater relative deficiencies to be able to face emergency care on days of demonstrations.

3 METHODOLOGY

To study the effect of social movement, demonstrations and subsequent public disorder over the number of admissions to public health services, the difference-in-differences model, also known as double difference, was used. This model assumes that the occurrence of the treatment, which in this case is the social movement event, is an exogenous variation and that it is not related to other events that occur in the affected area. It requires the comparison of two groups, one that has been affected by exogenous variation, known as the treatment group or affected area, and the other group that has not been affected, known as the control group or unaffected area. The technique compares two groups during the same time period, and the groups are observed before and after the change. The estimator measures the difference that exists between both groups, and in both periods of time.

Considering the above, the two groups to be compared were selected. The first of them corresponds to the municipality of Santiago, capital of the province of Santiago, established in 1928. It is known as Santiago Centro because it is located in the central sector of the city. It houses the historic centre and the main state agencies. This area was selected as a Treatment group since it was exposed to daily demonstrations that mostly ended with disorders during the period after the social movement. Second, the municipality of La Florida, located in the south-eastern sector of the city of Santiago, is a residential zone made up of families of medium socioeconomic level. During the 1980s and part of the 1990s, it was the most populated municipality in the country and is currently the fourth nationwide [18]. This municipality was selected as a Control group since no major disorders were recorded in the period after the occurrence of the treatment. Fig. 3 shows a map that refers to the location of both municipalities:

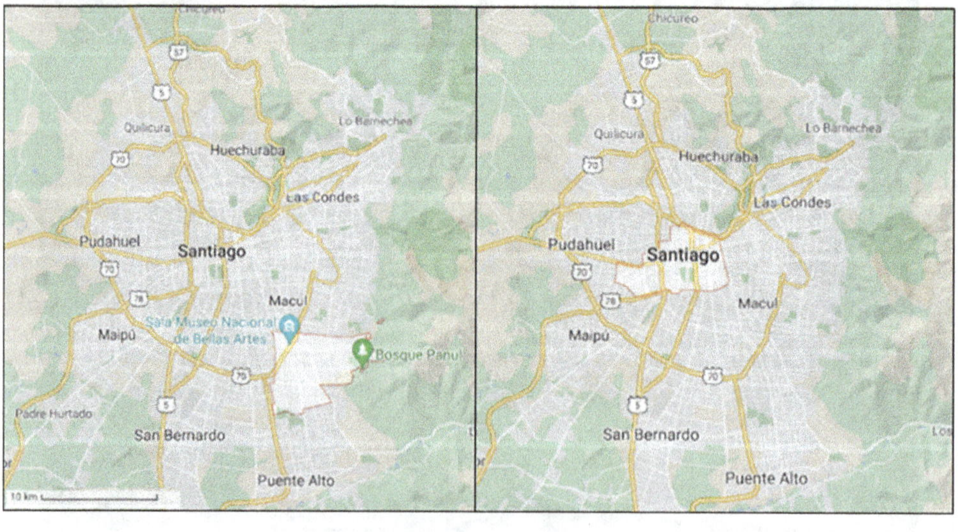

Figure 3: Location of the municipalities of Santiago (right) and La Florida (left).

Two models were estimated where the variable of interest for the first estimate is the total urgent care (1), while in the second the explained variable is the total respiratory care (2). In both cases, the estimated model corresponds to a difference-in-difference specification that aims to determine the effect of disorders on public emergency services, more explicitly:

$$Emergency_{it} = \beta_{it} + \beta_1 Affected_i + \beta_2 Post_t + \beta_3 Affected_i * Post_t + \epsilon_{it}, \quad (1)$$

$$Respiratory_{it} = \beta_{it} + \beta_1 Affected_i + \beta_2 Post_t + \beta_3 Affected_i * Post_t + \epsilon_{it}. \quad (2)$$

The dependent variable in the first model (1) is the number of emergency care services worked in logarithmic scale, *i* corresponds to the municipality and *t* to the day of the week. *Affected* is a binary variable that indicates whether the municipality was exposed to demonstrations or not. *Post* corresponds to a binary variable that takes the value 1 if it corresponds to the period of time after 18 October 2019 and 0 otherwise. Finally, the coefficient β_3 represents the difference between the affected and unaffected area in the period after the event, that is, it shows the impact to be measured. The estimation includes a fixed effect per day and centre of departure, thus controlling for all time- and space-varying differences. In the second model (2), the dependent variable is the total number of respiratory visits worked on a logarithmic scale, and the independent variables are the same as in the previous model.

To validate the results obtained, subsequently, a forgery analysis was performed. To carry out this test, the event was instrumentally moved to a previous moment in time. The data used were obtained from the website of the Department of Statistics of the Ministry of Health of Chile (known as DEIS), corresponding to the total number of emergency care visits and the number of care visits for respiratory symptoms of the 18 public health centres that exist in the municipalities under study. This information was downloaded from 1 January 2019 to 31 December of the same year, and was worked on a daily level, which obtained a data set that contains 5,421 records for the total urgent care and 5,045 observations for total respiratory care.

4 RESULTS

The model estimated according to eqn (1) shows that, during the period after 14 October 2019, in the municipality of Santiago, the number of emergency services actually increased due to the demonstrations. This can be seen in the coefficient $\beta_3$that accompanies the $Affected \times Post$ interaction with a positive sign. Taking this coefficient, it is possible to estimate that the disturbances in the municipality of Santiago increases the percentage of emergency attention by 5% in relation to the municipality of La Florida, ceteris paribus. However, this increase is not statistically significant (see in the first column of Table 1).

By including the control variables, the estimated model shows a better fit in its output, increasing from $R^2 = 0.1409$ to $R^2 = 0.7917$ and, on the other hand, the variable of interest remains positive and turns significant. This allows a conclusion that the occurrence of demonstrations in the municipality of Santiago generates an increase in the percentage of emergency attention of 4% concerning the municipality where there were less or no impacts due to demonstrations.

The second estimated model (eqn 2) shows that, after the occurrence of the social movement, demonstrations and subsequent disorders in the municipality of Santiago, the number of care visits for respiratory symptoms cases increased, that is, the $Affected \times Post$ interaction variable has a positive sign, although not significant (see in the third column

Table 1: Differences in differences results.

Variables	(1) Total emergency	(2) Total emergency	(3) Total respiratory	(4) Total respiratory
Affected	0.386172*** (0.0158139)	0.486920*** (.0200296)	−0.029347 (.0263038)	−3.71503*** (0.4216872)
Post	−0.18881*** (0.0222594)	0.145206** (0.1005198)	−0.642340*** (0.0350571)	0.168266 (0.1633352)
Affected x Post	0.0528024 (0.0349906)	0.042850** (0.0178686)	0.067580 (.0582259)	0.071099** (0.0295997)
R^2	0.1409	0.7917	0.0884	0.7823
No. of observations	5,421	5,421	5,045	5,045
F.E. Day	No	Yes	No	Yes
E.F. Health Centre	No	Yes	No	Yes

Note: Robust standard errors in parentheses.
* Significant at 10%; ** significant at 5%; *** significant at 1%.

of Table 1). The goodness of fit of the model improves from $R^2 = 0.0884$ to $R^2 = 0.7823$ when the control variables health centres and date were included, and the output turns significant, which allows for the conclusion that the occurrence of demonstrations in the municipality of Santiago increases the percentage of respiratory attention by 7% concerning the municipality of La Florida. Results of the estimates made are shown in Table 1.

Finally, to obtain a robust analysis, a falsification test was included. Hereby, among the health establishments of the municipality not affected by movements, a comparison of the number of emergency care cases and the number of care for respiratory symptoms cases were divided into two groups, the north zone and the south zone. In Table 2 it can be observed that the interaction variable is not significant when making the comparison. In other words, there are no differences between the health centres in the northern and southern areas, before and after the social outbreak.

Table 2: Falsification test.

Variables	(1) Total emergency	(2) Total emergency	(3) Total respiratory	(4) Total respiratory
Affected	0.163919*** (0.0235045)	−0.67495*** (0.0197622)	−0.581594*** (0.0304632)	−0.60509*** (0.0318757)
Post	−0.172530*** (0.0301096)	−0.288423** (0.1278949)	−0.660001*** (0.0388876)	−0.494822** (0.2134879)
Affected x Post	−0.0479809 (0.0521603)	−0.0277499* (0.0237027)	0.0488766 (0.067663)	0.0339966 (0.0384034)
R^2	0.0340	0.8009	0.2082	0.7459
No. of observations	3,233	3,233	3,221	3,221
F.E. Day	No	Yes	No	Yes
E.F. Health Centre	No	Yes	No	Yes

Note: Robust standard errors in parentheses.
* Significant at 10%; ** significant at 5%; *** significant at 1%.

In summary, the obtained output shows that social demonstrations do have an impact on emergency care and respiratory care in the public health network. The effect obtained in the municipalities exposed to demonstrations is positive and significant. Specifically, it increases the total emergency care by 4% and the total respiratory care by 7%.

5 CONCLUSIONS

The study carried out allows us to identify that there is an impact of the social demonstrations that occurred after October 2019 in the emergency public services. The results of the adjusted models allow us to estimate that the increase in care is significant and may reach 4% in the number of emergency care in the municipality exposed to demonstrations, concerning the municipality not exposed. Similarly, the second model shows a significant 7% increase in the number of respiratory care visits in the municipality exposed to demonstrations compared to the unexposed municipality.

The significant increase in the demand for care, in days of mobilization and protest that end in confrontations, causes an increase in work for personnel in the public health area and also a greater use of facilities, infrastructure, medicines and supplies for the care of patients. Recognizing these figures is not only relevant to have the information and increase the resources to attend and prevent unexpected or momentary situations, but it is also essential to generate public policies that allow having specialized personnel, better infrastructure and equipment and, in this way, manage and distribute public health resources and move towards equity in access to public health benefits. It is possible to see that the assumption is fulfilled that the increase in respiratory care is due to the harmful effects of riot control agents because it not only affects people exposed to direct action but also those who are confronted passively. With this, it is important that the containment policies carried out by the police take into account the damage they generate in the population in terms of health.

The results obtained can be useful to dialogue, discuss and legislate on "the protocols of police action in Chile" and take into account the health-related effects on citizens and their consequences on other public services.

REFERENCES

[1] Cociña, M., Desiguales: Orígenes, cambios y desafíos de la brecha social en Chile. Programa de las Naciones Unidas para el Desarrollo (PNUD), 2017.

[2] Sanhueza, C., Atria, J., Flores, I. & Mayer, R., Top incomes in Chile: A historical perspective on income inequality, 1964–2017. *Income and Wealth,* **66**(4), pp. 850–874, 2020.

[3] OECD, Inequality: Income inequality – OECD data, 2021. https://data.oecd.org/inequality/income-inequality.htm. Accessed on: 12 May 2021.

[4] Fuentes, A., Caracterización del desempleo en Chile. Puntos de Referencia Edición Digital MR, 2019.

[5] Banco Central, Tasa de desempleo, Serie Histórica, 2021. https://si3.bcentral.cl/Bdemovil/BDE/Series/MOV_SC_ML3. Accessed on: 2 Jun. 2021.

[6] González, A., El costo de la vida en Chile, una de las chispas del estallido social. Santiago, 2019.

[7] McDonald, B., A bullet to the eye is the price of protesting in Chile. *New York Times,* 19 Nov., 2019.

[8] Rodríguez, Á., et al., Ocular trauma by kinetic impact projectiles during civil unrest in Chile. *Nature Eye,* **35**, pp. 1666–1672, 2021.

[9] Gajardo Polanco, S., Región Metropolitana de Santiago: Índice de prioridad social de comunas 2019. Seremi de Desarrollo Social y Familia Metropolitana, Santiago, Chile, 2019.

[10] INE, Resultados Censo 2017: Web Diseminación Censo 2017, Instituto Nacional de Estadísticas, Santiago, Chile, 2017.

[11] MINSAL, Balance de Salud de la Red Pública (18 de octubre al 18 de diciembre 2019). https://www.minsal.cl/wp-content/uploads/2019/12/2019.12.20_BALANCE-ATENCIONES-DE-SALUD_04.pdf. Accessed on: 3 Jun. 2021.

[12] El Periodista, A un año del estallido social: más de 30 muertos y 460 personas con lesiones oculares, Diario El Periodista, Santiago, 2020.

[13] OHCHR, Report of the mission to Chile 30 October–22 November 2019, Office of the High Commissioner for Human Rights. https://www.ohchr.org/Documents/Countries/CL/Report_Chile_2019_EN.pdf. Accessed on: 12 Apr. 2021.

[14] La Tercera, Con turnos extra y a cuatro pabellones, la UTO del Hospital del Salvador atiende a cientos de pacientes con traumas oculares que dejan los balines, La Tercera, Santiago, Chile, 2019.

[15] La Tercera, Balance de atención de urgencias: Desde el estallido social se han registrado casi 11 mil consultas vinculadas a episodios de violencia, Diario La Tercera, Santiago, Chile, 2019.

[16] Ledur, J. & Levine, A.J., Chile woke up. *Thomson Reuters.* www.graphics.reuters.com/CHILE-PROTESTS/0100B32527X/index.html. Accessed on: 12 Apr. 2021.

[17] Lucas, D., Chile declares state of emergency amid riots. *Thomson Reuters.* www.reuters.com/news/picture/chile-declares-state-of-emergency-amid-r-idUSRTS2S1BS. Accessed on: 12 Apr. 2021.

[18] Gajardo, S., Índice de prioridad social de comunas 2019. http://www.desarrollosocialyfamilia.gob.cl/storage/docs/INDICE._DE_PRIORIDAD_SOCIAL_2019.pdf. Accessed on: 16 May 2021.

ECONOMIC RISKS FACING AFGHANISTAN: FAILED STATES AND BUREAUCRATIC RESISTANCE

J. RYAN TESTON, MEGAN E. KONYNDYK,
ROBERT C. FERGUSON & AUDREY HEFFRON-CASSERLEIGH
Florida State University, USA

ABSTRACT

The future of Afghan geopolitics is difficult to predict. Given the current state of infrastructure as well as pending currency collapse, and considering rapid inflation in the country, Afghanistan faces the risk of becoming a failed state if it is unable to secure stable relationships with bordering countries such as Pakistan, China, and Russia. The Taliban now have control of natural resources in the country that were last valued at $3 trillion. Some of these resources are in extremely high demand globally and could point to a large profit source for Afghanistan, or a conflict hotbed, should the Taliban be unable to maintain stability and security. Pakistani officials are speculating Afghanistan could join the China–Pakistan Economic Corridor (CPEC) and senior Taliban officials have expressed a desire for positive relations with China, as well as openness to China's interest in potential investment in mining in Afghanistan. Lack of recognition by the United States for the Taliban as the official leader of Afghanistan, as well as US and United Nations Security Council Sanctions on Afghanistan have had major impacts on investment and flow of aid to the country. The threat of terrorist organizations within Afghanistan will impact its geopolitical relations and ability to develop economic partnerships.

Keywords: Afghanistan, China, Pakistan, economy, risk, Taliban, terrorism, opium, mining.

1 INTRODUCTION

Failed states are unable to act as a service provider while, "strong states unquestionably control their territories and deliver a full range and a high quality of political goods to their citizens" (Rotberg [1]). Failed or "fragile" state indices easily apply to Afghanistan, but the implication of this label is more nuanced in a region fraught with failed states. Afghanistan, a failed state with a long history of armed conflict, may cause shock to neighboring countries that enjoy "strong state" attributes, generally defined as a state capable of providing political goods to its citizens. Research indicates that failed states influence their region's stability and "states which are generated by political instability and armed conflict, in turn generate subsequent unrest, instability, interstate, and civil war in their neighbors and near regions" (Iqbal and Starr [2]). This diffusion of failure can be compounded by the presence of violent conflicts. As resources diminish, the possibility of civil unrest or interstate conflict increases. Markers of unrest can be difficult to discern in a weak state and are impossible to measure in a failed state.

Through utilizing the state failure as an independent variable, Iqbal and Starr examine unrest, instability, civil war, state failure, and international war as measured dependent variables. Ultimately, it is determined that state failure is unlikely to spread to neighboring countries but there are still increased risks of unrest, instability, and civil war to states with close proximity, and more so to those that share borders with the failed state where research suggest "state failure in a contiguous state increases the odds of civil war by 45% and the odds of international conflict by 115% … [presenting] an especially attractive target for invasion, thus providing opportunities and incentives for interstate armed conflict" (Iqbal and Starr [2]).

Bureaucratic resistance is present when there is a conflict between elected or appointed leaders and the civil servants who work within the organization. Extensive research on

WIT Transactions on The Built Environment, Vol 207, © 2021 WIT Press
www.witpress.com, ISSN 1743-3509 (on-line)
doi:10.2495/DMAN210091

bureaucratic resistance and the national security state suggest that bureaucratic resistance can extend into the national security arena and effect the military and the security of the nation (Ingber [3]). Combined with Afghanistan's failed state status, the presence of bureaucratic resistance also contributes to the instability of its future.

While interstate conflict between Afghanistan and her neighbors seems unlikely, timely interventions could be a factor in disrupting the diffusion of state failure. Neighbors, such as China, which exhibit strong state attributes, may be more resistant to the effects of a failed state neighbor. Neighbors with weak state attributes, including political unrest or the inability to interact as a full member of the international community, may have increased impacts (Iqbal and Starr [2]).

2 CURRENT ECONOMIC CONDITION

If the Taliban can effectively maintain control in the face of the looming civil crisis, they face tremendous roadblocks towards successfully legitimizing their governance and advancing Afghanistan away from the world's lowest life expectancy at 53.25. Afghanistan also faces unique challenges including being one of the world's poorest populations with an average Gross Domestic Product (GDP) per capita of $2,065.00, terrorist organizations opposed to Taliban ideology, and the emergence of the National Resistance Front of Afghanistan (CIA [4]). The withdrawal of U.S. forces from Afghanistan, while controversial, coupled with the lightning-fast Taliban takeover, predictable yet unplanned for, raises more questions and concerns regarding the troubled state than it initially solved. While it is impossible to predict, numerous economic avenues, partnerships, and global statuses can be explored to examine what may lie ahead.

Entering the last quarter of 2021, the economy of Afghanistan is struggling with slow economic growth caused by deteriorating security conditions, intense drought negatively impacting agriculture, and a third wave of COVID-19 that began in April. Underserviced and isolated, the Afghanis enter winter with record high infection rates in a population with less than 5% fully vaccinated (World Bank [5]). This trend is inevitable given the gradual increase in inflation in the first 6 months of 2021, followed by an additional acceleration of inflation as the Afghani depreciated against the US dollar by approximately 4% after the Taliban took control in August (World Bank [5]). The World Bank notes that the Taliban is only getting about 50–60% of the daily average customs collections (based on 2020 data). The banking system in Afghanistan is in crisis. Despite banks reopening, firms and individuals are hitting withdrawal limits and not able to access higher amounts. This has resulted in limited ability to conduct business, since firms are unable to send funds overseas, as well as long lines of people waiting to withdraw funds. These are crushing variables and it's not possible to accurately assess changes to the exchange rate due to fragmentation of the financial market (World Bank [5]). This leaves Afghanistan in a dire economic and development situation.

The decrease in international aid and grants has worsened an already grim situation. Foreign aid made up 43% of Afghanistan's economy in 2020 (World Bank [5]). The US has blocked Afghanistan's access to 9.4 billion dollars in international reserves and the International Monetary Fund has paused release of 400 million dollars to Afghanistan (Rappeport [6]). Citing a variety of factors, the World Bank ceased distribution of aid on 24 August (Boghani [7]). The European Union, Germany, and Finland also halted funding for development on August 17, 2021. The World Food Program has stated that 1 in 3 people – over 12 million – in Afghanistan are going hungry, and they need 200 million USD to sustain operation to the year's end (WFP [8]). The stoppage of foreign aid is posing huge risks to not only the Afghanistan economy and development, but its people. Despite pledges of humanitarian aid from the US, EU, China, and others, Rina Amira, director of the

Afghanistan and Regional Policy Initiative at The Center on International Cooperation at NYU, stated "Humanitarian aid will not be sufficient to prevent the collapse of the Afghan economy" (Boghani [7]). For reference, UN member states have pledged more than 1.2 billion in humanitarian aid, but this is contingent on the Taliban abiding by certain behavior (Putz [9]). Compared to the 4.15 billion in Official Development Aid Afghanistan received in 2019, even if this pledged aid is fully dispersed, it will fall drastically short of what is needed. Return of development aid is also contingent on international recognition of the Taliban-run Afghan government. This is itself contingent upon, among other things, the Taliban upholding promises regarding the rights of women (Boghani [7]). These contingencies regarding aid present further obstacles to an economy teetering on the brink of collapse. Ultimately, the cost will be measured in human lives.

3 GEOPOLITICAL RELATIONSHIPS

How the Taliban will finance their government will be heavily influenced by relationships with China, Pakistan, and Russia. The takeover of the country by the Taliban has been supported by Pakistani Prime Minister Imran Khan (Pakistani Country Risk Report [10]). Pakistani senator Mushahid Hussain Sayed said "The best way forward and the immediately available alternative option for Afghanistan's economic development is CPEC, which includes Pakistan and China" (Greenfield [11]). The China–Pakistan Economic Corridor is part of China's Belt and Road Initiative, which has promised more than $60 billion to Pakistan for infrastructure projects. Listed among CPECs various projects are the planned power stations in the north of Pakistan, where several stations have been shuttered (CPEC [12]). There is low probability the Taliban could partake in the agreements CPEC required of Pakistan including, but not limited to, the TPPA. The Taliban, as of 8 September 2021, has officially "expressed the group's interest in joining the China–Pakistan Economic Corridor" (CPEC [12]). There would be substantial benefits were this to occur, such as "increased trade of goods, including oil via pipelines among China, Pakistan and Afghanistan" (Pakistan Country Risk Report [10]).

However, there are risks to Pakistan and China involving the Taliban's ability to prevent Afghanistan from becoming a haven to extremists. The very threats posed by a potential surge in extremists in Afghanistan could very well encourage Chinese and Pakistani involvement in an effort to control the situation. In a meeting on 25 October 2021 between Chinese Foreign Minister Wang Yi and acting deputy prime minister of Afghanistan Mullah Abdul Ghani Baradar, Wang conveyed that "the East Turkestan Islamic Movement (ETIM), an international terrorist organization listed by the UN Security Council, not only poses a real threat to China's national security and territorial integrity but also jeopardizes domestic stability and long-term stability in Afghanistan" (The People's Republic of China [13]). In response, Baradar stated "The Afghan Taliban, which attaches great importance to China's security concerns, will resolutely honor its promise and never allow anyone or any force to use the Afghan territory to harm China" (The People's Republic of China [13]). Afghanistan's official message continues following a path which would encourage Chinese support. Chinese officials also urged G20 members to lift economic sanctions on Afghanistan (The People's Republic of China [14]) and officials from China, Russian, Pakistan, and Iran have been in communication regarding Afghanistan (The People's Republic of China [15]). Chinese Foreign Minister Yi presented a "five-point proposal" on Afghanistan, including a proposal to, "help Afghanistan join in regional cooperation" (The People's Republic of China [15]). China has also promised $31 million in aid in the form of food and health supplies, including 3 million COVID-19 vaccines (Al Jazeera [16]). Given the population of nearly 39 million (World Bank [17]), this is enough for approximately 3.8% of the population to get fully vaccinated. For Afghanistan, it is clear that any path to avoiding failed state status must involve significant support from its neighbors.

4 UNTAPPED RESOURCES, CASH CROP ALTERNATIVES AND ILLICIT REVENUE

Prior to Taliban control, the governmental operations of Afghanistan predominantly relied upon international assistance to maintain effectiveness. At the same time, the Taliban insurgency from 2001 until August 2021, financed their operations through criminal activities both within Afghanistan and internationally (Azami [18]). While kidnappings, taxation of drug cultivation, and exploiting the natural resources of vulnerable communities may be efficient at financing an insurgency, to continue these practices will further isolate the newly reinstituted Taliban government. Currently the Taliban have control of natural resources in the country that were last valued at $3 trillion (Daly and Singh [19]). Some of these resources are in extremely high demand globally, and could point to a large profit source for Afghanistan, or a conflict hotbed, should the Taliban be unable to maintain stability and security. Roadblocks in building a stable mining economy include a history of corruption, the poor infrastructure of Afghanistan, issues of occupational safety, and questionable security of mining facilities. Copper, rare-earth metals, and gemstones are mentioned as the most sought-after resources abundant within Afghani borders, but also present are valuable commodities such as marble, natural gas, lithium, zinc, and many more (Daly and Singh [19]). Prior to the collapse of the previous Afghan government, it was reported that officials had control of 281 of the 709 identified mining zones. Of the remaining mines, 280 were under direct Taliban control, with 148 reportedly under the control of local warlords. In all, revenue from illegal mining operations were estimated in 2020 to finance the insurgency at approximately $464 million (UNODC [20]). In the history of Afghanistan, large scale investment in mining has failed, and warlords have profited from illegal mining royalties. Previous governments inability to ensure safe working practices by monitoring hours worked, proper training, and child labor, will also be a challenge to the Taliban (Pikulicka-Wilczewska [21]). While much has changed since 2016, estimates upwards of 10,000 deposits were believed to be outside of the previous government's control, and at the time of the reporting, were becoming an increasing avenue for Taliban financing (Kakar [22]).

Security is always the greatest risk in Afghanistan and there is low optimism the Taliban will be able to turn mining operations into a prosperous Afghan economy. The internal risk of attack at the hands of Islamic State–Khorasan (IS-K) continues to be a primary security threat, and potential investors remain cautious of the threat towards infrastructural developments and workers. In addition to the tremendously successful IS-K attack during evacuation operations, the terrorist organization has carried out 47 attacks aimed primarily at Taliban fighters (George et al. [23]).

Along with this vast collection of untapped natural resources, and potential cash crop alternatives, Afghanistan has largely controlled the opium market for decades, and while illicit, it has been a primary source of funding for the Taliban since their removal at the hands of U.S. and NATO forces in 2001 (UNODC [20]). Should the Taliban choose to once again adopt policies against opium cultivation, without exploring alternative crops that can ensure the financial security of rural communities as they did during their previous reign, this could prove disastrous for rural Afghanis who will be faced with the upheaval of their livelihood. Previously, at the hands of historically violent rule of law by the Taliban government, farmers had to choose to either adhere to the changing landscape or violate enacted policies.

As of September 2021, annual income for the Taliban reportedly ranged from $300 to $1.6 billion, with a large portion of annual revenue coming from opium poppy cultivation, and additional supplemental financial needs met through mining and donations (Maizland [24]). Should the Taliban re-establish bans upon opium poppy cultivation, it will further escalate the economic disparity of rural Afghanis who rely upon the crop for their income, and the Taliban who have relied upon generating tax income from its cultivation. By examining the most recent Afghanistan Opium Survey, conducted by the UN Office of Drugs and Crime, the 2020 opium production in the country is estimated at 224,000 hectares, with only 12 of Afghanistan's 36 Provinces not involved in substantial agricultural growth of the crop. Total USD value of the farm-gate price in 2020 totals $350 million, and while this pales in comparison to the pricing in 2011 estimated at $1.4 billion, this represents a tremendous source of income for rural Afghanis (UNODC [20]). If the Taliban remove this crop, without an efficient alternative crop that can compete with the economic return generated for these communities, it will further escalate the crisis in Afghanistan. While unlikely to end the large-scale narcotics industry within Afghanistan, alternative cash crops such as saffron can present a crop alternative. Afghan grown saffron is valued between $1,200 and 1,800 per kilogram, and during the coalition campaign was seen by some investors as a viable option to work towards lessening opium cultivation in the country (Razayee [25]).

5 SECURITY RISKS TOWARDS ECONOMIC PARTNERSHIP

The Taliban's current relationship with recognized Foreign Terrorist Organizations (FTOs) remains a primary concern in diplomatic relations. This exacerbates already fraught bureaucratic relationships in country. There are serving Taliban members with links to Al Qaeda, Al Qaeda in the Indian Subcontinent, The Haqqani Network, and additionally to some extent with the Pakistani Taliban and the Islamic Movement of Uzbekistan (IMU) (Congressional Research Service and Thomas [26]). Any idea the Taliban would sever ties with Al Qaeda negates their long and complex history – it is a relationship they have never broken. (Elias [27]). The ability of the Taliban to maintain consistent security for its citizens, as well as potential governmental partners and investors will be a chief economic risk facing the new regime. With so many historic ties to terrorist organizations operating and finding safe haven in Afghanistan, it will be interesting to see how the Taliban maintains control. Having the dual threat from IS-K, and the newly emerged National Resistance Front (NRF), the Taliban currently face major security and leadership challenges to effectively administering the country, its people, and resources.

While accurate numbers regarding the members of the foreign terrorist organizations are difficult to assess, a June 2021 publication from the UN Security Council estimated members of these organizations to be collectively in the thousands. Al Qaeda and Al Qaeda in the Indian Subcontinent are believed to collectively have upwards of 500 members in the border regions of Afghanistan/Pakistan. The Haqqani Network, seen as the liaison between Al Qaeda and the Taliban, are estimated to make up a major portion of identified Taliban fighters and specialists in the neighborhood of 3–10,000. The Pakistani Taliban have upwards of 6,000 fighters. The IMU have a strength estimated to total roughly 700, including the families of the fighters. IS-K fighters are estimated between 1,500 and 2,200 in total. ETIM numbers are estimated as several hundred, potentially as many as 400. Additionally, the UN report identifies fighters in northern Afghanistan from Khatiba Imam al-Bukhari (KIB) between 25 and 150 in total, and the Islamic Jihad Group with upwards of 100 fighters (Tirumurti [28]). All of these groups seek to challenge or exploit the current administrative apparatus of Afghanistan.

The Taliban have claimed public victory over NRF fighters, a claim disputed by NRF leadership (Robertson and Regan [29]). The exact numbers of NRF membership is not known, and it may be as many as 10,000, but the resistance fighters are reportedly not engaged in offensive operations as they prepare for winter (Coffey [30]). This suggests a possible surge in the spring, if survival through the coming winter can happen.

6 RECOMMENDATIONS

Given the state of crisis facing the people of Afghanistan, a path to stabilization would involve recognizing the Taliban officially and providing support to them as they transition into the role of official government. The Taliban – the new government of Afghanistan – will need outside support to return government functions for the people of Afghanistan. They will also need to be monitored to ensure any aid travelling into the country is utilized appropriately. With official recognition, Afghanistan could see the return of billions in development aid. This would provide a unique opportunity to leverage this aid with the Taliban and install a multinational monitoring group consisting of major and regional powers in Afghanistan to work with the government and individual aid agencies, ensuring its proper dispersal.

The emergence of extremists in Afghanistan would pose a huge threat to regional and even global stability. China is in a unique position to monitor any emerging extremist strongholds. China is motivated to do this by the threat extremists in Afghanistan have to their own stability in the Xinjiang region. It would be to the benefit of global anti-terrorism efforts to encourage China to step into a terrorism monitoring role in Afghanistan. China already has one unofficial military base in Tajikistan near the Afghanistan border and has just begun work on a second official base in Tajikistan (Standish [31]). This is a move on China's part to increase its physical security presence in Central Asia and provide a means for countering extremism in Afghanistan that could spill over into the Xinjiang region. It also provides a stepping point for China to spearhead an official, multinational anti-terrorism watch post within Afghanistan. Given that the Taliban has expressed in clear terms its wish to cooperate with Beijing on anti-terrorism, as well as their desire to receive the financial support that would come with China's Belt and Road initiative, there is real possibility for the Taliban agreeing to such an arrangement.

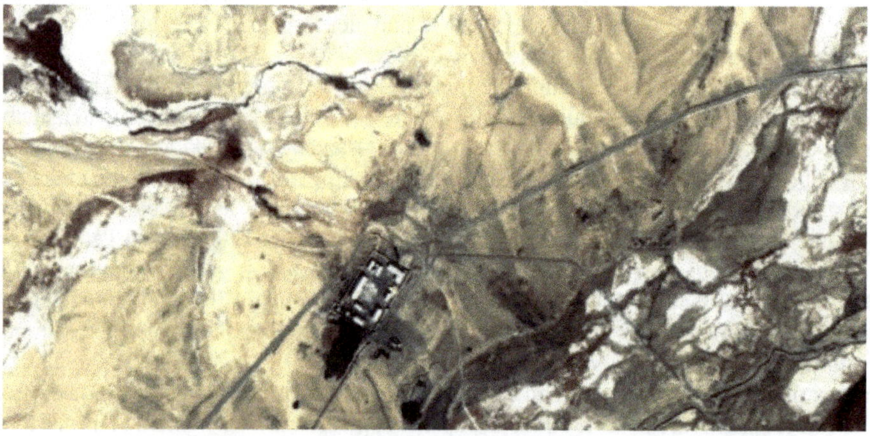

Figure 1: Military site near Shaymak before China's arrival [31].

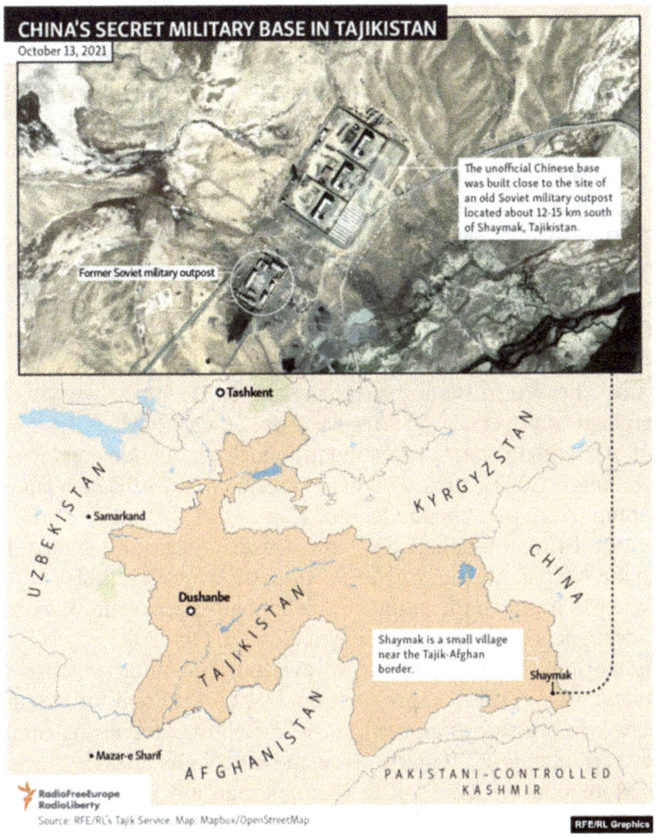

Figure 2: Base as of October 2021 after China's arrival [31].

7 CONCLUSION

Afghanistan is unlikely to stabilize without the reinstitution of international aid, the unfreezing of assets allocated towards governmental operations, or internal economic growth. All of these possibilities seem remote should the Taliban continue to go against the wishes of the international community. Long term economic sustainability is dependent upon their ability to establish cooperative relationships with their regional neighbors, principally with China, Pakistan, and Russia through incorporation into CPEC developments within China's Belt and Road Initiative, and their ability to effectively establish a reliable and safe infrastructure towards natural resource mining operations. The Taliban have expressed the desire to rid Afghanistan of opium cultivation, amongst other illegal narcotics, a decision which would further deteriorate the economic conditions of rural Afghanis. Additional concerns exist with the security of the region as IS-K initiates more violent attacks, and the NRF prepares for long term resistance to Taliban rule, along with the numerous terrorist organizations which continue to operate from Afghanistan.

REFERENCES

[1] Rotberg, R., Failed states, collapsed states, weak states: Causes and indicators, 2003. https://www.brookings.edu/wp-content/uploads/2016/07/statefailureandstate weaknessinatimeofterror_chapter.pdf. Accessed on: 1 Nov. 2021.

[2] Iqbal, Z. & Starr, H., Bad neighbors: Failed states and their consequences. *Conflict Management and Peace Science*, **25**(4), pp. 315–331, 2008. DOI: 10.1080/07388940802397400.

[3] Ingber, R., Bureaucratic resistance and the national security state. 2 Oct. 2018. https://www.americanbar.org/content/dam/aba/events/administrative_law/2018 meetings/11/016-future-federal-civil-service.pdf. Accessed on: 4 Nov. 2021.

[4] CIA, Afghanistan: The world factbook. https://www.cia.gov/the-world-factbook/ countries/afghanistan/. Accessed on: 22 Oct. 2021.

[5] World Bank, The World Bank in Afghanistan. 2021. https://www.worldbank.org/en/ country/afghanistan/overview#1. Accessed on: 26 Oct 2021.

[6] Rappeport, A., Afghanistan faces economic shock as sanctions replace foreign aid. *The New York Times*, 2021. https://www.nytimes.com/2021/08/21/business/afghanistan-economy.html. Accessed on: 25 Oct. 2021.

[7] Boghani, P., "Brink of collapse": How frozen assets and halted foreign aid are impacting the Afghan people. *Public Broadcasting Service*, 2021. https://www.pbs.org/wgbh/frontline/article/taliban-takeover-how-frozen-assets-foreign-aid-impacts-afghanistan/. Accessed on: 26 Oct. 2021.

[8] WFP, Afghanistan: WFP committed to averting humanitarian crisis as one in three people go hungry: World food programme. UN World Food Programme. https://www.wfp.org/stories/afghanistan-wfp-committed-averting-humanitarian-crisis-one-three-people-go-hungry. Accessed on: 24 Oct. 2021.

[9] Putz, C., More than $1.2 billion in humanitarian aid pledged to Afghanistan: The Diplomat for The Diplomat, 2021. https://thediplomat.com/2021/09/more-than-1-2-billion-in-humanitarian-aid-pledged-to-afghanistan/. Accessed on 25 Oct. 2021.

[10] Pakistan Country Risk Report, Q1 2022. Fitch Solutions Risk Reports 2022, First: pp. 1–80.

[11] Greenfield, C., Analysis: As west ponders aid for Afghanistan, China and Pakistan quick to provide relief. Reuters, 2021. https://www.reuters.com/world/middle-east/ west-ponders-aid-afghanistan-china-pakistan-quick-provide-relief-2021-09-12/. Accessed on: 26 Oct. 2021.

[12] China–Pakistan Economic Corridor (CPEC), Progress. http://cpec.gov.pk/progress-update. Accessed on: 26 Oct. 2021.

[13] The People's Republic of China, FM meets with acting deputy PM of Afghan Taliban's interim government, 2021. http://english.www.gov.cn/statecouncil/wangyi/202110/ 26/content_WS61774e4ac6d0df57f98e3ccc.html. Accessed on: 26 Oct. 2021.

[14] The People's Republic of China, FM urges lifting economic sanctions on Afghanistan at G20 meeting, 2021. http://english.www.gov.cn/statecouncil/wangyi/202109/23/ content_WS614c5c49c6d0df57f98e0ba2.html. Accessed on: 27 Oct. 2021.

[15] The People's Republic of China, China, Russia, Pakistan, Iran to strengthen coordination on Afghan issue, 2021. http://english.www.gov.cn/statecouncil/wangyi/ 202109/18/content_WS614534a4c6d0df57f98e0681.html. Accessed on: 27 Oct. 2021.

[16] Al Jazeera, China extends $31M in emergency aid to Afghanistan. *Taliban News, Al Jazeera*, 2021. https://www.aljazeera.com/news/2021/9/9/china-offers-31m-in-aid-3-million-covid-shots-to-afghanistan. Accessed on: 27 Oct. 2021

[17] World Bank, Population, total: Afghanistan. https://data.worldbank.org/indicator/SP.POP.TOTL?locations=AF. Accessed on: 27 Oct. 2021.

[18] Azami, D., Afghanistan: How do the Taliban make money. *BBC News*, 2021. https://www.bbc.com/news/world-46554097. Accessed on: 27 Oct. 2021.

[19] Daly, T. & Singh, S., Factbox: What are Afghanistan's untapped minerals and resources? *Thomson Reuters*, 2021. https://www.reuters.com/world/asia-pacific/what-are-afghanistans-untapped-minerals-resources-2021-08-19/. Accessed on: 27 Oct. 2021.

[20] UNODC, Afghanistan Opium Survey 2020, 2021. https://www.unodc.org/documents/crop-monitoring/Afghanistan/20210503_Executive_summary_Opium_Survey_2020_SMALL.pdf. Accessed on: 27 Oct. 2021.

[21] Pikulicka-Wilczewska, A., Why is Afghanistan unable to extract its vast mineral wealth? *Business and Economy, Al Jazeera*, 2019. https://www.aljazeera.com/features/2019/5/28/why-is-afghanistan-unable-to-extract-its-vast-mineral-wealth. Accessed on: 27 Oct. 2021.

[22] Kakar, J., Illegal mining, weak government help Taliban expand in Afghan north. *Thomson Reuters*, 2016. https://www.reuters.com/article/us-afghanistan-taliban-mining/illegal-mining-weak-government-help-taliban-expand-in-afghan-north-idUSKBN13114J. Accessed on: 27 Oct. 2021.

[23] George, S., Warrick, J. & De Young, K., Pakistan using informal intelligence channels to prop up Taliban fight against Isis. *The Washington Post.* https://www.washingtonpost.com/world/2021/10/23/afghanistan-isis-pakistan-intelligence/. Accessed on: 27 Oct. 2021.

[24] Maizland, L., What is the Taliban? Council on Foreign Relations, 2021. https://www.cfr.org/backgrounder/taliban-afghanistan. Accessed on: 27 Oct. 2021.

[25] Razayee, J.A., Spice of life: Saffron harvest offers jobs, opportunity in Afghanistan. *Thomson Reuters*, 2016. https://www.reuters.com/article/us-afghanistan-saffron-idUSKBN13I0CL. Accessed on: 27 Oct. 2021.

[26] Congressional Research Service & Thomas, C., *Terrorist Groups in Afghanistan*, 13th ed., Library of Congress: Washington, DC, pp. 1–3, 2021.

[27] Elias, B., Why the Taliban won't quit Al Qaeda. *Foreign Policy*, 2021. https://foreignpolicy.com/2021/09/21/taliban-al-qaeda-afghanistan-ties-terrorism/. Accessed on: 27 Oct. 2021.

[28] Tirumurti, T.S., Twelfth report of the Analytical Support and Sanctions Monitoring Team submitted pursuant to resolution 2557 (2020) concerning the Taliban and other associated individuals and entities constituting a threat to the peace stability and security of Afghanistan, 2021.

[29] Robertson, N. & Regan, H., Taliban claim to have "completely conquered" panjshir. *Cable News Network*, 2021. https://www.cnn.com/2021/09/06/asia/afghanistan-monday-intl-hnk/index.html. Accessed on: 27 Oct. 2021.

[30] Coffey, L., In Afghanistan, a resistance arises. The Heritage Foundation, 2021. https://www.heritage.org/terrorism/commentary/afghanistan-resistance-arises. Accessed on: 27 Oct. 2021.

[31] Standish, R., From a secret base in Tajikistan, China's war on terror adjusts to a new reality. 14 October, 2021. https://gandhara.rferl.org/a/tajikistan-china-war-on-terror-afghan/31509370.html. Accessed on: 4 Nov. 2021.

SECTION 3
ENVIRONMENTAL ISSUES

ENVIRONMENTAL REMEDIATION AND POSSIBLE USE OF TERRESTRIAL MICROBIAL FUEL CELLS

GIORGIA AIMOLA[1], GABRIELE GAGLIARDI[2], ANDREA PIETRELLI[3,4], VALERIA ANCONA[1], ANNA BARRA CARACCIOLO[1], DOMENICO BORELLO[2], VINCENZO FERRARA[5] & PAOLA GRENNI[1]
[1]Water Research Institute (IRSA), National Research Council (CNR), Italy
[2]Department of Mechanical and Aerospace Engineering, Sapienza University of Rome, Italy
[3]Univ Lyon, INSA Lyon, Université Claude Bernard Lyon 1, Ecole Centrale de Lyon, France
[4]ECP Lab, Université Lumière Lyon 2, France
[5]Department of Information Engineering, Electronics and Telecommunications, Sapienza University of Rome, Italy

ABSTRACT

Many soils contain a wide number of organic and inorganic chemicals and potential toxic elements due to industrial, agricultural and numerous anthropogenic activities. The recovery of polluted sites is an urgent need to be addressed and the development of innovative remediation technologies, which exploit nature-based solutions, is strongly encouraged, in line with the new EU Circular Economy Action Plan. Terrestrial microbial fuel cells (TMFCs) can be a valuable tool for recovering soils polluted by various organic and inorganic contaminants. TMFCs benefit from capabilities of microbial biofilms developed on the electrodes, which use the terminals as catalysts for metabolic activities, including contaminant degradation. This process produces energy, thanks to conversion of chemical bond energy (stored in the bonds of organic compounds) into electrical ones. This work describes construction materials, remediation capabilities and technology of the TMFCs, reporting the last advances in TMFCs such as soil-based reactors or in combination with plants (plant microbial fuel cell (PMFC)). Some aspects related to microbiological activities for pollutant biodegradation, plant-microbial interactions, energy production, and fields of application will be shown. Finally, abiotic factors which can improve bioremediation activities are also considered. Furthermore, limitations and issues for large-scale applications, as well as for stacking and scaling-up are discussed.

Keywords: electrogenic bacteria, microbial degradation, green technology, bioremediation, microbial fuel cell.

1 INTRODUCTION

In recent years, great attention has been provided to develop new sustainable strategies for soil recovery from contamination. Remediation strategies can be in situ (without excavation of soil or with excavation of soil and treated in the site) or ex situ (if contaminants or contaminated matrices are taken from a site and brought elsewhere for their treatment). In the following paragraphs bioremediation and plant-assisted bioremediation, in line with the Green Economy and the new EU Circular Economy Action Plan (https://ec.europa.eu/environment/circular-economy/pdf/new_circular_economy_action_plan.pdf) combined with terrestrial microbial fuel cells are described as possible technologies for recovery soils from organic and inorganic contaminants.

2 REMEDIATION PROGRESSES

2.1 Bioremediation

Bioremediation relies on the use of bacteria, fungi or plants (including in some cases their enzymes or growth stimulants) for degrading, transforming, sequestering, mobilizing contaminants. It exploits natural recovery abilities of microorganisms to degrade organic contaminants. Biotransformation is the transformation of a contaminant into another less

www.witpress.com, ISSN 1743-3509 (on-line)
doi:10.2495/DMAN210101

toxic or non-toxic molecule. Biodegradation is the breakdown of a parental molecule into smaller (less complex) or inorganic molecules (mineralization). Only microorganisms can mineralize organic compounds into inorganic constituents (CO_2 and H_2O). In some cases, the organic contaminant is used as carbon or nitrogen source for microbial community (metabolism), but in other cases degradation occurs by co-metabolism, i.e. the oxidation of a substance without its energy utilization and microbial growth.

Biodegradation can occur in soil, because natural microbial communities (and in particular *Bacteria*) have a great physiological versatility and catabolic potential, leading to breakdown of an enormous number of organic molecules, including persistent organic pollutants [1]–[3]. Bioremediation includes bioaugmentation (addition of previously selected microorganisms for removing specific contaminants) and biostimulation (addition of nutrients, oxygen, electron donors or acceptors).

Bioremediation effectiveness depends on site-specific abiotic and biotic conditions of soil and type and amount of contaminants. For example, contaminant concentrations above the toxicity tolerance level of microorganisms (which can vary with the compound bioavailability) will make biodegradation ineffective. Moreover, some microbes need specific conditions for being effective (e.g. specific acid/basic conditions, saline presence, organic carbon or water content, presence/absence of oxygen).

Terrestrial microbial fuel cells (TMFCs) require for their activity (electricity production), a high soil moisture in order to permit microorganisms to develop a biofilm on the anode (which is used as an electron acceptor in anaerobic conditions) and to promote the flows of protons through soil directly to the cathode. The latter is the site in which, in aerobic conditions, electrons and protons react with oxygen producing water. The oxidation of organic contaminants by electroactive bacteria is performed in anaerobic conditions, thanks to a high soil moisture. A soil moisture close to the maximum water holding capacity of soil and a constant environmental temperature favour microbial activity and promote TMFC performance (in terms of electricity production and organic contaminant degradation) [1], [3]. Even if in the case of TMFCs used for organic contaminant biodegradation the electricity outputs can be relatively low, this technology can be effective for bioremediation purposes. Mathematical models for MFCs are very useful for their design, control and optimization [4]. These models (e.g. Monod and Monod-type models) can describe microbial growth kinetics. If microbial gene regulatory networks are known and modelled, a more efficient MFC could be developed in order to scale up this technology. For example, advanced bio-based models which connect gene regulation of specific metabolic pathways with microbial growth kinetics enables a more accurate estimation of substrate biodegradation, microbial growth kinetics, and necessary gene and enzyme expression. The prediction of gene and enzyme expression can also be used in synthetic and biological systems for process optimization.

2.2 Plant assisted bioremediation

Plant assisted bioremediation (PABR) is a green remediation tool that relies on synergistic relations between naturally occurring microorganisms of soil, water, sludge or sediments and plant roots [5], [6], resulting in various contaminant clean up processes (phytoextraction, phytostabilization, phytoaccumulation, rhizodegradation). Plants release through their roots several low-molecular-mass compounds (e.g. sugars and polymerized sugar, amino acids and organic acids) together with root border cells and dead root cap cells. These root exudates positively affect rhizosphere microbiota because they are used as carbon sources, stimulating it for an active decontamination process [7]. PABR technology can recover matrices polluted by both organic and inorganic contaminants [8] and it has been successfully tested in greenhouse and field studies for promoting polychlorinated biphenyls (PCBs) degradation and heavy metals phytocontainment [7], [9]–[11].

A recent study [12] confirmed the effectiveness of the poplar clone Monviso in stimulating the natural belowground microbial community, in removing PCBs and heavy metals and improving the overall soil quality. The biomass generated by this PABR can be also used for producing energy. A fluidized-bed gasification treatment was successfully conducted on poplar biomass pruning residues. However, deep investigations are needed in different site-specific scenarios for ensuring a zero impact on environment (e.g. for trapping possible pollutant residues in ashes during the gasification process) [13].

Adding organic amendments such as biochar or compost can increase plant and microbial biomass in rhizosphere, with positive effects on the PABR performance [2].

3 POLLUTION MANAGEMENT OF SOIL WITH MFC

3.1 TMFC for remediation of organic pollutants

TMFCs can generate electricity and, at the same time, recover soil from different kind of pollution [14]–[16]. Specific microorganisms naturally present in soils (exo-electrogens bacteria) can use anode and cathode as electron acceptors during their metabolic activities, inducing contaminant transformation or degradation. TMFCs have to be designed ad hoc taking into consideration biotic and abiotic (e.g., pH, temperature, moisture) conditions for each specific contaminant.

TMFCs have proved to promote degradation also of some persistent organic compounds (POPs) [17]. In fact, the condition of anoxia necessary to favour degradation of these compounds is possible. The anode electrode is placed in anoxic conditions in soil where the oxidation of organic compounds by microorganisms occurs [18]. Meanwhile, the cathode is in contact with an accessible source of oxygen (water or air) and it works as a terminal electron acceptor [19]. Dichloroethane (DCA) degradation was tested with TMFCs in aerobic and anaerobic conditions in the presence/absence of microbial communities grown on the anode. Each condition was also tested in open and closed circuits. The microbial presence showed a clear improvement in TMFC performance in terms of DCA degradation and energy production (98% and 93% for closed and open circuit, respectively). In the absence of bacteria, no DCA removal was found in open circuit, while the removal increased to 95% in closed circuit [20].

Other authors reported phenol removal using TMFCs [21] with a higher degradation and microbial activity (90.1%) in closed than in open circuit (27.6%) condition. Moreover, Yu et al. [22] tested the degradation chlorinated phenol by TMFCs, confirming that degradation of this molecule occurs through a dehalogenation at the anode [23].

PCB degradation in TMFCs was found higher in the presence of surfactants (Tween80, sodium dodecyl sulfate SDS), used for stimulating microbial metabolic activity [24]. Cao et al. [25] used TMFCs to remediate soil contaminated by the pesticide hexachlorobenzene (HBC). A higher HBC removal (71%) was found in the presence of an anaerobic sludge; lower degradation rates were found in the open circuit (52%) or the absence of anaerobic sludge (39%). The presence of anode promoted the exo-electrogenic microbial activity, increasing metabolic rates of anaerobic bacteria and enhancing HCB degradation via the reductive dechlorination pathway under anaerobic condition.

Polycyclic aromatic hydrocarbon (PAH) degradation was more efficient in anaerobic condition than in aerobic one [26]. Moreover, PAH degradation was higher adding sulphates and nitrates because bacteria used them in their metabolism. PAH (anthracene, phenanthrene, pyrene) degradation using MFCs was found with higher removal rates in the closed condition [22]. Zhao et al. [27] studied the phenanthrene and pyrene degradation using plant microbial fuel cells. The plant presence in TMFCs increased the crude-oil degradation rate by 45% compared to the system without plants due to an improvement in the electron-transfer efficiency. In fact, the charge-transfer resistance in the TMFCs was significantly reduced by 57.9% with the addition of plants.

Recently, Borello et al. [1] reported the removal of DDE (2,2-bis (dichlorodiphenyl-dichloroethylene or p-chlorophenyl)-1,1-dichloroethylene), the main toxic metabolite of the pesticide DDT (dichlorodiphenyl-trichloroethane). After 2 months from the beginning of the experiment, the results showed that the TMFCs promoted a significant degradation of the pesticide (39%), compared to conditions in absence of TMFCs. At the same time of the pesticide degradation, an increase in microbial activity was observed.

TMFCs were also effectively in petroleum hydrocarbon biodegradation [28]. A good removal of alkanes, aromatics, asphaltene and polar materials was found after 223 days. Moreover, adding different biochars enhanced both the removal efficiency and the electrical generation. Similar results were found by Lu et al. [29] in which tubular BESs with carbon cloth anode or biochar anode were used in a field experiment where raw water saturated soils contaminated by petroleum hydrocarbons were used. The removal rate of the total petroleum hydrocarbons (11.46 g/kg dry soil) almost doubled in soils close to the anode (63.5–78.7%) than that in the open circuit positive controls (37.6–43.4%) during a period of 64 days.

3.2 TMFC for remediation of inorganic pollutants

Heavy metal removal from soil is more difficult than organic pollutants because they can only be converted into less toxic forms or immobilized to reduce their bioavailability. Cathode reduction is one of the main mechanisms to remediate heavy metals [30]. The final electron acceptors can be oxygen or metal ions, depending on redox potential of a metal ion: ions with low redox potential tend to use oxygen as terminal acceptor of electrons, while ions with high redox potential in the cathode area can compete with oxygen as terminal electron acceptors. These ions, through a reduction, can be transformed from the ionic form to the elementary state and they can precipitate and be easily separated from the soil [17], [31], [32]. The reaction could not occur at the cathode side when the redox potential of heavy metal (e.g. Pb(II), Cd(II) and Zn(II)) is lower than the anode potential. To facilitate soil removal of such metals, the application of air cathode was proposed [33]. Wang et al. [34] investigated the electrodynamic effect on Cr and Cd degradation and the air-cathode MFCs were effectively. Moreover, the remediation performance was enhanced by decreasing pH and extending operation time. Remediation of different Cr(VI) concentrations in soil was also analysed using a plant-MFC [35]. The latter achieved a 99% Cr(VI) and 27.4% total Cr removals in soils, suggesting that plant uptake, along with the common electrochemical process of a MFC, enhanced the effect of decontamination.

Soil contamination with copper (Cu) is one of the most widespread types of pollution; this metal can migrate under the electric field of a soil MFC from the anode to the cathode [36], making this technology easily used to remove copper from soil. In this regard, Zhang et al. [37] found that the heavy metal migration in soil was mainly due to the electricity generation of MFC, and the migration efficiency increased with increasing electricity generation.

Cadmium and lead (Cd and Pb) migration in TMFCs from the anodic to the cathodic chamber was also found and it was due to the weak electric field [38]. During this migration, a decrease in pH and electric conductivity at the anode was observed; in the cathode area, instead, an increase in pH and electric conductivity were found. Although the migration was observed for both metals, the removal efficiency was different, probably due to their mobility and treatment times.

Since the removal efficiency of the MFC for heavy metals in soil depends on the intensity of the electric field, electron acceptors with high reaction kinetic constants could be used instead of oxygen, or the area of the cathode could be expanded to increase the intensity of the electric field inside the MFC, promoting an efficient heavy metal remediation [39].

3.3 Plant microbial fuel cell for soil recovery

TMFCs have been recently coupled with PABR technique. Plant microbial fuel cell (PMFC) is a promising TMFC variant, which exploits the close relationships and synergic interactions between bacteria and plant in the rhizosphere for degrading contaminants. The rhizosphere secretions in PMFCs can be directly used as electron donors for the biofilm microbial community [40]. PMFCs integrate the photosynthesis mechanism and the TMFCs, to convert solar energy to a carbon source available for bacteria. Thanks to root exudates, bacteria transform/degrade organic matter and contaminants in soil, producing bioenergy by microbial oxidation at the anode region [41]–[43]. Light is a key factor for the technology success because is fundamental for photosynthesis and producing plant biomass and root exudates [42]. The light intensity, photoperiodism and photosynthetic pathway can affect system performance [42], [44]–[46]. Several studies showed the importance of trying different types of plants based on their photosynthetic activity (CAM, C3, C4) [38], [45], [47]. In a PMFC, fundamental factors in determining power outputs are roots, their development and exudates production, which in turn affect the relationships between plant and microbial community and the photosynthetic rates. Although a better performance was observed in C4 plants (in which the CO_2 is first fixed into a compound containing four carbon atoms before entering the Calvin cycle of photosynthesis), it also depends on others factors, such as electrode materials and operating parameters (temperature, pH, conductivity, etc.) [42], [48]. Furthermore, the trend of voltage generation is strictly dependent on the biomass developed, root morphology, released exudates and the microbial activity [42], [48].

PMFC efficiency was tested in closed and open circuit systems to remediate Cr(VI) polluted soil, using two plant species (*Pennisetum alopecuroides* and *Phragmites communis*) selected for their fast growth, high biomass, root development and a great capability to remediate contaminants in plant-assisted bioremediation [35]. Different types of electrode materials were also tested. The results showed that more than 90% of Cr(VI) could be removed in PMFCs; in the open circuit system with plants the removal rate was about 20–30%, proving that plants had an important role in contaminant uptake. The contaminant removal efficiency in the plant absence was approximately 7% due to the absorption onto electrodes and/or the reduction by Cr-reducing microorganisms. Furthermore, some abiotic characteristics of soil changed (pH: from slightly acidic to neutral; reduction of electrical conductivity).

PMFCs were used to degrade petroleum and PAHs (phenanthrene and pyrene), using *Aglaonema commutatum* cv. Silver queen and surfactants (β-cyclodextrin, Tween80 and glucose) [27]. The latter are in fact able to enhance the hydrophobic pollutant solubility in soil [49], [50]. The results showed that the co-presence of the plant and surfactants significantly enhanced the PMFC efficiency in the petroleum removal from soil. Moreover, the degradation rate of the two PAHs significantly increased with increasing surfactant concentrations and in presence of the plants. The β-cyclodextrin addition more strongly enhanced PAH remediation. The charge-transfer resistance values in the PMFCs were significantly reduced with the two surfactants β-cyclodextrin and glucose.

Until now, few works have been reported the role of PMFCs in bioremediation in field studies. There are some aspects to be evaluated and studied: plant viability, long term operation and rhizosphere mechanisms. Anyway, PMFCs have demonstrated their ability to remove contaminants and this technology is promising thanks to both the bioelectrochemical processes and the plant uptake.

4 TECHNOLOGY OF MFC: ELECTRODES AND STRUCTURES

In recent years, MFCs have been designed simplifying their structure and using more efficient and less expensive materials, making MFCs more sustainable electrical generation systems. For example, a reduction from two chambers to a single chamber, with the elimination of additional components such as the proton exchange membrane (PEM) was performed.

Anaerobic oxidation and extracellular electron transfer by specific microbes (e.g. *Geobacteria*), occur at the anode electrode, which works in anaerobic conditions. The cathode electrode is exposed to air, possibly also through an air-diffusion layer, and allows the transfer of oxygen for reduction reactions. Moreover, the oxygen reduction at the cathode electrode is advantageously supported with the aid of a catalytic binder layer [51].

The choice of materials and structure of the electrodes is very important to improve the electrical performance, in particular for electrical energy production. This is more evident for MFC, where output voltage values and intensity of current generated are low, normally insufficient to use MFC as power source directly connected to an electronic system to be powered.

Together with basic characteristics, such as a high conductivity obtained using pyrolyzed polymers, the best electrodes are nanostructured for increasing contact surface. In a fuel cell, a large surface area results, for example, in high reactivity with low material usage, useful for better catalysts [52]. A large surface area corresponds to a higher area of microbial activity, but the first priority is the use of materials showing biocompatibility for working bacteria. The opportunity for microorganisms to work in colonies and to produce conditions in which separate actions cooperate to produce enormous constructive results in terms of electricity generation. Further, a careful study on the geometry, composition, properties, and structure of the electrodes must be dedicated. Moreover, the biofilm development at the anode is crucial for the microbial cells activation. MFC without biofilm catalyst generates power density that is lower 2% than a mature biofilm [53]. The effect of biofilm growth is to decrease the anode impedance and facilitate electrochemical reactions: it also improves, in absence of mediators, the transfer of extracellular electrons from bacteria to anode electrode [54], increasing the generation of electricity. The growth of the biofilm itself is an indicator of better performance in terms of energy and potential clean-up capacity of the environment.

4.1 Research to increase electrical performance

The electrical performance depends on several factors that need to be controlled and optimized: extracellular electron transfer, biofilm growth and microbial adhesion to the surface of the anode substrate, reaction rate of oxygen reduction with oxygenation of the cathode substrate and ion transport. Many of these characteristics are related to the electrode properties. In recent years, the physical and chemical properties of the materials have been studied to be conveniently used as electrodes for MFCs. In particular, the choice was as objectives: biocompatibility of materials for promoting bacterial development, comprising a porous structure to increase the microbial adhesion area; chemical stability to avoid variations in functional conditions and keep the intensity of electricity generation constant; conductivity which must be as low as possible. For example, gold was adopted for its high conductivity and biocompatibility, eventually replaced by other metals with similar properties but with lower costs and simplified manufacturing. In particular, transition metals such as aluminium, copper, nickel, stainless steel and titanium have been used. The disadvantage of using metals for MFC electrodes is their passivation which hinders biofilm formation and microbial growth. Consequently, to obtain a better surface from the point of view of biocompatibility and electron transfer, materials based on carbon material were used. Since 2010, countless variants of electrodes based on this material have been proposed: carbon paper, graphite foil, carbon fibre well, polycrystalline carbon rod, carbon nanotube-textile, 3D carbon material, layered corrugated carbon, interlacing carbon yarn, activated carbon granules, granular activated carbon (GAC) particles, porous carbon with defined pore size, and so on [55]. In particular, the characteristic porosity of carbon-based materials naturally produces more power per geometric surface than the same material with a smooth surface. In addition to the properties of carbon-based materials that increase microbial adhesion to the anode substrate, conductivity, large effective surface area along with pore structure, techniques have been developed to form heteroatom-doped carbon that appear to improve performance of oxygen reduction. But the cost of such processes are high and a valid alternative low-cost has been recently studied that uses biomass-derived carbon [56]. Different solutions have been proposed to improve specific performance. For example, as regards the anode electrode, starting from a fabrication of the carbon cloth electrode, the adhesion of bamboo-like on nitrogen-doped carbon nanotubes has been proposed to improve biocompatibility and promote microbial adhesion to electrode [57]. Modifying electrodes with metal (or metal oxide) nanocomposite, such as titanium (Ti), manganese (Mn), and iron (Fe), allows the extracellular electron transfer from higher distances than a standard MFC electrode. Another example, the polyaniline (PANI) nanocomposite shows good and stable conductivity and the anode electrode produced with this nanocomposite exhibits better microbial adhesion [58]. Use of Poly(3,4-ethylenedioxythiophene) (PEDOT) is preferable when pH working conditions are wide and continuous redox activity is expected at the anode electrode [59].

4.2 Research to improve remediation

The proposal of an electrode modification has been cited previously, which realizes the adhesion of bamboo-like on nitrogen-doped carbon nanotubes in order to improve biocompatibility and promote microbial adhesion to the electrode. However, the goal of this change is mainly the electrical performance of the MFC. Li et al. [60] had an analogous objective but, aiming at the construction of carbon fibre electrodes, demonstrated to obtain that "effective range of remediation and the bioelectricity recovery was extended from 6 to 20 cm with the same area of air-cathode", referring to the bioelectrochemical remediation of petroleum hydrocarbons.

Yu et al. [22] examined the influence of electrode distance on PAH remediation rate in soils and using open soil MFC; removal rates of anthracene, phenanthrene, and pyrene significantly increased (from 21%, 17% and 12%, to 54%, 43 and 27% respectively). This was obtained using activated carbon fibre felt for both electrodes, and decreasing distance between anode and cathode electrodes from 10 to 4 cm. Other studies focused on electrode modifications for improving contaminant degradation in the area where MFCs work [61].

An example of the methodology used by the authors is extensively described in a previous article of Borello et al. [1]. It refers to the application of MFC for degrading DDE, the main toxic metabolite of the pesticide DDT. In order to evaluate the effectiveness of the method, soil was sampled (30 cm depth) of an ex-agricultural area located in central Italy, with the following characteristics: pH 7, 1.35% organic carbon and the texture was 28% sand, 24%

silt and 48% clay. A municipal solid waste compost was mixed (3%) with the soil to stimulate microbial community activity. Finally, a stock solution of DDE in acetone was diluted in distilled water and used to obtain a final DDE concentration of 1 mg/kg for each MFC.

The MFC used was a single-chamber type with a simplified configuration that uses graphite-based electrodes (anode and cathode) and does not require the use of semipermeable membranes. In this terrestrial MFC, the soil act as an electrolyte for the drift of hydrogen ions towards the cathode. Moreover, it provides both the bacterial community together with the nutrients existing naturally in the matrix. The anode was placed over a 1 cm layer of soil. Subsequently each chamber was filled with soil (at least 5 cm), the soil was completely saturated with water in order to obtain anaerobic conditions; moreover the air bubbles removed. The cathode was then placed on the top of the soil and was exposed to air. An external circuit connected anode to cathode to make the electron movement. Using this scheme, the analysis of the percentage of DDE removal was carried out experimentally after 2, 6 and 12 months, obtaining values up to 46.6%. The data were also compared to batches (in absence of cathode or anode electrodes, in order to find if the TMFC was more effectively in contaminant degradation.

5 TMFCS USED IN LARGE SCALE AND OPEN FIELD APPLICATIONS

As a matter of fact, a single TMFC/PMFC can provide low voltage and low power to supply ultra-low power electronic devices and transmitters [62]. Anyway, stacking MFCs allows to improve the output performance for large-scale applications. Unfortunately, there are possible issues to manage while stacking MFCs, as an example voltage reversal can occur. Pamintuan et al. [63] studied terrestrial PMFC with a combination of series and parallel connections in a stack. The combination of parallel and series configurations was the most efficient technique. Moreover, the voltage and current of stacking a few cells in series or parallel had a reduction in performances while increasing the number of cells connected linearly. Stacking of many cells in pure series or pure parallel connections is discouraged as voltage reversal can hinder the stack from reaching its theoretical potential.

Other reasons can bring to voltage reversal as several factors diversify the output of individual MFCs. The internal resistance and the construction design of reactors can be a key factor. Another way to reduce the losses due to differences between reactors connected in stack is to isolate each cell putting at the output of each cell a step-up converter instead of connecting all the cells together before the power management unit, which includes a single converter. The converter is always necessary to adapt the level of production of an MFC to the electrical needs of electronics devices. The power management comprises multiple DC–DC converters for each MFC [64]. The energy harvesting in MFCs is performed by proper arrangement of DC–DC converters where every MFC is connected to DC–DC converters and the acquired energy is collected as high voltage in a single capacitor that supplies the load. In this system, MFCs are connected in series with a resistor to make the mean internal resistance.

In the last few years, some interesting new construction typology of MFC has risen up. Sudirjo et al. [65] tested PMFCs in both lab and in field studies. The practical configuration used was a tubular MFC used for large-scale applications without topsoil excavation. In this study, tubular plant-MFC (1 m length; width: anode: 19 cm; cathode: 10 cm) was installed in triplicate in a paddy field located in West Kalimantan, Indonesia. The plant-MFC reactors were installed in lines with a 30 cm distance between reactors. Interestingly, the spacer was made from non-conductive materials to prevent short circuiting, but was otherwise completely permeable. The tubular plant-MFCs generated electricity continuously during the crop season as long as the rice paddy is flooded. In one crop season, 9.5–15 Wh/m^2 electricity was continuously generated with an average of 0.4 ± 0.1 mW per meter tube. Between two crop seasons, plant-MFCs may be utilized in the form of biosensors as rain event indicators. A transparent silicon tube (VMQ silicone, 12/16 mm inner/outer diameter) was used to supply oxygen. Although this MFC configuration is very innovative, it needs improvements to be valuable as a depollution system for field studies, although it is a promising technology. The system also included a wireless sensor network based on the LoRa system that allowed long range of data transmitting, in this case the wireless transmission system was empowered with external batteries and solar panels.

6 CONCLUSION

Microbial fuel cells (MFCs) are considered as a promising system, which can generate electricity, mitigate greenhouse gas emissions and potentially recover soil from pollution. They have a low cost and are able to work for long periods. Terrestrial microbial fuel cells (TMFCs) are an alternative method for energy production using a large variety of substrates and materials for bioenergy production. TMFCs have been tested for removing pollutants (persistent organic contaminants and metals) from contaminated matrices thanks to the presence of microorganisms naturally present in soils that can use electrodes as electron acceptors during their metabolic activities.

TMFCs in field studies have a great potential for both soil recovery and develop of monitoring systems. The energy harvested can be useful to empower low power wireless sensor network nodes, permitting to collect and send data for environmental monitoring. TMFC efficiency have been usually limited in conventional MFC design, which submerges the anode in the soil and places the cathode above the soil surface. Especially in field studies, the conventional configuration allows the depollution of a little portion of the surface soil. Some limitations related to several construction characteristics of TMFCs such as the distance between the electrodes, the shape and the dimensions of the reactors have to be better investigate in order to improve the soil recovery.

ACKNOWLEDGEMENTS

We acknowledge the COST Action Horizon program and the COST Action PHOENIX CA19123: Protection, resilience, rehabilitation of damaged environment.

REFERENCES

[1] Borello, D. et al., Use of microbial fuel cells for soil remediation: A preliminary study on DDE. *Int. J. Hydrog. Energy*, **46**, pp. 10131–10142, 2021.

[2] Di Lenola, M. et al., Combined effects of compost and *Medicago sativa* in recovery a PCB contaminated soil. *Water*, **12**(3), p. 860, 2020.

[3] Ancona, V. et al., Microbial fuel cell: An energy harvesting technique for environmental remediation. *Int. J. Environ. Impacts*, **3**(2), pp. 168–179, 2020.

[4] Tsipa, A., Varnava, C.K., Grenni, P., Ferrara, V. & Pietrelli, A., Bio-electrochemical system depollution capabilities and monitoring applications: Models, applicability, advanced bio-based concept for predicting pollutant degradation and microbial growth kinetics via gene regulation modelling. *Processes*, **9**(6), p. 1038, 2021.

[5] Wenzel, W.W., Rhizosphere processes and management in plant-assisted bioremediation (phytoremediation) of soils. *Plant Soil*, **321**(1), pp. 385–408, 2009.

[6] Ramanjaneyulu, A.V., Neelima, T.L., Madhavi, A. & Ramprakash, T., Phytoremediation: An overview. *Applied Botany*, eds R.M. Humberto, G.R. Ashok, K. Thakur & N.C. Sarkar, American Academy Press, pp. 42–84, 2017.

[7] Ancona, V., et al. Plant-assisted bioremediation: An ecological approach for recovering multi-contaminated areas. *Soil Biological Communities and Ecosystem Resilience,* eds M. Lukac, P. Grenni & M. Gamboni, Springer, pp. 291–303, 2017.

[8] Ancona, V. et al., Plant assisted bioremediation: Soil recovery and energy from biomass. *Assisted Phytoremediation*, V. Pandey, Elsevier, 2021.

[9] Ancona, V. et al., Plant-assisted bioremediation of a historically PCB and heavy metal contaminated area in Southern Italy. *New Biotechnol.* **38**, pp. 65–73, 2017.

[10] Ancona, V. et al., Heavy metal phytoremediation of a poplar clone in a contaminated soil in southern Italy. *J. Chem. Technol. Biotechnol.*, **95**, pp. 940–949, 2020.

[11] Ancona, V. et al., Poplar-assisted bioremediation for recovering a PCB and heavy-metal-contaminated area. *Agriculture*, **11**, p. 689, 2021.

[12] Barra Caracciolo, A. et al., Characterization of the belowground microbial community in a poplar-phytoremediation strategy of a multi-contaminated soil. *Front Microbiol.*, **11**, p. 2073, 2020.

[13] Ancona, V. et al., Gasification treatment of poplar biomass produced in a contaminated area restored using plant assisted bioremediation. *J. Env. Manag.*, **239**, pp. 137–141, 2019.

[14] Du, Z., Li, H. & Gu, T., A state of the art review on microbial fuel cells: A promising technology for wastewater treatment and bioenergy. *Biotechnol. Adv.*, **25**, pp. 464–482, 2007.

[15] Logan, B.E., *Microbial Fuel Cells*, John Wiley, 2008.

[16] Casula, E., Kim, B., Chesson, H., Di Lorenzo, M. & Mascia, M., Modelling the influence of soil properties on performance and bioremediation ability of a pile of soil microbial fuel cells. *Electrochimica Acta*, **368**, 137568, 2021.

[17] Abbas, S.Z. & Rafatullah, M., Recent advances in soil microbial fuel cells for soil contaminants remediation. *Chemosphere* **272**, 129691, 2021.

[18] Xu, X., Zhao, Q., Wu, M., Ding, J. & Zhang, W., Biodegradation of organic matter and anodic microbial communities analysis in sediment microbial fuel cells with/without Fe (III) oxide addition. *Biores. Technol.* **225**, pp. 402–408, 2017.

[19] Gustave, W. et al., Soil organic matter amount determines the behavior of iron and arsenic in paddy soil with microbial fuel cells. *Chemosphere*, **237**, 124459, 2019.

[20] Pham, H., Boon, N., Marzorati, M. & Verstraete, W., Enhanced removal of 1,2-dichloroethane by anodophilic microbial consortia. *Water Res.*, **43**(11), pp. 2936–2946, 2009.

[21] Huang, D.Y., Zhou, S.G., Chen, Q., Zhao, B., Yuan, Y. & Zhuang, L., Enhanced anaerobic degradation of organic pollutants in a soil microbial fuel cell. *Chem. Eng. J.*, **172**, pp. 647–653, 2011.

[22] Yu, B., Tian, J. & Feng, L., Remediation of PAH polluted soils using a soil microbial fuel cell: influence of electrode interval and role of microbial community. *J. Haz. Mat.*, **336**, pp. 110–118, 2017.

[23] Miran, W., Nawaz, M., Jang, J. & Lee, D.S., Chlorinated phenol treatment and in situ hydrogen peroxide production in a sulfate-reducing bacteria enriched bioelectrochemical system. *Water Res.*, **117**, pp. 198–206, 2017.

[24] Xu, X., Zhao, Q.L. & Wu, M.S., Improved biodegradation of total organic carbon and polychlorinated biphenyls for electricity generation by sediment microbial fuel cell and surfactant addition. *RSC Adv.*, **5**, 62534, 2015.

[25] Cao, X., Song, H., Yu, C. & Li, X., Simultaneous degradation of toxic refractory organic pesticide and bioelectricity generation using a soil microbial fuel cell. *Biores. Technol.* **189**, pp. 87–93, 2015.

[26] Sherafatmand, M. & Ng, H.Y., Using sediment microbial fuel cells (SMFCs) for bioremediation of polycyclic aromatic hydrocarbons (PAHs). *Biores. Technol.*, **195**, pp. 122–130, 2015.

[27] Zhao, L., Deng, J., Hou, H., Li, J. & Yang, Y., Investigation of PAH and oil degradation along with electricity generation in soil using an enhanced plant-microbial fuel cell. *J. Cleaner Prod.*, **221**, pp. 678–683, 2019.

[28] Li, X. et al., Long-term effect of biochar amendment on the biodegradation of petroleum hydrocarbons in soil microbial fuel cells. *Sci. Total Env.*, **651**, pp. 796–806, 2019.

[29] Lu, L., Huggins, T., Jin, S., Zuo, Y. & Ren, Z.J., Microbial metabolism and community structure in response to bioelectrochemically enhanced remediation of petroleum hydrocarbon-contaminated soil. *Environ. Sci. Technol.*, **48**, pp. 4021–4029, 2014.

[30] Erable, B. et al., Application of electro-active biofilms. *Biofouling*, **26**, pp. 57–71, 2010.

[31] Wu, M., Xu, X., Zhao, Q. & Wang, Z., Simultaneous removal of heavy metals and biodegradation of organic matter with sediment microbial fuel cells. *RSC Adv.*, **7**, pp. 53433–53438, 2017.

[32] Mathuriya, A.S. & Yakhmi, J.V., Microbial fuel cells to recover heavy metals. *Environ. Chem. Lett.*, **12**, pp. 483–494, 2014.

[33] Habibul, N., Hu, Y. & Sheng, G.P., Microbial fuel cell driving electrokinetic remediation of toxic metal contaminated soils. *J. Hazard. Mater.*, **318**, pp. 9–14, 2016.

[34] Wang, H. et al., Bioelectrochemical remediation of Cr(VI)/Cd(II)-contaminated soil in bipolar membrane microbial fuel cells. *Env. Res.*, **186**, 109582, 2020.

[35] Guana, C.-Y. et al., Wetland plant microbial fuel cells for remediation of hexavalent chromium contaminated soils and electricity production. *J. Haz. Mat.*, **365**, pp. 137–145, 2019.

[36] Wang, H. et al., New process for copper migration by bioelectricity generation in soil microbial fuel cells. *Environ. Sci. Pollut. Control Ser.*, **23**(13), pp. 13147–13154, 2016.

[37] Zhang, J. et al., Simultaneous enhancement of heavy metal removal and electricity generation in soil microbial fuel cell. *Ecotox. Environ. Saf.*, **192**, 110314, 2020.

[38] Habibul, N. et al., Bioelectrochemical chromium(VI) removal in plant-microbial fuel cells. *Environ. Sci. Technol.*, **50**, pp. 3882–3889, 2016.

[39] Zhang, J. et al., Effects of cathode/anode electron accumulation on soil microbial fuel cell power generation and heavy metal removal. *Environ. Res.*, **198**, 111217, 2021.

[40] Srivastava, P., Yadav, A.K. & Mishra, B.K., The effects of microbial fuel cell integration into constructed wetland on the performance of constructed wetland. *Bioresource Technol.*, **195**, pp. 223–230, 2015.

[41] Lu, L. et al., Microbial community structure accompanied with electricity production in a constructed wetland plant microbial fuel cell. *Biores. Technol.*, **195**, pp. 115–121, 2015.

[42] Nitisoravut, R. & Regmi, R., Plant microbial fuel cells: A promising biosystems engineering. *Ren. Sust. Energy Rev.*, **76**, pp. 81–89, 2017.

[43] Jadhav, D.A., Ghosal, D., Chendake, A.D., Pandit, S. & Sajana, T.K., Plant microbial fuel cell as a biomass conversion technology for sustainable development. *Catalysis for Clean Energy and Environmental Sustainability*, eds K.K. Pant, S.K. Gupta & E. Ahmad, Springer: Cham, 2021.

[44] Strik, D.P.B.T.B., Hamelers, H.V.M., Snel, J.F.H. & Buisman, C.J.N., Green electricity production with living plants and bacteria in a fuel cell. *Int. J. Energy Res.*, **32**, pp. 870–876, 2008.

[45] Helder, M. et al., Concurrent bioelectricity and biomass production in three plant-microbial fuel cells using *Spartina anglica, Arundinella anomala* and *Arundo donax. Biores. Technol.*, **101**, pp. 3541–3547, 2010.
[46] Xiao, L. & He, Z., Applications and perspectives of phototrophic microorganisms for electricity generation from organic compounds in microbial fuel cells. *Ren. Sust. Energy Rev.*, **37**, pp. 550–559, 2014.
[47] Yan, Z., Jiang, H., Cai, H., Zhou, Y. & Krumholz, L.R., Complex interactions between the macrophyte *Acorus calamus* and microbial fuel cells during pyrene and benzo[a] pyrene degradation in sediments. *Sci. Rep.*, **5**, 10709, 2015.
[48] Deng, H., Chen, Z. & Zhao, F., Energy from plants and microorganisms: Progress in plant–microbial fuel cells. *Chem. Sus. Chem.*, **5**, pp. 1006–1011, 2012.
[49] Li, X. et al., Sand amendment enhances bioelectrochemical remediation of petroleum hydrocarbon contaminated soil. *Chemosphere*, **141**, pp. 62–70, 2015.
[50] Bardi, L,. Mattei, A. & Steffan, S., Hydrocarbon degradation by a soil microbial population with b-cyclodextrin as surfactant to enhance bioavailability. *Enzyme Microb. Technol.*, **27**, pp. 709–713, 2000.
[51] Wang, Z., Cao, C., Zheng, Y., Chen, S. & Zhao, F., Abiotic oxygen reduction reaction catalysts used in microbial fuel cells. *Chem. Electro. Chem.*, **1**, pp. 1813–1821, 2014.
[52] Stamatin, I., Morozan, A. & Dumitru, A., Nanostructured materials used for design the components of fuel cells. *Rom. J. Phys.*, **49**, pp. 767–775, 2004.
[53] Hou, B., Hu, Y.Y., Sun, J. & Cao, Y.Q., Effect of anodic biofilm growth on the performance of the microbial fuel cell (MFC). *2010 4th International Conference Bioinforma. Biomed. Eng.*, pp. 1–4, 2010.
[54] Choudhury, P. et al., Performance improvement of microbial fuel cell (MFC) using suitable electrode and Bioengineered organisms: A review. *Bioengineered.*, **8**, pp. 471–487, 2017.
[55] Jayesh, M. et al., Recent advances in the development and utilization of modern anode materials for high performance microbial fuel cells. *Biosens. Bioelectr.*, **90**, pp. 558–576, 2017.
[56] Yang, W. & Chen, S., Biomass-derived carbon for electrode fabrication in microbial fuel cells: A review. *Ind. Eng. Chem. Res.*, **59**, pp. 6391−404, 2020.
[57] Ci, S., Wen, Z., Chen, J. & He, Z., Decorating anode with bamboo-like nitrogen-doped carbon nanotubes for microbial fuel cells. *Electrochem. Commun.*, **14**, pp. 71–74, 2012.
[58] Wang, P., Li, H. & Du, Z., Polyaniline synthesis by cyclic voltammetry for anodic modification in microbial fuel cells. *Int. J. Electrochem. Sci.*, **9**, pp. 2038–2046, 2014.
[59] Kang, Y.L., Ibrahim, S. & Pichiah, S., Sinergetic effect of conductive polymer poli(3,4-ethylenedioxythiophene) with different structural configuration of anode for microbial fuel cell application. *Bioresour. Technol.*, **189**, pp. 364–369, 2015.
[60] Li, X. et al., Carbon fiber enhanced bioelectricity generation in soil microbial fuel cells. *Biosens. Bioelectron.*, **85**, pp. 135–141, 2016.
[61] Logan, B.E., Rabaey, K., Conversion of wastes into bioelectricity and chemicals by using microbial electrochemical technologies. *Science*, 337, pp. 686–690, 2012.
[62] Rajeev, P. et al., Plug into a plant: Using a plant microbial fuel cell and a wake-up radio for an energy neutral sensing system. *Proceedings of 2017 IEEE 42nd Conference on Local Computer Networks Workshops (LCN workshops)*, IEEE Press, Singapore, pp. 18–25, 2017.

[63] Pamintuan, K. et al., An analysis of the stacking potential and efficiency of plant-microbial fuel cells growing green beans (*Vigna ungiculata* ssp. *sesquipedalis*). *Int. J. Renew. Energy Develop.*, **9**, pp. 439–447, 2020.
[64] Saravanakumar, K. & Rajeswari, R., Microbial fuel cell-based self-powered biosensor for environment monitoring in IoT cloud framework. *Concurrency Comput. Pract. Ex.*, **31**(15), e5165, 2019.
[65] Sudirjo, E. et al., Performance and long distance data acquisition via LoRa technology of a tubular plant microbial fuel cell located in a paddy field in West Kalimantan, Indonesia. *Sensors* **19**(21), p. 4647, 2019.

[63] Pamuncak, A. et al., An analysis of [illegible] stacking protocols and efficiency of [illegible] microbial fuel cells [illegible] (Vertical [illegible]) [illegible], pp. 429–447, 2020.
[64] [illegible] & Rameswari, K., Microbial fuel cell based self-powered biosensor for environmental monitoring in [illegible] 34(15), pp. 163, 2019.
[65] [illegible], [illegible], Performance and long distance data acquisition via LoRa technology of a tubular plant microbial fuel cell located in a paddy field in West Kalimantan, Indonesia. Sensors, 19(21), p. 4647, 2019.

WATER POLLUTION THREATENING MARINE, COASTAL AND ESTUARINE SYSTEMS: A REVIEW OF ENVIRONMENTAL-ECONOMIC APPROACHES FOR THE ASSESSMENT OF DEVELOPMENT STRATEGIES

MARIA I. BASTOS[1,2], PETER C. ROEBELING[2,3], FÁTIMA LOPES ALVES[2] & SEBASTIAN VILLASANTE[1,4]
[1]Campus Do*Mar, University of Vigo Campus, Spain
[2]CESAM and Departamento de Ambiente e Ordenamento, University of Aveiro, Portugal
[3]Wageningen Economic Research, Wageningen University and Research (WUR), The Netherlands
[4]Cross-Research in Environmental Technologies (CRETUS), Department of Applied Economics, University of Santiago de Compostela, Spain

ABSTRACT

High-risk industrial pollution is one of the most harmful events threatening natural ecosystems. Water pollution abatement strategies (WPAS) – ranging from mitigation to adaptation strategies – are key to deal with these threats. While WPAS have to be adequately valued and assessed, environmental-economic (EE) reasoning – resorting mainly to three analytic tools, namely cost effectiveness analysis (CEA), cost–benefit analysis (CBA) and efficiency analysis (EA) – is critical for developing regional development strategies. This study reviews the state of knowledge on environmental-economic water pollution abatement studies, using a semi-systematic literature review approach. Results show, first, that economic analysis can support the sustainable planning and management of marine, coastal, estuarine and freshwater socio-ecological systems (MCEF-SES), effectively dealing with environmental pollution and push WPAS forward. Second, the most used EE approach within the context of MCEF-SES is CEA (34%), followed by CBA (19%), EA (19%), cost assessment (18%) and benefit assessment (10%). Third, until now EE research focused mostly on: (i) the gathering and examination of data; (ii) determining how to assess anthropogenic-induced ecosystem damages; (iii) assessing pollution damages; and (iv) performing CEA of diffuse-source/non-point source pollution. Fourth, even though some works have already been conducted on CBA and spatially explicit dynamic economic models, produced research still concentrates mostly on continuous/diffuse source pollution and oil spills. Out of these, EE scientific research remains sparse. Fifth, CBA is a useful tool to support EE decisions and to explore alternative WPAS, thus supporting better social assessments. Sixth, EA allows for the simultaneous analysis of multiple variables, dimensions and scenarios and is viewed as the most comprehensive of all EE analysis tools. However, most often its complexity limits its usage.
Keywords: coastal socio-ecological systems, water pollution, ecosystem impacts, environmental-economic approach, mitigation and adaptation strategies.

1 INTRODUCTION

Changes in land use, industrialization and urbanization, intensify environmental vulnerability and expose marine, coastal, estuarine and freshwater socio-ecological systems (MCEF-SES) to several types and sources of pressure [1]–[4]. Water pollution (WP) types range from continuous-persistent pollution (CPP) to sudden-accidental pollution (SAP), and sources of WP range from point-source pollution (PSP) to diffuse-source pollution (DSP). CPP refers to the sustained, uninterrupted and lingering release of substances that remain in the environment and bio-accumulate through the food-chain [5]. SAP refers to "releases of pollutants that begin abruptly, are fortuitous, and arise without warning, or unexpectedly" [6, p. 617]. PSP is the pollution produced by a stationary site or fixed facility discharging pollutants into the environment, which can easily be traced-back to its origin [7]. Finally, DSP is the pollution coming from dispersed sources with no single source-point or pollution outlet, whose origin is difficult to trace-back to its source [8].

WIT Transactions on The Built Environment, Vol 207, © 2021 WIT Press
www.witpress.com, ISSN 1743-3509 (on-line)
doi:10.2495/DMAN210111

The most significant impacts of pollution on water quality include toxic inorganic compounds (heavy metals; toxic elements), toxic organic compounds (pesticides, paints, pharmaceuticals), infectious organisms, oxygen-demanding wastes (sewage, paper/pulp effluents, food-processing wastes), and eutrophication [9]. These, in turn, impact natural ecosystem functions, services and values with subsequent biological and socioeconomic consequences [1], [10], [11]. To halt this dilapidation of natural capital and attain sustainable development, public policies are needed that aim to reinforce ecosystem conservation, remedy persistent adverse consequences of human action on natural ecosystems, improve the way resources are used and mitigate ecological hazards [12].

Water pollution abatement strategies (WPAS) comprehend mitigation and/or adaptation strategies. Mitigation measures seek to decrease or completely stop the likelihood and severity of pollution [13] and, thus, aim at reducing the disposal of water pollutants into the environment [14]. Adaptation measures seek to minimize the impact of emitted pollution, and focus mainly on cleaning operations or remediation measures [15]–[18].

Sustainable economic development requires the valuation of the negative impacts of pollution as well as the estimation of abatement costs, abatement benefits and net welfare gains. However, most economic studies on the subject do not assess the monetary value of ecosystem services (ES) provided by nature, do not consider environmental values in the decision making processes [19] and do not incorporate sociocultural values and the economic impact of the ES on welfare [20]. Environmental-economics aids in the quantification and valuation of benefits provided by the environment, the evaluation of policy instruments and the estimation of welfare implications from sustainable natural resources management in MCEF-SES – thus establishing a bridge between underpinning bio-physical sciences on the one hand and environmental governance and policy on the other hand [21].

Taking the production and protection standpoint analysis of public policies formulation (which balances environmental protection targets and economic development goals), the objective of this paper is to review the current status of environmental-economic knowledge dealing with water pollution threatening MCEF-SES so that research gaps are identified. After presenting the methodology (Section 2), obtained results are summarized and organized in a Cartesian plane (Section 3). Finally, strengths and weaknesses of each environmental-economic approach are discussed (Section 4), and knowledge gaps identified (Section 5).

2 MATERIAL AND METHODS

This review study focusses on peer-reviewed scientific papers (published in English, between 2000 and 2016) on the economics of MCEF-SES anthropogenic pollution, the strategies to reduce this anthropogenic pollution and the environmental-economic (EE) tools to address them. To this end, the Web of Science (WoS) database and the WoS main collection were used as search tool and primary database.

2.1 Search method

A WoS multistep web search based on specific keywords was undertaken, which was completed by a supporting-tier of multiple snow-ball reviews (based on WoS pop-up results when inspecting WoS citations and on the scrutiny of WoS references from key-selected papers). The anchor point of this exploration is the economic approach applied to high-risk WP events threatening MCEF-SES. Using approximately 50 keywords and expressions (see Table 1), the process resulted in 94 selected papers focusing on MCEF water pollution issues, analyzed from an economic perspective.

Table 1: Web of Science used search topics.

Rounds	Used search keywords
1st	["spill", "industrial risk", "industrial disaster", "port risk", "port disaster", "port hazard"] AND costs AND benefits AND [marine, maritime, coastal, estuary, lagoon].
2nd	["environmental impact*", "pollution cost*", "point source pollution", "sudden pollution", "pollution hazard", contaminant] AND costs AND benefits AND [port, harbor, harbor, "freshwater ecosystems"].
3rd	["societal costs", "cost assessment", "abatement costs", "damage costs", "cost-benefit analysis", "comparing costs and benefits", "costs-benefits", costs] AND [estuary, ports, coast*, marine, "fresh water ecosystems"] AND [hazard, pollution, contamination, "risk assessment", risk, pollution, "environmental hazard*"].
4th	"EU Water Framework Directive" AND "cost-benefit analysis".

2.2 The analytical framework

Selected papers were organized in a Cartesian plane (Fig. 1), with the horizontal axis representing the type of pollution source (DSP vs. PSP) and the vertical axis representing the type of water pollution (SAP vs. CPP). This results in four quadrants: Q1 demarcated by the PSP and the CPP axes (labelled *Ominous Plume* quadrant); Q2 defined by the CPP and the DSP axes (labelled *Ominous Tide* quadrant); Q3 delineated by the DSP and the SAP axes (labelled *Toxic Tide* quadrant); and Q4 defined by the SAP and the PSP axes (labelled *Toxic Plume* quadrant). The center of the Cartesian plane is labelled *Common-mire*.

3 RESULTS

3.1 Synthesis of results

The 94 selected papers represent studies developed worldwide, with 51 studies focusing on European, 18 on Australian, 13 on North American, eight on Chinese and four on World systems. Fig. 1 shows that the majority of considered papers (N = 52 (55%)) deal with CPP/DSP events (*Ominous Tide*; Q2), with most of them focusing on nutrient enrichment (20 out of 52) and pollution produced by agriculture activities (15 out of 52) while only few focusing on other types of pollution events (diversified anthropogenic activities, water scarcity and climate change). The second biggest group of papers (N = 15 (16%)) focuses on SAP/PSP events (*Toxic Plume*; Q4), with nearly all of these papers studying oil spills while only two focusing on industrial hazards. The third group of papers (N = 12 (13%)) focuses on PSP/CPP events (*Ominous Plume*; Q1), with studies focusing on a variety of topics, like shipping, water treatment and aquaculture. The fourth group of papers (N = 13(14%)) is generic or covers multiple types of pollution events and, as such, overlaps all four quadrants and sits in the center of the Cartesian plane. Finally, the smallest group of papers (N = 2 (2%)) focusses on SAP/DSP events (*Toxic Tide*; Q3), mostly originated by flooding and associated pollution.

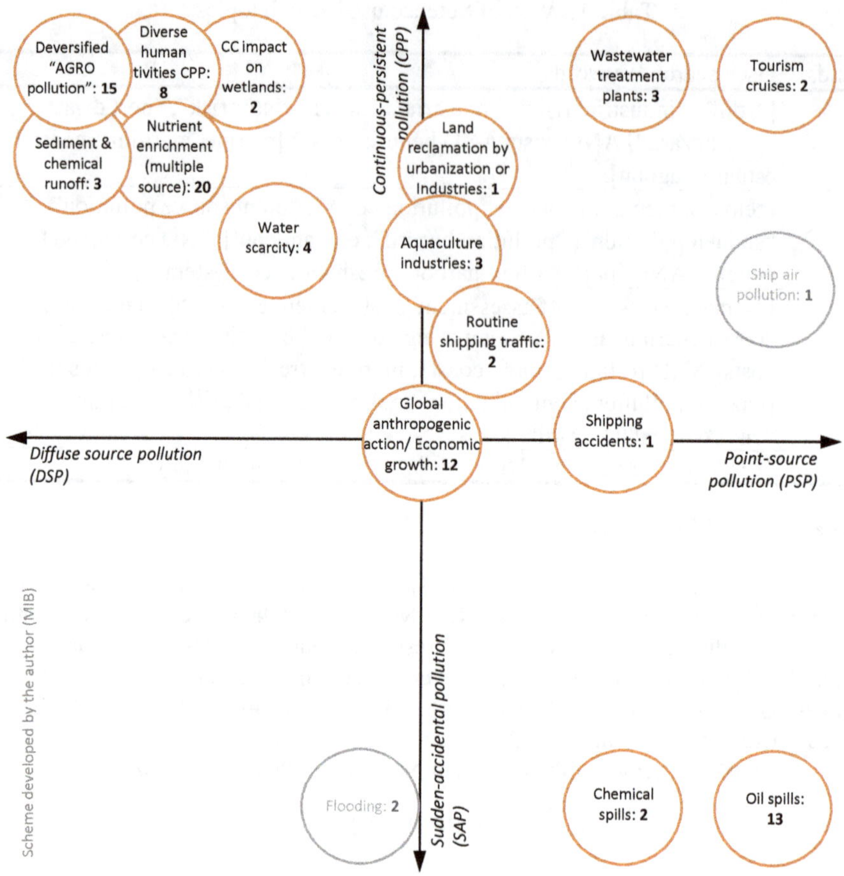

Figure 1: Water pollution types threatening marine, coastal, estuarine and freshwater socio-ecological systems.

3.2 Description of results per quadrant

In the *Ominous Plume* quadrant (PSP/CPP; Fig. 1 and Table 2), one study uses a cost-effectiveness (CEA) approach, five studies use a cost–benefit assessment (CBA) approach, five studies use a partial CBA approach (three cost-assessments (CA) and two benefit assessments (BA)) and, finally, one study uses an efficiency assessment (EA) approach. In the *Ominous Tide* quadrant (CPP/DSP), 28 studies use a CEA approach, 15 studies use an EA approach, five studies use a partial CBA approach (focusing on BA) and four studies use a CBA approach. In the *Toxic Tide* quadrant (DSP/SAP), the two studies use a CBA approach. In the *Toxic Plume* quadrant (SAP/PSP), 11 studies use a partial CBA analysis (focusing on CA), two studies use a CBA approach, one study uses a CEA approach and, finally, one study uses an EA approach. Finally, in the *Common-mire* (center), five studies use a CBA approach, five studies use a partial CBA approach (three focusing on CA and two focusing on BA), two studies uses a CEA approach and one study uses a EA approach.

Table 2: Assessed measures and used methods in considered studies per quadrant.

Quadrant	Study	Measures	Methods
Ominous Plume (Q1)	**Cost-effectiveness analysis (CEA): 1 study**		
	[3]	M/A	Valuation methods
	Cost–benefit analysis (CBA): 5 studies		
	…	…	…
	[23]	Mitigation	Valuation methods
	…	…	…
	…	…	…
	Cost Assessment (CA): 3 studies		
	[30]	Adaptation	Valuation methods
	…	…	…
	[25]	M/A	Valuation methods
	Benefit Assessment (BA): 2 studies		
	[33]	Mitigation	Valuation methods
	[34]	Mitigation	Valuation methods
	Efficiency analysis (EA): 1 study		
	[35]	M/A	Optimal control
Q3	**Cost–benefit analysis (CBA): 2 studies**		
	[37]	Mitigation	LP + VM
	[38]	Mitigation	Valuation methods
Common-mire (Q1, Q2, Q3, Q4)	**Cost-effectiveness analysis (CEA): 1 study**		
	[39]	M/A	LP
	[11]	Mitigation	Valuation methods
	Cost–benefit analysis (CBA): 5 studies		
	[41]	M/A	Valuation methods
	[42]	M/A	Valuation methods
	…	…	…
	[44]	M/A	BBN
	…	…	…
	Cost assessment (CA): 3 studies		
	[46]	M/A	Valuation methods
	[47]	Mitigation	Meta-analysis
	[48]	M/A	LP
	Benefit assessment (BA): 2 studies		
	[49]	Mitigation	Valuation methods
	[50]	M/A	LP
	Efficiency analysis (EA): 1 study		
	[51]	M/A	Optimal control

Quadrant	Study	Measures	Methods
Ominous Tide (Q2)	**Cost-effectiveness analysis (CEA): 28 st.**		
	[22]	M/A	LP
	…	…	…
	[21]	Mitigation	Meta-modelling
	[28]	M/A	LP
	…	…	…
	Cost–benefit analysis (CBA): 4 studies		
	[29]	Adaptation	Valuation methods
	…	…	…
	[31]	M/A	Valuation methods
	[32]	M/A	Valuation methods
	Benefit Assessment (BA): 5 studies		
	[24]	Adaptation	Valuation methods
	…	…	…
	…	…	…
	…	…	…
	[36]	Mitigation	Valuation methods
	Efficiency analysis (EA): 15 studies		
	[27]	Mitigation	Optimal control
	…	…	…
	[26]	M/A	Optimal control
	…	…	…
	[40]	Mitigation	Optimal control
	…	…	…
Toxic Plume (Q4)	**Cost-effectiveness analysis (CEA): 1 study**		
	[43]	Mitigation	Meta-modelling
	Cost–benefit analysis (CBA): 2 studies		
	[45]	M/A	Probabilistic (BBN)
	[16]	Adaptation	Optimal control
	Cost assessment (CA): 11 studies		
	[1]	Adaptation	Valuation methods
	…	…	…
	[17]	Adaptation	Probabilistic (BBN)
	…	…	…
	[18]	Adaptation	RA
	Efficiency analysis (EA): 1 study		
	[52]	Mitigation	Optimal control

Notes: M/A = Mitigation/adaptation; BBN = Bayesian Believe Network; LP = Linear programming; CP = Compromising programming; RA = Regression analysis; VM = Valuation methods.

CEA studies (32 out of 94 studies) aim to assess which pollution abatement measures achieve pollution reduction targets and/or environmental improvements at least cost [3], [21], [22]. CBA studies (18 out of 94 studies) aim to identify the abatement measure scenario offering the largest net welfare (i.e. difference between costs and benefits [23]). BA studies (nine out of 94 studies) aim to assess the willingness to pay for pollution adaptation measures or determine the benefits from pollution mitigation options [24]. CA studies (17 out of 94 studies) aim to assess the costs of water pollution or the costs of pollution abatement measures [1], [25]. Studies using the EA approach (18 out of 94 studies) aim to identify the portfolio of abatement measures that provide the largest total net welfare [26], [27].

Studied abatement measures in the *Ominous Plume* quadrant (PSP/CPP) include water pollution mitigation measures (such as improved port safety, shipping traffic and ship characteristics, in-site tertiary treatment of wastewater, and biomitigation) and adaptation measures (such as utilization of sulphur scrubbers and marine gas oil, treatment of emissions, waste collection, transportation and disposal). Studied abatement measures in the *Ominous Tide* quadrant (CPP/DSP) focus mainly on best management practices (such as land retirement, (re)construction of wetlands and riparian buffers) and adaptation measures (such as using coastal waters and wetlands to filter nutrients and care for endangered species). In the *Toxic Tide* quadrant (DSP/SAP), no studies focusing on water pollution were found. However, given that flooding cause noxious impacts on the quality of MCEF-SES, and flooding mitigation articles can provide valuable insights on EE analysis applied to MCE-SES, 2 studies focusing on flooding issues were included in the present analysis; both studies assess flood mitigation options by local authorities. Studied abatement measures in the *Toxic Plume* quadrant (SAP/PSP) include mitigation measures (such as technical improvement of oil tankers and accident prevention measures) and adaptation measures (such as clean-up, oil combating equipment and restoration of ecosystems).

4 DISCUSSION

4.1 Water pollution events and impacts on marine coastal, estuarine and freshwater socioecological systems

Research on EE approaches to MCEF-SES pollution shows that continuous diffuse source (*Ominous Tide* quadrant) and sudden point source (*Toxic Plume* quadrant) water pollution constitute the most investigated types of events. Investigated continuous point source water pollution (*Ominous Plume* quadrant) remains miscellaneous and there are no studies focusing explicitly on sudden diffuse source WP; both studies included in this analysis assess flood mitigation options by local authorities (*Toxic Tide* quadrant; Fig. 1 and Table 2).

In the *Ominous Plume* quadrant only few EE studies investigate each of the diverse WP events (e.g. shipping traffic, wastewater treatment plants, aquaculture pollution or land reclamation by industry). In the *Ominous Tide* quadrant most of the EE studies investigate eutrophication induced by sediment and nutrient enrichment of water systems, produced by multiple sources and, specifically, agriculture-related activities; other EE studies in this quadrant focus on a variety of anthropogenic pressures (such as land use change and urbanization), water scarcity, contaminated runoffs and climate change induced impacts. In the *Toxic Plume* quadrant, the frequency of oil tanker accidents over the last decades and the resulting EE impacts attracted great public and scientific attention, giving rise to an international research effort directed at assessing leakage damages and devising risk-minimizing measures [43]. Thus, in this quadrant, EE studies include mostly oil spill accidents, followed by very few chemical industry and processing industry hazards.

4.2 MCEF-SES pollution abatement strategies

Apparent in all examined EE papers is the notion that water quality improvement strategies are flexible portfolios of abatement measures, whose composition is defined according to the relative policy priority given by local authorities to environmental protection and economic development. Over the four quadrants, the most investigated abatement strategies are the mixed mitigation and adaptation measures (predominant in the central all-source/all-type *Common-mire* pollution group, in the *Ominous Tide* quadrant and the *Ominous Plume* quadrant; 73%, 52% and 50% of the considered papers), followed by mitigation measures (18%, 26% and 43% of papers). Adaptation strategies are extensively investigated in the *Ominous Tide* quadrant, the *Toxic Plume* quadrant and the *Common-mire* pollution group (22%, 13% and 9% of papers). The majority of the *Toxic Plume* quadrant studies (60%) do not focus on any specific abatement strategy and opt for remaining generic (the same holds for 7% of the papers in the *Ominous Plume* quadrant). Studies in the *Toxic Tide* quadrant investigate mitigation strategies (Fig. 2(a)).

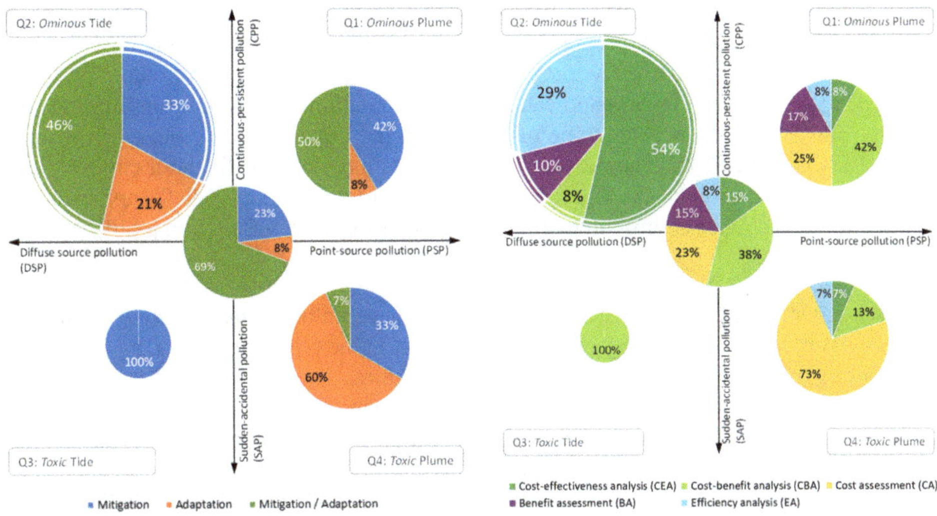

(a) (b)

Figure 2: (a) Identified pollution abatement measures; and (b) EE approaches.

4.3 Environmental-economic approaches to assess production and protection strategies

Environmental-economic (EE) approaches draw mainly on cost-effectiveness analysis (CEA), cost–benefit analysis (CBA) and efficiency analysis (EA). Examined scientific papers cover all EE approaches: cost effectiveness analysis (CEA), full cost–benefit analysis (CBA), partial CBAs (CA or BA) and efficiency analysis (EA) (Fig. 2(b)). *Ominous Plume* quadrant studies resort to all the EE approaches in a quite balanced way: the full CBA (5 (36%)), the partial CBAs (CA (3 (21%)) and BA (2 (14%))), the CEA (2 (14%)) and the EA (2 (14%)). In the *Ominous Tide* quadrant, the CEA is the most used approach (27 (54%)), followed by the EA (14 (28%)), the BA (5 (10%)) and the full CBA (4 (8%)). *Toxic Plume* studies focus on CA (11 (74%)), followed by the full CBA (2 (13%)), the CEA (1 (7%)) and the EA (1 (7%)). Both *Toxic Tide* studies opt for CBA analysis.

EE approaches constitute an important support to the process of policy-formulation, decision-making and the assessment of their premises. Robust EE multi-criteria decision tools, like CBA or EA, can help devising the optimal mix of strategies that most adequately balance environmental protection and economic development objectives, especially if all relevant information is put together in a spatially explicit analysis [45]. Just like traditional economic efficiency requires the whole value of resources to be established and incorporated into private and public decision-making processes, EE efficiency requires a comprehensive evaluation of both environmental costs and benefits [39]. Pollution accidents are rather varied, but it is crucial to control case-specificities and implement appropriate abatement strategies [18]. Thus, ecosystem values and pollution damage costs ought to be integrated into EE analysis [53]. Besides the determinants of pollution accidents and their consequences, pollution damage costs are strongly dependent on issues like politics, compensation policies and media coverage of pollution accidents which influence public awareness [1], [13], [18]. Moreover, in the *Ominous Tide* quadrant, some papers indicate that WP mitigation may come at a private benefit when nutrient emission reductions are low (due to cost savings that outweigh production losses); yet private WP mitigation costs increase exponentially when nutrient emission reductions become larger (due to production losses, see e.g. [21], [28]). Other authors indicate that WP adaptation always comes at a private cost (see e.g. [26]). Regardless of the type of environmental impacts and the lack of consensus among EE experts and stakeholders on how to value and account for both environmental benefits and environmental damages, EE assessment tools prove especially appropriate to estimate these and other environmental impacts (and ecosystem benefits) [1], [30].

5 CONCLUSIONS

Negative environmental impacts and ecosystem service value losses resulting from water pollution (WP) events can be large and vary directly with the type of pollutant, the geographic location of pollution occurrence, the geological features and biological characteristics of impacted area, climatological circumstances and the way contaminants spread through the water system. However, as large parts of natural capital services and values remain "outside the market", pollution losses remain widely underestimated.

The objective of this review study is to assess the current knowledge on the environmental-economics (EE) applied to maritime, coastal, estuarine and freshwater socioecological systems (MCEF-SES), focusing on WP pressures. EE studies were mapped in a four quadrant Cartesian plane, having on one axis the type of WP events (sudden to continuous pollution) and on the other axis the type of pollution sources (point to diffuse sources). Five groups of EE studies were mapped: the *Ominous Plume* quadrant (Q1), the *Ominous Tide* quadrant (Q2), the *Toxic Tide* quadrant (Q3), the *Toxic Plume* quadrant (Q4) and, finally, the *Common-mire* quadrant (dwelling on all four quadrants).

Results show that, hitherto EE water pollution research focused mostly on (1) the gathering of data and the building of knowledge on how to assess damages induced by human action on ecosystems; (2) the assessment of damages produced by pollution; and (3) the cost-effectiveness analysis of diffuse-source/non-point source pollution. Also, WP events associated with sediment and nutrient enrichment from point source pollution (e.g., industrial and residential) to diffuse-source pollution (e.g., agricultural) remain in the spotlight. Overall, the pollution sources considered most troublesome are oil spills, diffuse-source pollution and routine shipping operations, while wastewater, aquaculture, routine cruise tourism, chemical industries and land reclamation also constitute important sources of WP.

The most investigated type of MCEF-SES pollution events, taking an EE perspective, are CPP/DSP events (*Ominous Tide* quadrant) and oil spills (*Toxic Plume* quadrant). Further EE research focusing on the *Toxic Plume* and *Ominous Plume* quadrants is needed. Besides oil spills, *Toxic Plume* pollution events have been sparsely investigated until now, namely industrial spills. In the *Ominous Plume* quadrant, studies tend to focus on wastewater and aquaculture pollution, leaving without attention pollution produced by big industrial compounds. In the *Ominous Tide* quadrant, most studies investigate eutrophication induced by sediment and nutrient enrichment of water systems produced by multiple sources (mainly agriculture) and a variety of anthropogenic pressures (e.g. land use change, urbanization, water scarcity and contaminated runoffs).

The most used EE approaches to investigate MCEF-SES WP issues are the cost-effectiveness analysis (CEA) approach (34%), followed by the cost–benefit analysis (CBA) approach (19%) and the efficiency analysis (EA) approach (19%). Partial-CBAs are also used, such as cost-assessment (18%) and benefit assessment (10%) studies. Most used methods within the context of WP investigation are valuation methods (35%), linear programming (26%) and optimal control (17%). Probabilistic Bayesian-belief-network, non-linear programming, meta-modelling, regression analysis, meta-analysis and compromise programming methods remain marginal within the context of EE research. Static-deterministic techniques congregate preferences (50%), followed by dynamic-deterministic ones (30%).

In an analysis by quadrant, CEA is the most used EE tool to investigate CPP/DSP pollution events (*Ominous Tide* quadrant) and helps assuring economic efficiency of environmental management measures, by including in the analysis both the costs of pollution abatement and corresponding environmental impacts. The most used EE tool to investigate PSP/CPP pollution events (*Ominous Plume* quadrant) is CBA that aims to assess the welfare benefits from abatement strategy scenarios. Hence, CBA is useful in assessing trade-offs between the economic benefits from environmental protection and economic development. The second most used EE tool to investigate CPP/DSP pollution events (*Ominous Tide* quadrant) is EA that allows for the simultaneous analysis of multiple variables, dimensions and scenarios and is viewed as the most comprehensive and flexible of all EE analysis tools. However, its complexity may limit its usage.

When environmental impacts of human activities are not properly valued, incorporated into cost structures or counted-in the final market prices of products and services, market distortions occur and environment policies are ineffective. Research shows that profit seekers tend to opt for defensive and cost-effective pollution remedying measures, instead of large (long-term) investments in costly clean and pollution-preventing technologies. Thus, to avoid further costs from externalities, current and future impacts of environmental pollution produced by industries, infrastructures and urban settlements ought to be evaluated, together with a comprehensive valuation of ecosystems and services (also including non-market values) so that both the value of biodiversity and ecosystem services and the costs of pollution are taken into account in decision-making.

In conclusion, environmental-economic analysis can aid and support the implementation of ecosystem-based management of MCEF-SES and is viewed as an effective approach to deal with environmental degradation and help push forward relevant pollution abatement strategies. However, it is evident from the literature that context, spatially and temporally explicit assessments, at scales meaningful for policy formulation, require the development of spatially explicit models that consider bio-physical and socio-economic heterogeneity.

Some caveats remain and some important research knowledge gaps have been identified. First, most of natural capital and ecosystem service values remain "outside the market" and its consumption remains unvalued, allowing for environmental depletion and negative externalities to persist. Thus, the value of biodiversity and ecosystem services ought to be comprehensively valued, assessed and taken into account; moreover, pollution costs ought to be fully internalized. Specifically, there is a particular need to: (i) to capture the full range of ecosystem services and values; (ii) to develop bio-physical measures and indicators of biodiversity and ecosystems, suitably related to socio-economic attributes of ecosystem valuation; (iii) to develop economic assessments that are context, spatially and temporally explicit at scales meaningful for policy formulation; (iv) to integrate analyses of risks, uncertainties, irreversibility and resilience in environmental-economic assessments and valuation studies; and (v) to develop robust complex systems modelling approaches that integrate linkages between biodiversity, ecosystem services and human well-being.

ACKNOWLEDGEMENTS

This study is part of the thesis of a PhD Programme on Marine Science, Technology and Management (Do*Mar). Thanks are due, for the financial support, to CESAM (UIDB/50017/2020 and UIDP/50017/2020), to FCT/MCTES through national funds, and to the co-funding by European funds when applicable. Thanks are also due to the UNaLab project, financed by the EU Horizon 2020 research and innovation program under grant agreement No. 730052: Topic: SCC-2-2016-2017: Smart Cities and Communities Nature based solutions. Finally, thanks are due to the financial support of the Xunta de Galicia for financial aid through project GPC 2013-045, the ICES Science Fund to the project "Social transformations of marine social-ecological systems" and the EC "Ocean governance for sustainability challenges, options and the role of science" (COST Action CA15217).

REFERENCES

[1] Depellegrin. D. & N. Blazauskas, N., Integrating ecosystem service values into oil spill impact assessment. *J. Coast. Res.*, **29**(4), pp. 836–846, 2012.

[2] Roebeling, P.C., Hendrix, E.M.T. & Van Grieken, M.E., Exploring industry specific social welfare maximizing rates of water pollution abatement in linked terrestrial and marine ecosystems. *J. Coast. Res.*, **56**, pp. 1681–1685, 2009.

[3] Xue, M. & Luo, Y., Dynamic variations in ecosystem service value and sustainability of urban system: A case study for Tianjin city, China. *Cities*, **46**, pp. 85–93, 2015.

[4] Wen, Z. & Chen, J., A cost-benefit analysis for the economic growth in China. *Ecol. Econ.*, **65**(2), pp. 356–366, 2008.

[5] GreenFacts, Persistent organic pollutants, 2018. https://www.greenfacts.org/glossary/pqrs/persistent-organic-pollutants-POPs.htm. Accessed on: 19 Jan. 2018.

[6] Ballard, N. & Manus, P.M., Clearing muddy waters: Anatomy of the comprehensive general liability pollution exclusion. *Cornell Law Rev.*, **75**(3), pp. 610–643, 1990.

[7] Aquarium of the Pacific, Pollution and Health, 2016. https://www.aquariumofpacific.org/mcri/pollution_and_health.

[8] EPA, Nonpoint source (NPS) pollution, 2016. https://www.epa.gov/nps/basic-information-about-nonpoint-source-nps-pollution.

[9] Walmsley, J.J., Framework for measuring sustainable development in catchment systems. *Environ. Manage.*, **29**(2), pp. 195–206, 2002.

[10] Stam, G.J., Bottelberghs, P.H. & Post, J.G., Environmental risk: Towards an integrated assessment of industrial activities. *J. Hazard. Mater.*, **61**(1–3), pp. 371–374, 1998.

[11] Walker, W.E., POLSSS: Overview and cost-effectiveness analysis. *Saf. Sci.*, **35**, 2000.

[12] Islam, M.S. & Tanaka, M., Impacts of pollution on coastal and marine ecosystems including coastal and marine fisheries and approach for management: A review and synthesis. *Mar. Pollut. Bull.*, **48**(7–8), pp. 624–649, 2004.

[13] Psarros, G., Skjong, R. & Vanem, E., Risk acceptance criterion for tanker oil spill risk reduction measures. *Mar. Pollut. Bull.*, **62**(1), pp. 116–127, 2011.

[14] Gren, I., Adaptation and mitigation strategies for controlling stochastic water pollution: An application to the Baltic Sea. **6**, 2008.

[15] Kontovas, C.A., Psaraftis, H.N. & Ventikos, N.P., An empirical analysis of IOPCF oil spill cost data. *Mar. Pollut. Bull.*, **60**(9), pp. 1455–1466, 2010.

[16] Monte, L., Brittain, J.E., Gallego, E., Håkanson, L., Hofman, D. & Jiménez-Martin, A., MOIRA-PLUS: A decision support system for the management of complex fresh water ecosystems contaminated by radionuclides and heavy metals. *Comput. Geosci.*, **35**(5), pp. 880–896, 2009.

[17] Montewka, J., Weckström, M. & Kujala, P., A probabilistic model estimating oil spill clean-up costs: A case study for the Gulf of Finland. *Mar. Pollut. Bull.*, **76**(1–2), pp. 61–71, 2013.

[18] Ventikos, N.P. & Sotiropoulos, F.S., Disutility analysis of oil spills: Graphs and trends. *Mar. Pollut. Bull.*, **81**(1), pp. 116–123, 2014.

[19] Villasante, M., Sumaila, S. & Antelo, R., Why cooperation is better: The gains to cooperative management of the Argentine shortfin squid fishery in South America. *Environment and Development Economics: Essays in Honour of Sir Partha Dasgupta*, Oxford University Press, pp. 271–294, 2014.

[20] Mach, M.E., Martone, R.g. & Chan, K.M.A., Human impacts and ES: Insufficient research for trade-off evaluation. *Ecosyst. Serv.*, **16**, pp. 112–120, 2015.

[21] Roebeling, P., Rocha, J., Nunes, J.P., Fidélis, T., Alves, H. & Fonseca, S., Using the soil and water assessment tool to estimate dissolved inorganic nitrogen water pollution abatement cost functions in central portugal. *J. Environ. Qual.*, **43**(1), pp. 168–176, 2014.

[22] Ahlvik, L., Ekholm, P., Hyytiäinen, K. & Pitkänen, H., An economic-ecological model to evaluate impacts of nutrient abatement in the Baltic Sea. *Environ. Model. Softw.*, **55**, pp. 164–175, 2014.

[23] Djukic, M., Jovanoski, I., Munitlak, O. & Lazic, M., Cost-benefit analysis of an infrastructure project and a cost-reflective tariff: A case study for investment in wastewater treatment plant in Serbia. **59**, pp. 1419–1425, 2016.

[24] Faccioli, M., Riera Font, A. & Torres Figuerola, C.M., Valuing the recreational benefits of wetland adaptation to climate change: A trade-off between species' abundance and diversity. *Environ. Manage.*, **55**(3), pp. 550–563, 2015.

[25] Ng, A.K.Y. & Song, S., The environmental impacts of pollutants generated by routine shipping operations on ports. *Ocean Coast. Manag.*, **53**(5–6), pp. 301–311, 2010.

[26] Roebeling, P., Cunha, M.C., Arroja, L. & Van Grieken, M.E., Abatement vs. treatment for efficient diffuse source water pollution management in terrestrial-marine systems. *Water Sci. Technol.*, **72**(5), pp. 730–737, 2015.

[27] Doole, G.J. & Romera, A.J., Implications of a nitrogen leaching efficiency metric for pasture-based dairy farms. *Agric. Water Manag.*, **142**, pp. 10–18, 2014.

[28] Roebeling, P.C., Van Grieken, M.E., Webster, A.J. & Biggs, J., Cost-effective water quality improvement in linked terrestrial and marine ecosystems: a spatial environmental–economic modelling approach. *Mar. Freshw. Res.*, pp. 1150–1158, 2009.

[29] Birol, E., Koundouri, P. & Kountouris, Y., Assessing the economic viability of alternative water resources in water-scarce regions: Combining economic valuation, cost-benefit analysis and discounting. **69**, pp. 839–847, 2010.

[30] Caric, H., Challenges and prospects of valuation: Cruise ship pollution case. **111**, 2016.

[31] Lei, L., Jian, L., Yutao, W., Nvjie, W. & Renqing, W., Cost-benefit analysis and payments for watershed-scale wetland rehabilitation: A case study in Shandong Province, China. **5**(3), pp. 787–796, 2011.

[32] Martínez-Paz, J.M., Perni, A. & Martínez-Carrasco, F., Assessment of the programme of measures for coastal lagoon environmental restoration using cost–benefit analysis. *Eur. Plan. Stud.*, **21**(2), pp. 131–148, 2013.

[33] Martínez-Espiñeira, R., Chopin, T., Robinson, S., Noce, A., Knowler, D. & Yip, W., Estimating the biomitigation benefits of integrated multi-trophic aquaculture: A contingent behavior analysis. *Aquaculture*, **437**, pp. 182–194, 2015.

[34] Martinez-Espiñeira, R., Chopin, T., Robinson, S., Noce, A., Knowler, D. & Yip, W., A contingent valuation of the biomitigation benefits of integrated multi-trophic aquaculture in Canada. *Aquac. Econ. Manag.*, **20**(1), pp. 1–23, 2016.
[35] Doyle, M.W., Patterson, L.A., Chen, Y., Schnier, K.E. & Yates, A.J., Optimizing the scale of markets for water quality trading. *AGU Publ. – Water Resour. Res.*, pp. 7231–7244, 2014.
[36] Zander, K.K., Garnett, S.T. & Straton, A., Trade-offs between development, culture and conservation: Willingness to pay for tropical river management among urban Australians. *J. Environ. Manage.*, **91**(12), pp. 2519–2528, 2010.
[37] Broekx, S., Smets, S., Liekens, I., Bulckaen, D. & de Nocker, L., Designing a long-term flood risk management plan for the Scheldt estuary using a risk-based approach. *Nat. Hazards*, **57**(2), pp. 245–266, 2011.
[38] de Nocker, L. et al., Cost-benefit analysis to select the optimal flood protection strategy along the Scheldt. *WIT Trans. Ecol. Environ.*, vol. 91, WIT Press: Southampton and Boston, pp. 271–277, 2006.
[39] Oinonen, S. et al., Cost-effective marine protection: A pragmatic approach. *PLoS One*, **11**(1), pp. 1–19, 2016.
[40] Roebeling, P., Alves, H., Rocha, J., Brito, A., Almeida, P. & Mamede, J., Gains from trans-boundary water quality management in linked catchment and coastal socio-ecological systems: A case study for the Minho region. *Water Resour. Econ.*, **8**, pp. 32–42, 2014.
[41] Bertram, C. & Rehdanz, K., On the environmental effectiveness of the EU Marine Strategy Framework Directive. *Mar. Policy*, **38**, pp. 25–40, 2013.
[42] Bertram, C., Dworak, T., Görlitz, S., Interwies, E. & Rehdanz, K., Cost-benefit analysis in the context of the EU Marine Strategy Framework Directive: The case of Germany. *Mar. Policy*, **43**, pp. 307–312, 2014.
[43] Vanem, E., Endresen, Ø. & Skjong, R., Cost-effectiveness criteria for marine oil spill preventive measures. *Reliab. Eng. Syst. Saf.*, **93**(9), pp. 1354–1368, 2008.
[44] Scharin, H. et al., Processes for the sustainable stewardship of marine environments. *Ecol. Econ.*, **128**, pp. 55–67, 2016.
[45] Helle, I., Ahtiainen, H., Luoma, E., Hanninen, M. & Kuikka, S., A probabilistic approach for a cost-benefit analysis of oil spill management under uncertainty: A BN model for the Gulf of Finland. *J Env. Manag.*, **158**, pp. 122–132, 2015.
[46] Levrel, H. et al., The maintenance costs of marine natural capital: A case study from the initial assessment of the Marine Strategy Framework Directive in France. *Mar. Policy*, **49**, pp. 37–47, 2014.
[47] Sobota, D.J., Compton, J.E., Mccrackin, M.L. & Singh, S., Cost of reactive nitrogen release from human activities to the environment in the United States Cost of reactive nitrogen release from human activities to the environment in the US. 2015.
[48] Zhang, K., Li, Y.P., Huang, G.H., You, L. & Jin, S.W., Modeling for regional ecosystem sustainable development under uncertainty: A case study of Dongying, China. **533**, pp. 462–475, 2015.
[49] Blau, J. & Green, L., Assessing the impact of a new approach to ocean management: Evidence to date from five ocean plans. *Mar. Policy*, **56**, pp. 1–8, 2015.
[50] Wang, W., Guo, H., Chuai, X. & Dai, C., The impact of land use change on the temporospatial variations of ecosystems services value in China and an optimized land use solution. *Environ. Sci. Policy*, **44**(5), pp. 62–72, 2014.

[51] You, L., Li, Y.P., Huang, G.H. & Zhang, J.L., Modeling regional ecosystem development under uncertainty: A case study for New Binhai District of Tianjin. **288**, pp. 127–142, 2014.

[52] Di Jin, H., Kite-Powell, L. & Broadus, J., Dynamic economic analysis of marine pollution prevention technologies: An application to double hulls and electronic charts. *Environ. Resour. Econ.*, **4**, pp. 555–580, 1994.

[53] Loureiro, M.L., Loomis, J.B. & Vázquez, M.X., Economic valuation of environmental damages due to the prestige oil spill in Spain. *Environ. Resour. Econ.*, **44**(4), pp. 537–553, 2009.

SECTION 4
EMERGENCY PREPAREDNESS AND RISK MITIGATION

ROLE OF COMMUNITY-LED PLANNING FOR EFFECTIVE WIDE-AREA REGIONAL EVACUATION

TAKEYASU SUZUKI
Disaster and Environmentally Sustainable Administration Research Center, University of Yamanashi, Japan

ABSTRACT

Abnormal weather caused by global warming has brought about unprecedented torrential rain, causing heavy rain disasters in various parts of Japan. Since recent river flooding caused inundation damage over a wide area, area-wide evacuation, that is, evacuation to a safe place beyond municipalities' boundaries is necessary to save lives. This paper first introduces the community disaster management planning system that was established in Japan in the wake of the Great East Japan Earthquake in 2011. Next, the necessity of wide-area evacuation that has become apparent due to the recent flooding of large-scale rivers and a few problems that have arisen in the government-led wide-area evacuation plan are introduced, because residents' concrete evacuation behaviour was not sufficiently reflected. The author discusses efforts of a community in the Kofu basin, Yamanashi Prefecture which has been working to formulate a community disaster management plan for wide-area evacuation and city and prefectural efforts that support this district. The author also describes the risk communication that was undertaken to clarify the role of central ministries, local governments and community residents in the formulation of wide-area evacuation plans. Finally in wide-area evacuation drills, planned and conducted by the community in 2020 during the COVID-19 pandemic, the community disaster management plan is further reviewed. In this article, the author addresses how such a community-led wide-area evacuation plan is indispensable for administrations' effective wide-area evacuation planning.

Keywords: risk communication, wide-area evacuation, community-led, community disaster management plan, evacuation drill, flood countermeasures.

1 INTRODUCTION

Extreme weather caused by global warming has resulted in increased intensity and numbers of heavy rain disasters worldwide. In Japan since 2014, for example, four to six typhoons have made annual landfall, doubling from 2.7, the average number of landings during the 30-year period from 1981 to 2010 [1]. The highest rainfall in the history of observations has been updated nationwide, and use of past meteorological observation data to predict stochastic disaster events is no longer meaningful. In Japan, river management is determined based on degree of importance, and river facilities are designed based on design discharge. However, compared to flood control in Europe and the United States, Japan still has room for improvement in flood control safety and achievement [2].

Considering physical measures' limitations in strengthening river facilities to protect against floods (i.e., measures to protect), a drastic change was made to more intangible measures to protect lives by evacuation through amendment of the Flood Control Law in 2005 (i.e., measures to evacuate) [3]. However, these measures cannot now match the level of abnormal weather events because floods due to the greatest expected rainfall have actually occurred and social systems have collapsed. Damage caused by the 2018 heavy rainfall disaster in western Japan and the 2019 typhoon No. 19 has indicated the need to build resilient cities able to recover quickly from flooding. Therefore, in addition to "protecting" and "evacuating", developing cities that can escape or overcome floods and minimize damage (i.e., measures to "dodge" flood hazards) is crucial. Thus, the countermeasure, "dodge",

WIT Transactions on The Built Environment, Vol 207, © 2021 WIT Press
www.witpress.com, ISSN 1743-3509 (on-line)
doi:10.2495/DMAN210121

which involves building cities that can maintain their function and minimize damage to inhabitants' property and lives, is essential for urban sustainability [4].

The southern Kofu basin, Yamanashi Prefecture, Japan, is a region with high risk of flood damage. The region has no safe upland, and severe inundation during heavy rainfall is expected. Furthermore, a new station for the Linear Motor Chuo Shinkansen Line is planned for construction as a center for the neighbourhood's urban development. To build a sustainable, flood-resilient city in such a hazardous area, the author established an urban development study group [5], which has proposed a future sustainable Kofu basin under knowledge brokers' initiative and with experts' cooperation. The group applied the concept of sustainable cities presented by the author and proposed building detention/retention basins, a nature museum, stay-type leisure facilities, raised ground created by banking, artificial land and high-rise buildings, and a transportation network – thus forming constituent elements for a sustainable, flood-resilient Kofu basin [4].

However, sustainable city development requires sufficient discussion and time-consuming consensus building. After that, residential land will be constructed by raising ground, and for residents, moving to safe from hazardous areas will take 20 or 30 years. Therefore, protecting lives from flood damage is essential – most likely via wide-area evacuation – that is, evacuation to a safe place beyond municipalities' boundaries – until residences' relocation to safe areas is completed.

Here, the author discusses efforts of the Riverside District, Chuo City in the Kofu basin, which has been working to formulate a community flood plan [6], [7]. Via risk communication conducted by the author, this article introduces city and prefectural efforts that support this district. As a result of wide-area evacuation drills, planned and conducted in 2020 during the COVID-19 pandemic, the community disaster management plan was further reviewed. In the article, the author addresses how such a community-led wide-area evacuation plan is indispensable for administrations' effective wide-area evacuation planning.

2 NECESSITY OF FORMULATING COMMUNITY-LED WIDE-AREA EVACUATION PLAN

2.1 Community disaster management plan and wide-area evacuation

Japan has established the Disaster Countermeasure Basic Act as a comprehensive and long-term national and regional disaster management plan for prefectural and municipal governments, implementing disaster management activities at each level.

However, immediately after the 2011 Great East Japan earthquake, many deeply recognized that disaster countermeasures after a large-scale wide-area disaster work well, especially with mutual cooperation through self-help, mutual help, and public help [8]. On the basis of lessons learned, in 2013, some provisions concerning self-help and mutual help were added to the Disaster Countermeasure Basic Act [7]. At that time, from the perspective of the local community's mutual help disaster management activity, a community disaster management plan was established. This involved voluntary disaster management activities conducted by residents and businesses (community residents, etc.) in certain municipal districts [8].

The 2015 Kanto and Tohoku heavy rains resulted in massive flooding of the Kinugawa River and 40 km^2 or one-third of the area of Joso City in Ibaraki Prefecture, causing two human casualties. At the flood's height, more than 4,000 people were stranded and later rescued with helicopters and boats [9]. The Irie survey [10] indicated that the greatest number of evacuees – 35% – headed out of Joso City via car (89% of all related evacuation means).

In terms of motivation, 41% of evacuees were "family and friends prompted". Moreover, most of those affected indicated that they remained indoors due to normality bias, believing that the heavy rain would not inundate or that moving was difficult "considering the elderly". The reason "neighbours are not evacuating" also rated high, accounting for 12% of survey responses [10].

Based on lessons from Joso City flooding due to the 2015 Kanto and Tohoku heavy rain disaster and Mabi-cho in Kurashiki City, Okayama Prefecture, due to the western Japan heavy rain disaster, a wide-area evacuation plan is necessary to facilitate early evacuation across municipal boundaries before an incoming disaster [9], [11]. Such a plan should prioritize support for persons needing evacuation help that requires municipal transportation to welfare facilities, along with evacuation support by the neighbourhood [12]. Hence, residents and the administration must work together to support those who cannot evacuate without assistance. To achieve a community's wide-area evacuation, residents must have the intention of evading and evacuating; a system is created that encourages them to do so, with encouragement not only from their families and friends but also their neighbours. The administration's role in wide-area evacuation is controlling traffic, providing guidance to enable evacuation, and securing vehicle evacuation routes. Particularly important is securing public shelters, including welfare facilities; therefore, prior coordination among municipalities is indispensable [12].

2.2 Cooperation between community- and government-led wide-area evacuation plans

The Cabinet Office of Japan has summarized the basic concept of wide-area evacuation from floods [13], created guidelines [14], and is promoting local governments' formulation of wide-area evacuation plans. Some government-led plans are ineffective, however. Kazo City, Saitama Prefecture, and Sakai Town in Ibaraki Prefecture, with the support of the Ministry of Land, Infrastructure, Transport and Tourism (MLIT) and prefectures, conducted surveys and research on wide-area evacuation [15] (for example, Sakai Town, Local Autonomy Research Organization, 2018) and formulated wide-area evacuation plans, adding them to their regional disaster management plans. At the time of typhoon No. 19 in 2019, both cities and towns conducted wide-area evacuation according to the plan. In Kazo City, however, the following problems were highlighted: the parking lot of the evacuation destination became full, the road became heavily congested, and the residents were confused due to a delay in issuing evacuation information [16].

In addition, Kazo City's bulletin (August 2020) reported that residents were surprised that the list of evacuation destinations outside the prefecture disappeared from the evacuation route map [17]. Even in Sakai Town, Ibaraki Prefecture, a line hundreds of meters long of residents evacuating by car formed at the evacuation center. Moreover, residents complained of anxiety because of the time it took to enter the shelter [18]. Finally, immediately after the town issued an evacuation order, residents began moving all at once, causing traffic congestion. These problems are considered to have arisen because residents' concrete evacuation behaviour was not sufficiently reflected in the plan. After that, Sakai Town increased the number of evacuation sites outside of town and negotiated with the city to secure four parking lots of the city's parks and temples where townspeople who evacuated by car could spend time [18].

In other words, if a wide-area evacuation plan is formulated under residents' initiative, the local government can consider evacuation support according to the actual situation. In government-led plans, however, residents often act according to their judgment while also following government instructions, causing a gap between response actions. Especially in large-scale and wide-area evacuation, top-down or government-led plans tend to be emphasized. However, it should go without saying that creating a bottom-up or community-led wide-area evacuation plan is important in district disaster prevention. Also important is enhancing the plan's effectiveness by coordinating mutual and public assistance.

3 WIDE-AREA EVACUATION PLAN AND RISK COMMUNICATION AT THE RIVERSIDE DISTRICT

3.1 Outline of the Riverside District

Fig. 1 illustrates an assumed inundation map of the Kofu Basin and the River-side District, which takes the name of the community of "Riverside Town" in Chuo City, Yamanashi Prefecture. Situated on the Kamanashi River's left bank, the Riverside District was once a water reservoir sandwiched between a dike and the river. After the Kamanashi River flooded in 1907, the Usuinuma swamp was formed. Reclamation of the swamp began in 1977, and residential construction has created the largest residential district in the Kofu basin, with nearly 1,400 households, 4,000 residents, and a potential population of 7,000. Residential land was first sold from the north area, represented by three residents' associations: the 1st Association of the North, the 2nd Association of the Central, and the 3rd Association of the South [7].

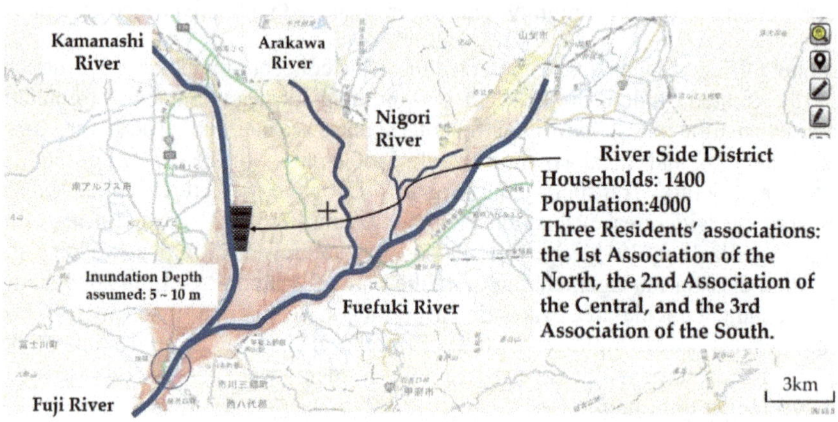

Figure 1: Assumed inundation map of the Kofu Basin and the location of the Riverside District.

3.2 Risk communication for a wide-area evacuation plan

To evacuate community residents, the author proposed implementation of a risk communication approach using the "creating Confidence, generating Awareness, enhancing Understanding, gaining Satisfaction, and motivating Enactment (CAUSE) model" proposed by Rowan [19], Rowan et al. [20] and Suzuki [21]. In the Riverside District, Chuo City, the author implemented these processes twice for wide-area evacuation plan development. The first process covered the 3rd Residents' Association evacuation plan, and the second expanded it to the entire Riverside District. Consequently, in May 2019, the whole district's (1st through 3rd residents' associations with 1,400 households and 4,000 people) community disaster management plan was formulated as a wide-area evacuation plan at the final stage of "Enactment" (see Fig. 2 for a schematic) [22].

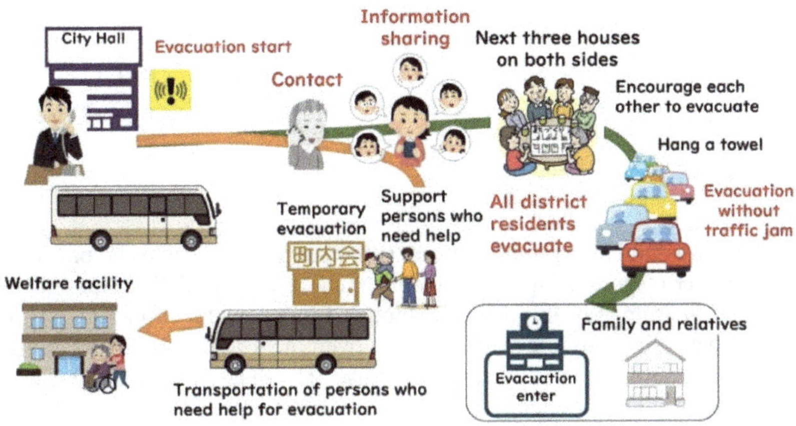

Figure 2: Schematic representation of the Riverside District's evacuation plan [22].

Fig. 3 is the "My Timeline" of evacuation (preparation and) behaviour distributed to each household, but when the authors presented this timeline in 10 stages, residents added some ideas to create the timeline shown in Fig. 3. Finally, the timeline was printed on B5-size glossy paper in black and red two-colour printing and distributed to all households. In 2018, the timeline had a red border, but in 2019, a blue one. The 2020 version is described below. The timeline is now published annually, and residents update such information as evacuation sites.

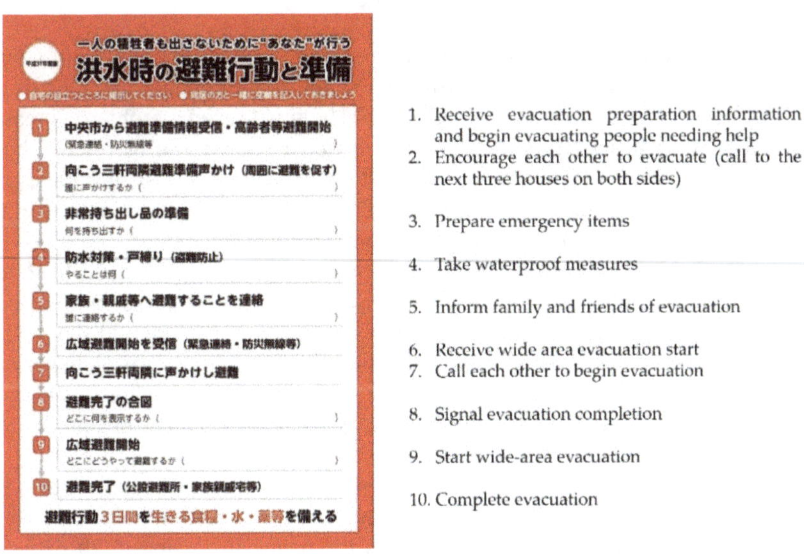

Figure 3: "My Timeline" of evacuation preparation and behaviour distributed to each household.

Through three years of risk communication activities, local residents became well aware of and understood the community's flood risk, found a solution on their own, accepted proposals for wide-area evacuation, and finally formulated the community disaster management plan for wide-area evacuation.

4 RISK COMMUNICATION TO CLARIFY EACH STAKEHOLDER'S ROLE IN WIDE-AREA EVACUATION

4.1 BECAUSE model

The author developed the BECAUSE model as a training process that fosters local government officials' ability to communicate with organizations inside and outside agencies to facilitate disaster response. Therefore, the meaning of each CAUSE steps in BECAUSE differed somewhat from Rawan's CAUSE model. Moreover, the S in BECAUSE means "Satisfaction", which includes the stage of local government staff proposing a "Solution" on their own or with a new method of disaster response [23]. In addition, BE refers to the stage of preparation for training in which an environment must be built where local government staff can participate; BE is therefore a stage for gaining familiarity with the local administration's governor, mayor, and senior staff. In 2013, the BECAUSE model was applied to construction of a regional cooperation system for wide-area evacuation, having been classified into three sections: a disaster-affected municipality; neighbouring municipalities supporting it; and a supporting organization of disaster management and construction departments of the prefectural government, prefectural police headquarters, and river administrators of MLIT. Later, a disaster response exercise was conducted as the final step, E, of the BECAUSE model to verify the constructed regional cooperation system's effectiveness [24]. However, wide-area evacuation at this time involved residents from one municipality evacuating to another nearby municipality.

4.2 Each stakeholder's role in wide-area evacuation

The author individually interviewed and exchanged opinions with the mayor of Chuo City, the director of the Disaster Management Bureau, Yamanashi Prefecture, and the director of the Kofu Regional River and Road Office, MLIT. The most serious problem is that people do not evacuate even when an evacuation information is issued. Furthermore, the author explained that such stakeholders as local and national governments should have a role in wide-area evacuation, as summarized in Fig. 4, if all basin residents are to implement the plan as the risk communication stage of Understanding and Satisfaction [25].

Risk communication's purposes are for residents of Chuo City, Yamanashi Prefecture, and MLIT to declare the role they should play in the Kofu basin's large-scale flood evacuation and to recognize one another's roles. Essentially, citizens should share wide-area evacuation plans as much as possible. Apart from the wide-area evacuation plan, various non-structural measures such as a legal-based system and structural measures such as raising residential lands, roads, and high-rise buildings as water-proofing measures might be future targets to improve the Kofu basin's flood resilience.

On 25 December 2018, the author held a disaster management symposium and planned "Make use of lessons learned from Mabi-cho for wide-area evacuation in the Kofu basin" as a panel discussion of each stakeholder's role. In the panel discussion, both panelists and citizens should discuss appropriate actions. Besides the author, the panel included the

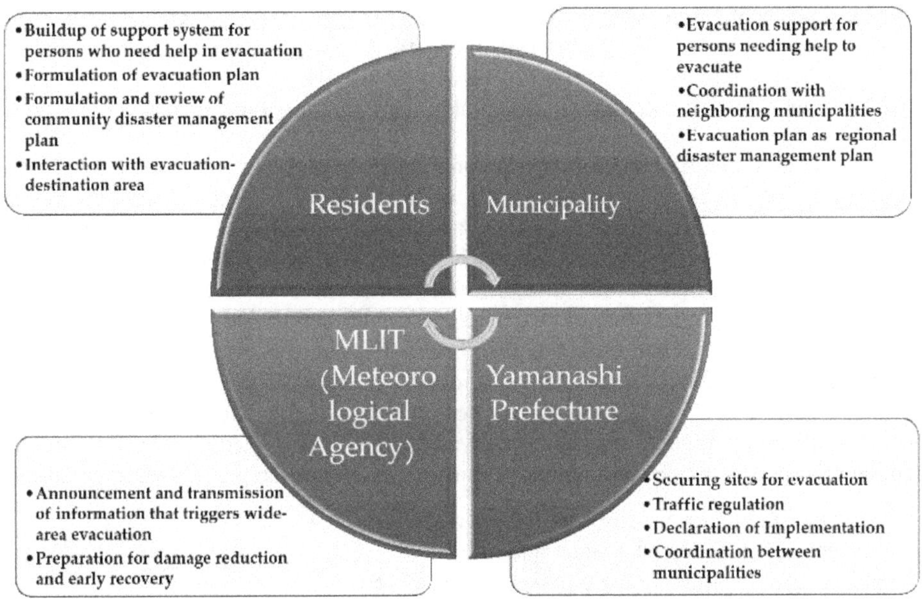

Figure 4: Roles of Ministry of Land, Infrastructure, Transport and Tourism (MLIT), prefectures, municipalities, and residents in wide-area evacuation.

chairman of the Riverside District's 3rd Residents' Association, the mayor of Chuo City, the director of the Disaster Management Bureau in Yamanashi Prefecture, and the director of the Kofu Regional River and Road Office of the MLIT. The chairman of the Riverside District 3rd Residents' Association stated that to evacuate a wide area without casualties, evacuation support is necessary for those who need help during evacuation, and so is development of an wide-area evacuation plan by the 3rd Residents' Association for the entire Riverside District. Additionally, the chairman declared that in the next stage, Chuo City should develop such a plan for the entire city and further requested that the city issue early evacuation information for citizens to initiate the evacuation plan and secure evacuation destinations.

The mayor of Chuo City stated the essence of assessing transportation means for evacuation, given the wide-scale plan. He further mentioned that the city was initiating an interregional evacuation study in cooperation with the neighbouring Showa Town, supported by MLIT, within the framework of the Disaster Mitigation Council in the Fuji River basin. Moreover, the mayor expressed interest in continuous assessment of the wide-area evacuation plan with cooperation of Yamanashi Prefecture and MLIT.

The director of the Disaster Prevention Bureau, Yamanashi Prefecture stated that the prefecture's role is actively to advise and support municipalities' development of an evacuation plan, establish a forum for consultation among related organizations, and ensure consistency among municipalities' implementation of evacuation plans. Moreover, the director stressed the prefecture's responsibility to inform residents of the wide-area evacuation plan and implement an evacuation exercise accordingly.

The director of Kofu Regional River and Road Office stated that the office's role is to advise Yamanashi Prefecture on criteria for wide-area evacuation. Therefore, criteria that trigger wide-area evacuation are to be decided at this examination stage. In the event of a flood, MLIT should provide information when such criteria are reached.

5 MEASURES AGAINST COVID-19 AND WIDE-AREA EVACUATION DRILLS IN THE RIVERSIDE DISTRICT

5.1 Wide-area evacuation in a safe parking lot in neighbouring municipalities

In the wide-area evacuation plan, evacuation itself does not mean "sheltering" at an evacuation center but escaping dangerous places and saving lives. In the event of heavy rain when the Riverside District conducts wide-area evacuation, neighbouring municipalities should be evacuation destinations. Therefore, designated evacuation centers in areas not directly affected by heavy rain disasters are accustomed to accepting evacuees, so neighbouring municipalities cannot be expected to open designated evacuation centers for the Riverside District. Therefore, the district's wide-area evacuation plan requires that residents secure themselves with relatives and in acquaintances' homes and hotels, etc. as evacuation destinations. Residents who cannot do so may choose to evacuate in a car and park on a hill in a neighbouring municipality. The author examined a parking lot that is a candidate for an evacuation site in nearby Nirasaki City, based on the following conditions:

1. Large parking lot or parking space (20 cars or more);
2. Existence of facilities with water and toilets;
3. Existence of facilities where elderly people can rest;
4. Wide access road from public roads to facilities;
5. Existence of residents' association (community residents) that can support evacuees.

The fifth condition above is necessary because evacuees who do not know the community cannot well deal with unforeseen problems occur and they need help. Yamamura also emphasizes the need for cooperation with neighbouring residents' associations while evacuees stay in their cars [25].

Examination of such places on the map and in the field survey revealed 23 facilities: public halls, welfare facilities, closed elementary schools, municipal facilities/parks, and parking lots of private companies with space for 700–800 cars. Thus, if approximately half of the households in the Riverside District can secure evacuation destinations with relatives and in acquaintances' homes and hotels, the remaining households can evacuate in their cars to a safe parking lot. Elderly people who cannot secure shelter or travel to a parking lot on their own will be asked to register as residents who need assistance and be evacuated to the welfare center by city bus or car with their neighbours' support.

5.2 Wide-area evacuation action considering measures against COVID-19 infection

At the 15 July 2020 district leaders' meeting, the Riverside District representative proposed the third-year (2020) version of "Evacuation Action and Preparation in the Event of a Flood Disaster", which is the My Timeline scheduled for annual updating. The 2020 My Timeline included measures to prevent COVID-19 infection. In addition, distributed evacuation by each household was added while responding to the five levels of alert for heavy rain disasters, corresponding to municipalities' evacuation information.

Therefore, preparation (preparedness), evacuation behaviour (evacuation), and staying at an evacuation center (sheltering) are mixed. Furthermore, coronavirus infection control was added, so the most important wide-area evacuation is actually inconspicuous. Since the district representative is quite familiar with water disaster prevention, he attempted to squeeze every action into the timeline: advance preparation, wide-area evacuation 24 hours before a disaster, and evacuation of residents late for the beginning of wide-area evacuation.

As a result, the My Timeline was somewhat lacking in cohesion and organization. Some leaders at the meeting commented that the timeline should be simpler and that residents could not write on it; even so, a concrete amendment was not finalized.

Therefore, the author proposed to examine the timeline and to present an amended one at a later date; this proposal was approved. In the amended timeline, the preparatory stage was first described and next the timeline for wide-area evacuation. After describing how to ensure emergency safety in case of being late for wide-area evacuation, actions prior to beginning evacuation and those during evacuation implementation/completion are described separately. For the Riverside District, the most notable evacuation behaviour is to call to three houses on either side before beginning evacuation. Finally, the timeline's evacuation implementation/completion column describes COVID-19 countermeasures for evacuation sites.

At the next board meeting, leaders welcomed and approved the author's amended timeline. By August 19, the My Timeline for "evacuation action and preparation" was printed on a green background (Fig. 5) and presented to the board prior to distribution to all district households.

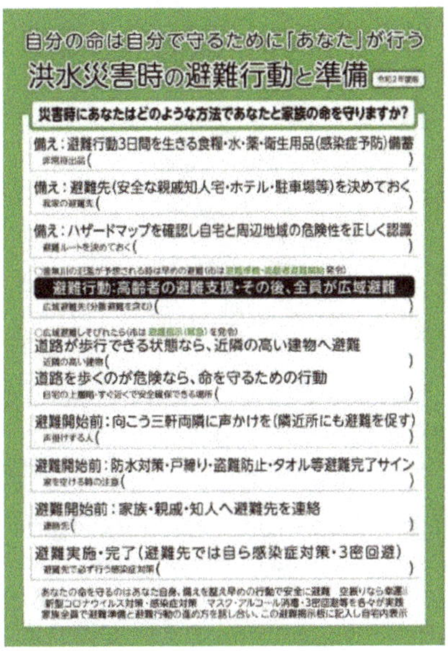

Evacuation Action and Preparation in the Event of a Flood Disaster

How do you save your family's life?
Preparedness: Stockpile of food, water, medicine, and sanitary goods (COVID-19 infection countermeasures) to survive for 3 days /Emergency carry-out ( )
Preparedness: Determine where to evacuate (secure evacuation destinations such as relatives, acquaintances' homes, and hotels, others) /evacuation destination ( )
Preparedness: Check the hazard map and correctly recognize the danger around my home /Evacuation route ( )

Evacuation action
Wide-are evacuation: Support persons requiring assistance and evacuate all / Wide-are evacuation destination ( )
Wide-are evacuation destination ( )
Evacuation of residents who have been late for the start of wide-area evacuation
Safe place (building) nearby ( )
A safe place on the upper floors of your home or nearby ( )
Before wide-are evacuation
Persons you will call out ( )
Signal evacuation completion ( )
Family and relatives who you will inform ( )
Implement and complete evacuation
COVID-19 infection countermeasures ( )

Figure 5: 2020 My Timeline for evacuation behaviour and preparation.

5.3 Cooperation from Chuo City and Yamanashi Prefecture

For the Riverside District's wide-area evacuation, evacuation by bus of those requiring assistance was vital. Therefore, Chuo City positioned this wide-area evacuation drill as training for transferring people requiring assistance. The city requested cooperation from Santo Kotsu Co. Ltd., and in March 2020, Chuo City signed an agreement that the company's bus could be used for evacuation. Chuo City arranged minitrucks, tents, questionnaires, rubbing alcohol, chairs, cardboard beds, portable toilets, etc., along with buses for transporting people needing evacuation support. Chuo City's Crisis Management Division attended the Riverside District's board of directors meeting to share information from the training's preparatory stage.

The Riverside District's wide-area plan assumes evacuation to the hills of Nirasaki City, and in the 2020 evacuation drill, residents' association leaders planned to evacuate to Nirasaki City. However, Nirasaki City refused to provide an evacuation site it had once offered because it could not obtain nearby residents' cooperation due to the COVID-19 pandemic. In response, Chuo City consulted with the Disaster Management Bureau of Yamanashi Prefecture, which proposed the parking lots of two prefectural facilities as alternatives. From the two candidate sites, the Riverside District selected the Prefectural Clean Energy Center in Kai City located on a hill between Nirasaki City and Chuo City (Fig. 6). Since the center's toilet facilities can be rented along with the parking lot, conducting wide-area evacuation drills has become possible. Yamanashi Prefecture not only cooperated by providing evacuation shelters but also dispatched two staff members on the day of the drill.

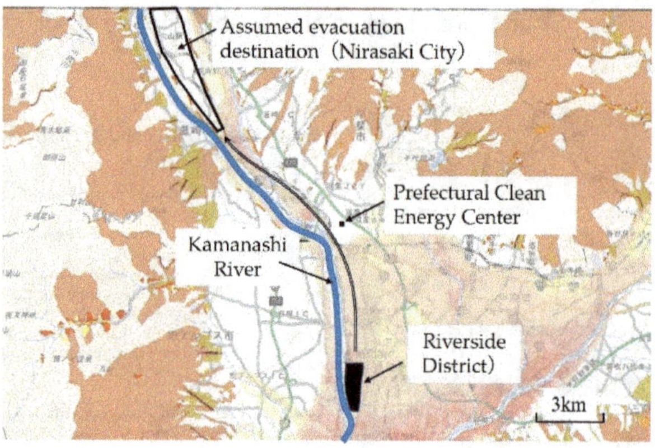

Figure 6: Positional relationship among the Riverside District, the Prefectural Clean Energy Center, and the hills of Nirasaki City.

On the day of the drill, Santo Kotsu Co., Ltd. provided two buses. In the district, two residents needing evacuation assistance and four supporters had been selected in advance, and these six residents experienced bus boarding training. The drill provided Santo Kotsu an opportunity to notice points to consider when transporting people requiring evacuation assistance (Fig. 7(a)). As a measure to prevent COVID-19 infection, each inhabitant wore a self-provided mask, disinfected the hands with alcohol, and maintained an interval of 1.5 m. Having provided 5 l of alcohol for hand disinfection (given to participants in small containers) and a thermometer, district officials took participants' temperatures. Fig. 7(b) shows the training led by residents.

6 RISK COMMUNICATION FOR FORMULATION OF WIDE-AREA EVACUATION PLANS FOR YAMANASHI PREFECTURE AND CHUO CITY

6.1 Meeting between Riverside District and Chuo City

Riverside District community leaders visited Chuo City's mayor and deputy mayor to thank them for the city's cooperation in wide-area evacuation drills (Fig. 8). In addition, drills' results were summarized and reported. Chuo City was also requested to formulate a wide-

(a) (b)

Figure 7: Photographs of wide-area evacuation drills' implementation. (a) A resident needing assistance is helped to board a bus; and (b) Community leaders explain training content.

Figure 8: Meeting between community residents and the city at Chuo City Hall.

area evacuation plan, secure wide-area evacuation destinations for the district at an early stage, and develop a community disaster management plan for other areas.

Disaster management drills sponsored by the city have been held close to September 1st every year, the day the Great Kanto Earthquake struck in 1923. However, disaster management drills for flood control are more effective around June, before the flood season. In addition, holding a disaster management drill in September, a period of high temperatures and high humidity, might harm participating residents' health. For these reasons, Riverside District community leaders have offered to hold a disaster drill for Chuo City the following year in June, and Chuo City promised to respond to this offer in early December. At a later date, the drill scheduled for June was postponed due to lack of preparation, but the city's cooperative attitude was recognized.

The author also attended this meeting and advised Chuo City on the following three points:

1. Resume negotiations with Nirasaki City on acceptance of evacuees.

2. Visit Kazo City, Saitama Prefecture to investigate the actual situation of the wide-area evacuation from Kazo City during typhoon No. 19 in 2019. Reports have indicated that typhoon evacuation exposed the disadvantage of wide-area evacuation formulated by the government as a back-up for the city. As a result, Kazo City is reportedly putting a brake on wide-area evacuation. Since it seems that wide-area evacuation's effectiveness cannot be guaranteed without bottom-up efforts by the Riverside District, the author proposed that Chuo City officials visit Kazo City to interview city officials and residents who conducted wide-area evacuation.
3. Expand formulation of community disaster management plans for wide-area evacuation and begin formulating wide-area evacuation plans for the city. It has already been confirmed that the MLIT and Yamanashi Prefecture are ready to support formulation of a wide-area evacuation plan for Chuo City.

6.2 Risk communication with stakeholders

As previously mentioned, the author exchanged opinions on each organization's role in conducting wide-area evacuation with the Kofu Regional River and Road Office of the MLIT that manages the Kamanashi River, Yamanashi Prefecture, and Chuo City. As a result, a symposium was held with representatives of these organizations, residents in the Riverside District, and the author as an expert in disaster management, to clarify each stakeholder's role [25].

To ensure the effectiveness of the Riverside District and Chuo City's wide-area evacuation, the plan should specify each institution's role as agreed upon during the symposium. Although Chuo City's mayor remains in office, since the symposium, personnel shifts have brought new directors to the Kofu Regional River and Road Office of the MLIT and to the Disaster Management Bureau of Yamanashi Prefecture. Therefore, the author exchanged opinions about Chuo City's wide-area evacuation plan with the Kofu Regional River and Road Office of the MLIT and confirmed that the office understands its role and is ready to support the plan as much as possible. In fact, the MLIT is strengthening its stance of actively supporting formulation of regional disaster management plans for municipalities against flooding by rivers under the ministry's control. The author also exchanged opinions with the Disaster Management Bureau of Yamanashi Prefecture, which is also preparing to formulate a wide-area evacuation plan and therefore promised to provide as much support as possible to improve mutual plans' consistency.

After confirming these organizations' intentions, the author urged Chuo City to consult with the Kofu Regional River and Road Office and Yamanashi Prefecture to confirm the content of support and then prepare for formulation of a wide-area evacuation plan. In response, Chuo City visited the Kofu Regional River and Road Office of the MLIT and Yamanashi Prefecture to discuss support for formulating a wide-area evacuation plan. After that, Chuo City informed the author that it would start formulating a wide-area evacuation plan in 2021.

7 CONCLUDING REMARKS

In the Riverside District of Chuo City, which has a high risk of flooding, a community disaster management plan for wide-area evacuation was formulated as a result of support applied by the author through risk communication. In this district, an evacuation support system for people needing evacuation assistance has been established, and wide-area evacuation actions for each household have been summarized in a timeline. However, because this evacuation plan contains measures dependent on administrative agencies –

municipalities, prefectures, and central ministries – for evacuating people requiring assistance, selection of evacuation sites, and traffic control of evacuation routes, it needs to be consistent with these agencies' wide-area evacuation plan.

When a community-led plan is formulated in this way, the municipality can formulate a concrete, highly effective wide-area evacuation plan to support residents' evacuation behaviour. In fact, Chuo City has decided to formulate a wide-area evacuation plan in 2021. Similarly, to enhance the effectiveness of municipalities' wide-area evacuation plans, issues that prefecture should address will be clarified, and the prefecture's wide-area evacuation plan's effectiveness will also increase. Taking this opportunity, Yamanashi Prefecture is expected to begin formulating a plan for the entire Kofu basin – not government-led, top-down but a highly effective wide-area evacuation plan that combines bottom-up and top-down.

ACKNOWLEDGEMENTS

The success of risk communication reported in this study is largely due to the great efforts of Mr. Hosokawa and other community leaders of the Riverside District and the active efforts of the residents of the district. I express my sense of gratitude to them all of the Riverside District. I would like to express our deepest gratitude to Mr. Kaneko, Deputy Director of the Kofu River National Highway Office for his great cooperation in risk communication with Yamanashi Prefecture and Chuo City based on the promotion measures of the basin flood control project of the Ministry of Land, Infrastructure, Transport and Tourism.

REFERENCES

[1] Japan Meteorological Agency (JMA), Number of typhoons landing. https://www.data.jma.go.jp/fcd/yoho/typhoon/statistics/landing/landing.html. Accessed on: 10 Sep. 2021.

[2] Japan Institute of Construction Engineering (JICE), Flood control project implementation system in European countries: Flood control project plan based on climate change. *JICE Report*, 111001, 2011. (In Japanese.)

[3] Suzuki, T., Legal systems. *Disaster Engineering*, Riko Tosho: Tokyo, pp. 287–299, 2019. (In Japanese.)

[4] Suzuki, T., Building up a common recognition of city development in the southern part of Kofu Basin under the initiative of knowledge brokers with the cooperation of experts. *Sustainability*, **12**(16), p. 6316, 2020. https://doi.org/10.3390/su12166631.

[5] Suzuki, T., Risk communication process applicable to make the Kofu Basin resilient to flood. *KASEN(River)*, **874**, pp. 23–27, 2019. (In Japanese.)

[6] Suzuki, T., Watanabe, T. & Okuyama, S., Community disaster management plan to implement extensive evacuation without a single victim against flooding. *Journal of Community Disaster Management Society* (*C-Bosai*), **13**, pp. 35–50, 2018. (In Japanese.)

[7] Suzuki, T., Risk communication in order to facilitate community resilience against a large-scale flood. *WIT Transaction on The Built Environment Disaster Management and Human Health Risk VI*, vol. 190, WIT Press: Southampton and Boston, pp. 145–155, 2019.

[8] Murosaki, Y. & Tominaga, Y., Chapter 5. *Mutual-Help in Creation of People to Confront Disasters*, Minerva Shobo: Kyoto, pp. 77–86, 2018. (In Japanese.)

[9] Joso City Flood Control Verification Committee, Verification report on disaster response to flood in Joso City due to Kinugawa River flood, 2015. (In Japanese.)

[10] Irie, S., Levee collapse of the Kinu River: How did Joso city residents evacuate? From a survey on residents' disaster information awareness and evacuation behaviour at the 2015 heavy rainfall in Kanto and Tohoku regions. *The NHK Monthly Report on Broadcast Research,* **66**(8), pp. 36–65, 2016. (In Japanese.)
[11] Verification Committee of the July 2018 Heavy Rain disaster, Okayama Prefecture, Verification Report on the July 2018 flood disaster. http://www.pref.okayama.jp/uploaded/life/601705_5031910_misc.pdf. Accessed on: 10 Sep. 2019.
[12] Suzuki, T., System creation and information sharing environment aiming for wide area cooperation at the time of large-scale disaster. *City Planning Review*, **64**(6), pp. 64–67, 2015. (In Japanese.)
[13] Cabinet Office of Japan, Guidelines for disaster information, May 2021. http://www.bousai.go.jp/oukyu/hinanjouhou/r3_hinanjouhou_guideline/pdf/hinan_guideline.pdf. Accessed on :10 Sep. 2021.
[14] Cabinet Office of Japan, Basic way of thinking about wide area evacuation from flood damage, May 2021. http://www.bousai.go.jp/fusuigai/suigai_kouikihinan/pdf/kangaekata.pdf. Accessed on: 10 Sep. 2021.
[15] Sakai Town & Local Government Research Organization, Survey and research on damage mitigation measures and wide-area evacuation action plan creation based on heavy rain disasters, 2018.
[16] *Mainichi Newspaper*, One month from wide-area evacuation due to the typhoon No. 19, 2019. https://mainichi.jp/articles/20191112/ddl/k11/040/118000c. Accessed on: 10 Sep. 2021.
[17] *Mainichi Newspaper*, Wide-area evacuation in Kazo City, Saitama. https://mainichi.jp/articles/20201010/k00/00m/040/038000c. Accessed on: 10 Sep. 2021.
[18] *Asahi Newspaper*, Search for wide-area evacuation, Sakai Town, Ibaraki. https://www.asahi.com/articles/ASNC97VFWNBDUJHB00S.html. Accessed on: 10 Sep. 2021.
[19] Rowan, K., Why rules for risk communication are not enough: A problem-solving approach to risk communication. *Risk Analysis*, **14**(3), pp. 365–374, 1994.
[20] Rowan, K., Botan, C.H., Kreps, G.L., Samoilenko, S. & Farnsworth, K., Risk communication education for local emergency managers: Using the CAUSE model for research, education, and outreach. *Handbook of Risk and Crisis Communication*, Taylor & Francis: London, pp. 168–191, 2009.
[21] Suzuki, T., *Wisdom, Technique and Mechanism to Protect Life from Great Disaster*, Shizuoka Academic Publishing: Shizuoka, pp. 154–196, 2014. (In Japanese.)
[22] Suzuki, T. Risk communication in the riverside district of Chuo City, Yamanashi Prefecture. *Journal of Community Disaster Management Society* (*C-Bosai*), **19**, pp. 58–70, 2020. (In Japanese.)
[23] Suzuki, T. & Uno, S., Development of disaster response management system with inter-organizational coordination. *Journal of Disaster Information Studies*, **10**, pp. 122–133, 2012. (In Japanese.)
[24] Suzuki, T., Hada, Y. & Miyamoto, T., Application of the BECAUSE model to establishment of a collaboration system with relevant administrative agencies for regional evacuation due to large river flooding capability and training process of its introduction. *Journal of Disaster Information Studies*, **14**, pp. 105–115, 2016. (In Japanese.)
[25] Suzuk, T., Watanabe T. & Okuyama, S., Facilitating risk communication for wide-area evacuation during large-scale floods. *International Journal of Environmental Research and Public Health,* **16**, p. 2466, 2019. DOI: 10.3390/ijerph16142466.

[26] Yamamura, T., *Practical Decentralized Evacuation and Shelter Management: Infectious Diseases X Large-scale Disasters*, Gyosei: Tokyo, pp. 190–204, 2020. (In Japanese.)

SPATIAL RELATIONSHIP BETWEEN PLACEMENT OF SHRINES AND RISK OF FLOODING IN LOW-LYING AREAS AT THE EASTERN TOKYO METROPOLITAN REGION

HIROSHI GOTOH
Department of Town Planning and Design, College of Science and Technology, Nihon University, Japan

ABSTRACT

Since ancient times, in Japan, shrines and temples have been a gathering place for local people to say prayers for their well-being. Also, shrines are commonly used by communities as evacuation centers in the event of a disaster at the present time. However, if the location of the shrine in the evacuation area is vulnerable to disasters, then the ramifications in terms of safety could be very serious. In this study, the author examined the spatial distribution of shrines in low-lying areas of eastern Tokyo. Recently, it was reported that shrines damaged by the tsunami that followed the Great East Japan Earthquake are related to the kind of main deities. In particular, since extensive ground subsidence has occurred in the low lying areas of eastern Tokyo, the author investigated the ground height before ground subsidence occurred based on land subsidence data, and considered the relationship between the ground height and the enshrined main deities at shrines taking into account with the danger of flood damage. The area where the shrine is located is thought to have existed as a town since ancient times. As a result, the area is generally considered to be less affected by floods. The information obtained from this survey will confirm the truth of this tradition, and will be useful for the management of urban planning from the viewpoint of disaster prevention.

Keywords: shrine, deities, low-lying areas of Tokyo, evacuation areas, subsidence, flood hazard.

1 INTRODUCTION

In the last decade, Japan has experienced several largescale natural disasters: the tsunami caused by the Great East Japan Earthquake that occurred on 11 March 2011, flooding of the Kinugawa River caused by the heavy rains in the Kanto and Tohoku regions that occurred on 11–12 September 2015, and flooding that occurred over a wide area of western Japan due to heavy rains on 3–8 July 2018 [1]. Of these events, the tsunami caused by the Great East Japan Earthquake caused the most extensive damage to shrines in the region. These shines, which are traditionally regarded as safe spaces by local communities, were the sites of extensive loss of life because people believed that they would be safe there [2]. As an example of recent river flooding, the Nagano Hachiman Shrine was damaged by the heavy rains caused by Typhoon No. 19 in 2020 [3], and the Aso Shrine, which is a national treasure, was damaged by the heavy rains in July 2021 [4].

Regarding the flood damage caused by the tsunami after the Great East Japan Earthquake, Uda et al. reported that there were many shrines near the periphery of the flooded area [2]. They therefore proposed that, historically, residents of areas that are susceptible to flooding may have built shrines in areas that were considered to be safe from tsunamis. Similarly, Hattori and Nii examined the spatial relationships among shrines, villages, and rivers in flood-prone areas of the Kurobe River alluvial fan [5]. Their findings showed that shrines are often located along the valley ridge line, which is the microtopography of the alluvial fan flood channel. Thus, they considered that the shrines played an important role in protecting local inhabitants against flood damage. Furthermore, Takada et al. [6] categorized the occurrence or absence of inundation damage to shrines caused by the Great East Japan

www.witpress.com, ISSN 1743-3509 (on-line)
doi:10.2495/DMAN210131

Earthquake tsunami by focusing on the main deity, and showed that there were differences in the extent of damage depending on the deity. Specifically, they found that damage to the Shinto shrines – an ancient religion practiced in Japan in which numerous deities have specific roles – differed depending upon the deity being glorified. Further, they showed that the extent of flood damage caused by the tsunami to shrines dedicated to the god of hydraulic control, "Susanoo" (e.g., the Kumano and the Hachiman shrines), was limited. The results of Takada's survey has potential that serve as a reference material for future infrastructure management, urban planning, and urban design in Japan. However, the application of this hypothesis, which Shrines dedicated to Susanoo are resistant to floods to flood control in other areas, is unknown. In order to test this hypothesis, it is necessary to investigate other areas.

In this study, the author targeted four low-lying wards (Taito, Sumida, Arakawa, and Koto wards) in the eastern part of Tokyo that are vulnerable to river floods and storm surges. The relationship between the spatial characteristics of the shrines and the main deity were analyzed, by referring to the study by Takada et al. [6] For comparison, the author conducted a survey on the location of churches, which serve the same purpose as shrines abroad, in low-lying areas near Amsterdam and in north Holland.

2 METHODOLOGY

The areas investigated were Arakawa Ward, Sumida Ward, Taito Ward, and Koto Ward, which are all low-lying areas (zero-meter zone, i.e., located at under the mean sea level) in the eastern part of Tokyo. First, the author investigated the history of flood damage during the Edo period, when the lowland areas of eastern Tokyo began to develop, by surveying the literature [7]. A general residential map was used to locate all of the shrines in the four wards [8]. Each site was then visited on foot to confirm the identity of the main deity to which the shrine was dedicated. In the absence of any information at the site of the shrine, the homepages of shrines were consulted [9], [10]. Next, in order to determine the altitude of each shrine, information on the current ground level was obtained from the website of the Geospatial Information Authority of Japan [11]. The predicted inundation depth at each shrine was then obtained from a hazard map issued by the local government, and compared with the ground level [12]–[15]. These estimates were necessary because considerable ground subsidence has occurred in the study area since the shrines were built, primarily due to excessive pumping of groundwater from the early Showa period to around 1970 [16]. In order to accurately determine the extent of land subsidence at the location of each shrine, the author obtained annual elevation data from the Tokyo Metropolitan Government Bureau of Construction. Since the spatial extent of the benchmarks measured in Tokyo has increased in number over time, some benchmarks were measured after severe land subsidence occurred. Therefore, in order to estimate the ground level as close to the Edo period as possible, the author selected and examined data for old benchmarks that were measured before the early Showa period (around 1935). The cumulative subsidence was then calculated and added to the current ground level by referring to the amount of ground subsidence for each year obtained from the old benchmark data nearest to each shrine. In this way, the ground level before land subsidence occurred was estimated, and the geographic locations of shrines dedicated to deities that offer protection from flood damage were examined. Some shrines in Japan have been relocated in the past, but it was sometimes difficult to investigate the relocation history; in such cases the author assumed that the shrine had been located at the current location for a long time. The author also examined Christian churches in this study because they fulfill a role similar to Shinto shrines in Japan, i.e., they serve as places of worship and protection. For reference, the author selected areas in north Holland and the city of Amsterdam in The Netherlands, as most of these areas are also located at or below sea level. The existence of churches was confirmed on Google Maps [17], and data on the ground level was determined using an electronic map published by the Geospatial Information Authority of Japan [11]. In this way, the ground level at each church location was measured.

3 RESULTS ON LITERATURE REVIEW ON SPATIAL LOCATIN OF SHRINE

3.1 History of flood damage in Edo area during Edo period

Fig. 1 shows a regional map of Edo, the old part of Tokyo. In this map, the locations of the three major floods in Edo are shown.

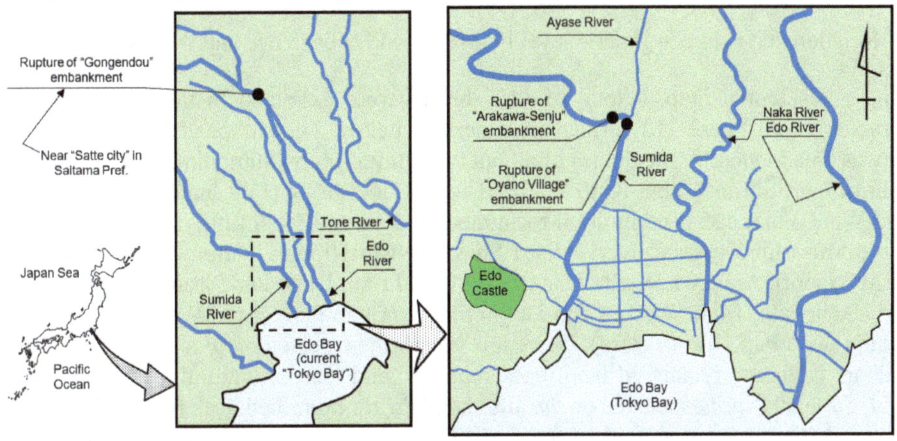

Figure 1: State of Edo (Tokyo) during the Edo era with insets showing rivers that experienced flooding and locations of three ruptured river embankments.

There were numerous floods in Edo (Fig. 1) during the Edo period (1603–1868), when many shrines are thought to have been built. It can be read from the literature that numerous floods occurred during this period and caused considerable damage [7]. It is likely that there were many more small and medium-sized flood disasters not mentioned in the literature [7]. Since the urban areas in the low-lying areas in the eastern part of Tokyo are built on a very large alluvial, these areas were greatly affected by both river floods and storm surges. According to the history of Katsushika Ward [18], the three major floods that occurred in Edo (Table 1) were caused by the collapse of the Kotanomura embankment in 1742 (Kanpo 2), the collapse of the Gongendo embankment in 1786 (Tenmei 6), and the collapse of the Arakawa Senju embankment in 1846 (Koka 3). Fig. 1 shows the locations of each breach. As can be seen, the levees broke near the center of Edo. Other literature sources corroborate the information shown in Table 1. According to the history of Katsushika Ward, in 1680 (Enpo 8), 1742 (Kanpo 2, one of the three major floods in Edo), 1791 (Kansei 3), 1823 (Bunsei 6) and 1856 (Ansei 3), the Katsushika area of Tokugawa Shogunate (currently, Chiba Prefecture, Saitama Prefecture, and northeastern Tokyo [19]) experienced severe damage due to high tides along the coast. It is therefore presumed that the residents in this area would have relied on deities to protect them from repeated water-related disasters and that this would be reflected in the spatial arrangement of shrines dedicated to deities that protected people from water-related disasters.

Table 1: Relationship between kind of deity and estimated maximum inundation depth.

Blessing	Estimated inundation depth					
	0 m	≤ 0.5 m	≤ 1 m	≤ 3 m	≤ 5 m	> 5 m
God of everything (main deity in Japanese mythology)				4	4	
Good luck		1	1	3	1	
Peace on Earth	1	1		17	11	
Good harvests (Inari system)		1	2	23	18	
Victory in war				8	5	1
Academic achievement				2		

3.2 Relationship between ground level before ground subsidence and deities

Using a residential map, a total of 113 shrines were extracted in the survey area. The characteristics of these 113 shrines are described below. The author visited all 113 shrines and was able to identify the deities of all but two shines: Inari Daimeijin (Eitai, Koto-ku) and Fushimi Inari Shrine (Eitai, Koto Ward). There is one shrine (Fuji Inari Shrine (Eda-gawa, Koto Ward)) that had been built on reclaimed land around 1920 [20]. There are six shrines on the Musashino terrace: Koryō Inari Shrine (Nishi Nippori, Taito Ward), Suwa Shrine (Nishi Nippori, Arakawa Ward), Hanazono Inari Shrine (Ueno Park, Taito Ward), Nanakura Inari (Ikenohata, Taito Ward), Gojo Tenjin Shrine (Ueno Park, Taito Ward), Ueno Toshogu Shrine (Ueno Park, Taito Ward). There are 104 shrines located on plains that were built on sandbars including reclaimed land, back marshes, and land on natural levees [21]. In this study, these 104 shrines located on the alluvial plain are examined. Referring to Takada et al. [6], the deities were classified as follows: (1) the main deity of Japanese mythology (the Shinto religion) called as Amaterasu Okami; (2) the deity that brings good luck to local residents; (3) the deity of peace on Earth (creator of the Earth and mountains, control of oceans and rivers, protection from fire and wind damage); (4) the deity of the Inari system (ensures good harvests, acquisition of wealth, business prosperity, good health of children); (5) the deity of victory in war (the god of warriors such as the Shogun and Yamato Takeru); and (6) the deity of academic learning. Table 1 shows the relationship between the main deity and the maximum inundation depth at the shrine estimated based on the inundation hazard map issued by the local government. It was difficult to find a statistically significant between the main god and the predicted inundation depth estimated using the current level of the ground. So, it was found that most shrines would be flooded in the event of a flood. As mentioned above, shrines are often not well suited for use as evacuation shelters during floods. In addition, it was difficult to determine the relationship on the spatial arrangement between the danger of flooding and the kind of main deity based on the current ground level. Consequently, the author decided to proceed again with the study taking into consideration the effects of severe land subsidence that occurred mainly in the 1970s from 1925 (early Showa period).

Specifically, the author estimated the ground level in the Edo era with reference to the fluctuation history of the nearest benchmark of each shrine measured from the beginning of the Meiji era (1868–1912), the Taisho era (1912–1926), and the Showa era (1929–1989). The author then reproduced the condition of the ground level before subsidence and tried to re-examine it. Fig. 2 shows the changes in land subsidence from 1929 (Showa 4) to 2018. As can be seen, the maximum subsidence in the study area since 1929 is at least 3 m. Indeed, based on benchmark levels measured since the Meiji era, there are some points where the ground level has changed by 4 m or more. Fig. 3 shows the relationships among the current ground level, the ground level at around the Edo period, and the main enshrined deities. The results show that during the Edo period, in this area, it was not the area that was not at under the sea level like it is today. Inari shrines and shrines for praying for happiness should be familiar to residents. Therefore, as shown in Fig. 3, it can be seen that the shrines are arranged in a wide range of ground heights so as to be close to the place where the inhabitants live in the Edo period. In addition, it was difficult to determine whether a shrine was dedicated to academic achievement because the number of samples was small, but such shrines were considered to be the same as the Inari system shrines because they are located close to residential areas. Furthermore, as reported by Takada et al. [6], shrines dedicated to Amaterasu were considered to be located in areas of high ground because Amaterasu has been the supreme deity related Japanese mythology and revered by Japanese people, and can therefore be regarded as having similar characteristics to Inari system shrines. On the other hand, shrines dedicated to peace on Earth and victory in war, as shown in Fig. 3, were built on ground that was slightly higher than the others(Amaterasu, Good luck, Good harvest, and Academic achievement). The locations of these(Peace on Earth and Victory in war) shrines are located seems to have been slightly higher than the surrounding area. The difference in ground height could not be determined clearly in the previous discussion, a one-way analysis of variance (ANOVA) was performed to confirm whether there is a significant difference in the mean value of ground height. Table 2 shows the calculation results. The results indicate that the hypothesis that the ground level of the shrine does not depend on the type of deity enshrined could not be rejected at a significance level of 5%. Therefore, there was no stochastic difference in the mean value of ground height. In other words, as concluded by Takada et al., it was not possible to demonstrate that there is a difference in the environment in which it is difficult to encounter flood damage depending on the kind of deity.

3.3 Spatial arrangement of churches in lowlands of northern Holland

The author extracted 25 churches in Amsterdam and 100 churches in the low-lying areas of north Holland along the transect indicated by A–B shown in Fig. 4.

Then, the author investigated the ground level at the location of each church and arranged the percentage for each ground level. Fig. 5 shows the frequency distribution of churches and their heights above sea level. In Europe, churches are typically located in center of towns. Fig. 6 shows the C–D transect through Amsterdam shown in Fig. 4. This is a cross section

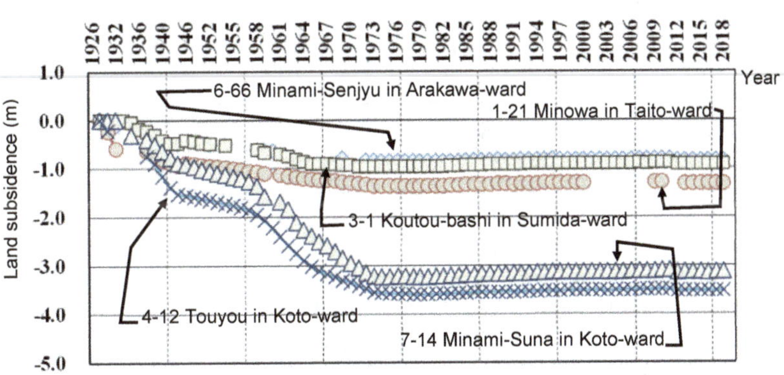

Figure 2: Land surface subsidence at several sites in the survey area over time.

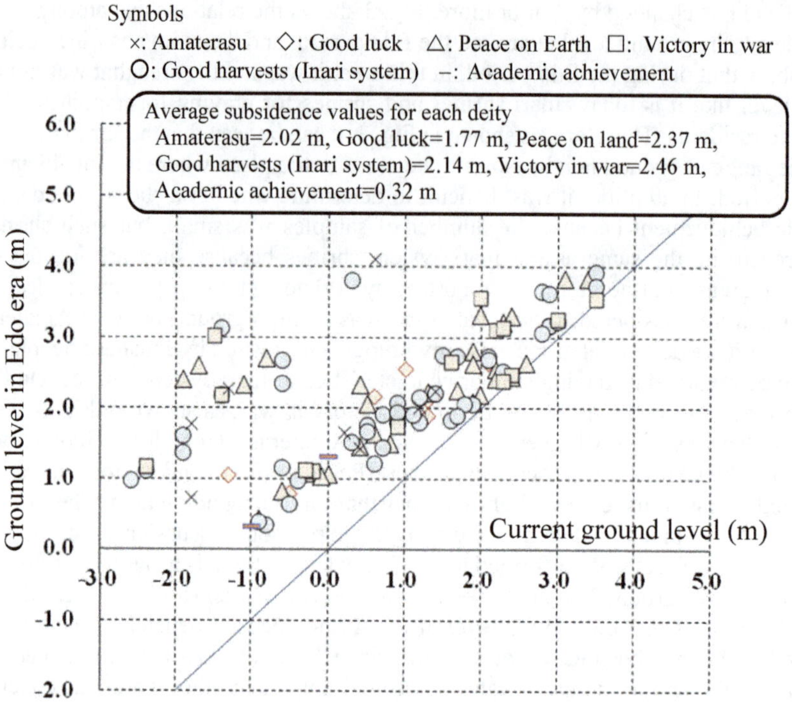

Figure 3: Relationship between deities and ground level.

Table 2: Results of one-way ANOVA.

Overview				
Group	Number of shrines	Total	Average	Variance
Amaterasu	8	16.161	2.020	0.950
Good luck	6	10.639	1.773	0.506
Peace on Earth	30	71.135	2.371	1.192
Good harvest	44	94.464	2.147	0.850
Victory in war	14	34.442	2.460	0.779
Academic achievement	2	1.643	0.822	0.500

Table of analysis of variance						
Fluctuation type	Fluctuation	Degrees of freedom	Variance	Observed variance ratio	P value	Boundary value for F
Between groups	7.101	5	1.4202	1.531	0.187	2.31
Within groups	90.923	98	0.928			
Total	98.024	103				

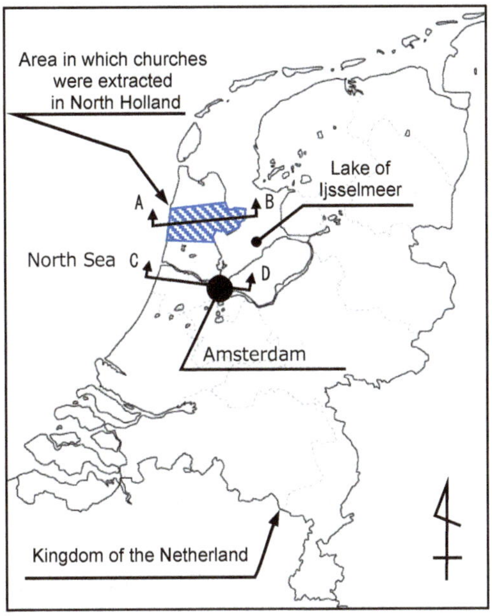

Figure 4: Area in which churches were extracted in The Netherlands.

obtained from the Geospatial Information Authority of Japan [11], and it can be seen that the city of Amsterdam was originally built on a hill. Most of the churches constructed in region are therefore built on high ground (Fig. 5). Fig. 7 shows a cross-sectional view of the A–B transect in north Holland shown in Fig. 4. Much of the land between the land protected by the coastal embankments at the North Sea and at Lake Ijsselmeer is at sea level. According to 1 m ground level data for The Netherlands obtained from the Geospatial Information Authority of Japan, no further details of the unevenness of the ground is known. However, as shown in Fig. 5, the ground level at the locations where some churches were built was below sea level, but in most cases, the ground level is above sea level. Therefore, although this area is lowland, it is considered that churches, which became the centers of towns, were established in areas that were slightly elevated. These churches then attracted residents, and the town developed and expanded. Thus, the density and placement of churches in elevated areas may have been to avoid damage due to inundation and, particularly, flood damage.

4 CONSIDERATIONS RELATED TO THE USE OF SHRINES AS EVACUATION SITES IN THE EVENT OF DISASTERS IN JAPAN

In the four wards targeted in the eastern lowlands of Tokyo, like the shrines on the Pacific coast of the Tohoku region, it was not found that shrines dedicated to specific deities were spatially arranged to avoid disasters. However, shrines are considered to function as centers of communities, where many people gather at local festivals, and as places to worship deities on a daily basis. The shrine grounds, although generally large in size, are safe from earthquakes as there are no tall buildings. They are also famous landmarks in the area, there is a possibility so each family member decide to use them as a meeting place in case of disaster. When the author surveyed the evacuation facilities for disaster countermeasures in

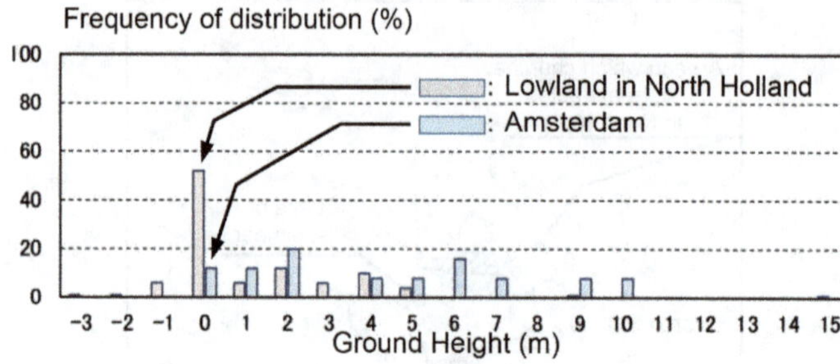

Figure 5: Distribution of ground height at church locations in different regions of The Netherlands.

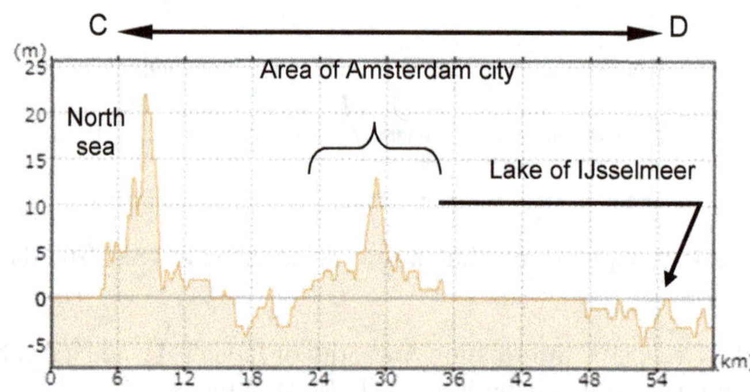

Figure 6: C–D transect through Amsterdam and surrounding area. *(Source: Modified from [11].)*

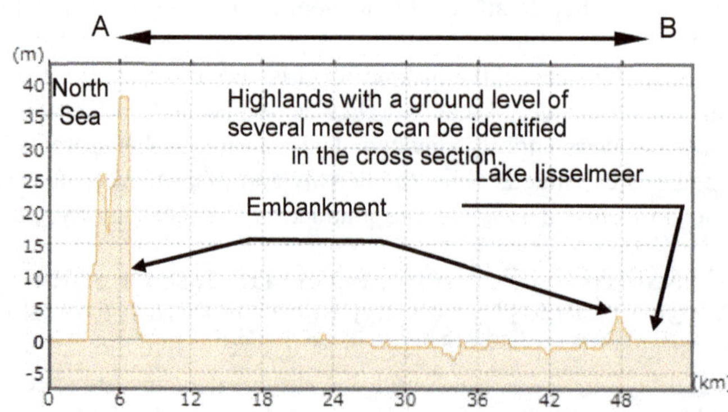

Figure 7: A–B transect in north Holland. *(Source: Modified from [11].)*

the four wards, most were public facilities such as schools; shrines were typically not designated as evacuation centers. Indeed, the survey showed that this is the case in many of the other wards in Tokyo. However, although the number was not large, there were some cases where shrines and temples were designated as temporary meeting places in the event of disasters caused by fire or earthquakes. It is considered that some residents cannot distinguish between the evacuation sites on disasters caused by fire/earthquakes and those due to flooding. Therefore, depending on the type of disaster(earth quake, fire and flooding), it is considered necessary to inform local residents of the locations of evacuation facilities and temporary meeting places.

5 CONCLUSION

The results of this study are summarized briefly. In the eastern lowland of Tokyo, there was no difference in the arrangement of shrines according to the type of deity, based on the present ground level. However, when cumulative ground subsidence was taken into account, the average ground height of shrines dedicated to deities responsible for peace on earth and victory in war was slightly higher than that of other shrines. However, a one-way analysis of variance (ANOVA) showed that there was no significant difference in ground height between the shrines.

Japanese people perceive areas with shrines as ancient towns, which would be safe against floods. In urbanized Tokyo, shrines have existed for a long time, but they are not always located on higher ground due to land alteration. Therefore, it was found that the areas where shrines are located may also be affected by floods. Therefore, the author believes that, from the perspective of disaster prevention in urban planning, shrines in Tokyo are not suitable as evacuation sites in the event of floods, and therefore, careful publicity should be promoted to residents.

In some areas of Japan, shrines are used as shelters. Therefore, I will conduct further research on the relationship between floods and the experience of damage at shrines and the actual situation of evacuation sites during floods.

As an example of the layout of religious sites outside Japan, The author have investigated the lowlands of The Netherlands and found that churches are located on slightly higher ground. In the future, the author would also like to focus on the lowlands of The Netherlands and consider classifying the layout of churches according to denomination, such as whether they are Catholic or Protestant.

ACKNOWLEDGEMENT

The author would like to express their gratitude to the Technical Support and Human Resource Development Center of the Tokyo Metropolitan Government Bureau of Construction who provided the benchmark data used in this study.

REFERENCES

[1] Japan Meteorological Agency, Meteorological cases that caused disasters. https://www.jma.go.jp/jma/index.html. Accessed on: 1 Mar. 2019.

[2] Uda, T. et al., Damage due to the 2011 Great Tsunami and shrines that escaped from being damaged. *Journal of Japan Society of Civil Engineers B3 (Ocean Engineering)*, **68**(2), pp. 43–48, 2012. (In Japanese.)

[3] *Mainichi Newspaper*, Inauguration ceremony of a new shrine in Nagano City, An area affected by Typhoon No. 19. https://mainichi.jp/articles/20201002/k00/00m/040/038000c. Accessed on: 18 Mar. 2021.

[4] *Yomiuri Newspaper*, Heavy rains cause flooding of national treasure shrine, damage to 61 cultural properties. https://www.yomiuri.co.jp/culture/20200717-OYT1T50146/. Accessed on: 18 Mar. 2021.

[5] Hattori. S. & Nii, A., Basic research on the location of shrines in flood-prone areas, targeting Kurobe River fan-shaped land and Nyuzen Town, Toyama Prefecture. *Proceedings of the 8th Landscape and Design Research Conference, Japan Society of Civil Engineers*, pp. 241–248, 2012. (In Japanese.)

[6] Takada, T. et al., A study on the deities and spatial arrangement of shrines in tsunami disaster caused by the Tohoku Earthquake. *Journal of Japan Society of Civil Engineers (Safety Problem (F6))*, **68**(2), pp. 167–174, 2012. (In Japanese.)

[7] Hatakeyama, N., *Meteorological Disasters*, Kyoritsu Shuppan Co., Ltd.

[8] Shobunsha Publications, Master of the City Tokyo 23 wards: Useful information map, 2018.

[9] Tokyo Shrine Office, Find a shrine. http://www.tokyo-jinjacho.or.jp/. Accessed on: 1 Mar. 2019.

[10] Each shrine in the five target wards: Homepage of each shrine. Accessed on: 1 Mar. 2019.

[11] Geospatial Information Authority of Japan, Electronic land WEB. http://www.gsi.go.jp/. Accessed on: 1 Mar. 2019.

[12] Arakawa Ward: Hazard map. https://www.city.arakawa.tokyo.jp/. Accessed on: 18 Mar. 2021.

[13] Sumida Ward: Hazard map. http://www.city.sumida.lg.jp/. Accessed on: 18 Mar. 2019.

[14] Taitou Ward: Hazard map, etc. https://www.city.taito.lg.jp/. Accessed on: 18 Mar. 2019.

[15] Koto Ward: Hazard map, etc. https://www.city.koto. lg.jp/. Accessed on: 1 Mar. 2019.

[16] Tokyo Metropolitan Government Bureau of Construction Koto Waterworks Office, Business summary of Koto Waterworks Office. http://www.kensetsu.metro.tokyo.jp/. Accessed on: 1 Mar. 2019.

[17] Google, Google Maps. https://www.google.com/. Accessed on: 1 Mar. 2019.

[18] Katsushika Ward: History of Katsushika Ward. http://www.city.katsushika.lg.jp/. Accessed on: 1 Mar. 2019.

[19] *Asahi Newspaper*, Kotobank. https://kotobank.jp/. Accessed on: 1 Mar. 2019.

[20] Tokyo Port Promotion Association, History of Tokyo harbor landfill. https://www.tokyoport.or.jp/43pdf_01.pdf. Accessed on: 1 Mar. 2019.

[21] Chuo-Kaihatsu Corporation, Ground Information Navigation. http://www.geonavi.net/georisknavi2/html/13_tokyo.html. Accessed on: 1 Mar. 2019.

CRITERION-REFERENCED ASSESSMENT INDEX FOR EVALUATING SOCIAL WELLBEING IN FLOOD RELIEF CAMPS

ADITYA S. NATH[1*], RITIKA SHARMA[1†], ARUNDAS KANDATHIL[1], CHINMAY KARAN[1] & CHAITALI BASU[2]
[1]Department of Architecture, School of Planning and Architecture, New Delhi
[2]Department of Building Engineering and Management, School of Planning and Architecture, New Delhi

ABSTRACT

The Sendai Framework for Disaster Risk Reduction advocates for reduction in disaster risk in the social assets of persons and communities. Well-being has been recognized by the World Health Organization as a significant determinant of health. Relief camps, being a vital community resource during the testing times of disaster, therefore, have a social obligation to enable expeditious recovery of displaced communities by offering social support. Despite the presence of numerous studies focusing on well-being in the built environment, very few explore social well-being in relief camps. The study aims to address this gap. The paper identifies social issues through mutual corroboration and interdependencies found by a mixed-review method involving three research approaches: comparative analyses of existing relief-camp guidelines, review of assessment reports of two prominent floods in India, and a structured interview of 255 inhabitants from 10 flood relief camps in the southern Indian state of Kerala. Recurrent issues were discerned from distinct sources, prompting the identification of assessment constructs and consequently, 50 assessment items falling under these constructs. The findings from this research can serve as an operative-index for evaluating social well-being in flood relief camps. The assessment index is fundamentally region-specific. However, the methodology employed may be adapted to develop similar indices. The originality of this research is in establishing the link between social well-being and relief-camps, and thereafter, formulating a method to assess social well-being in flood relief camps.
Keywords: social well-being, flooding, relief camp, resilience, community-centred, criterion-referenced assessment.

1 INTRODUCTION

Relief camps can play a significant role in developing psychological resilience of victims in the event of a disaster [1]. However, in the moment of urgency, all attention and resources are primarily focused on catering to critical physical requirements such as food, water and medical care. While those unquestionably remain the need of the hour, social well-being often gets compromised in the process due to uninformed decision-making at flood relief camps. The Sendai Framework for Disaster Risk Reduction 2015–2030 [2] advocates for reduction in disaster risk in the social assets of persons and communities. Even though social well-being is widely explored in the built-environment, it fails to have implications in relief camps, where it could play a vital role. The basic concepts relevant to this study are discussed in this section.

1.1 Background

Flood damage and resilience: The Inter-Governmental Panel on Climate Change (IPCC), in its *Fifth Assessment Report* predicts that "heavy precipitation events, which are very likely to increase in frequency in the future, will augment flood risk" [3]. In this context, studies

* *ORCID: https://orcid.org/0000-0003-3743-2643*
† *ORCID: https://orcid.org/0000-0003-0072-0414*

www.witpress.com, ISSN 1743-3509 (on-line)
doi:10.2495/DMAN210141

exploring various aspects of urban flood resilience recommend cities and villages to amplify focus on enhancing resilience rather than resistance [4].

Relief camps: Used as temporary accommodation for a displaced group of people until they move back to permanent homes, relief camps gain significance following disasters. Subsequent to the rescue period of 72 hours, victims are accommodated in camps during the recovery and reconstruction period lasting up to a year. Relief camps are considered vital for personal safety, climate protection, security, and resistance to diseases [5].

Social well-being in the built environment: World Health Organization defines health as "a state of complete physical and mental well-being and not merely the absence of disease or infirmity" [6]. Well-being is established as one of the major determinants of personal health as per this revised definition. Disasters have a chronic effect on the physical and mental health of the affected population and hostile conditions could foster psychological trauma due to loss of family, property or livelihood. Mental stress manifest as bodily complaints, highlighting need for psychological support.

Community frameworks: The United Nation International Strategy for Disaster Reduction defines disaster as a serious disruption of the functioning of a community, causing widespread losses, which exceed its ability to subsist using its own resources. The community is of significance since it is often the group that shares space in relief camps in case of an unprecedented event [7].

Criterion-referenced assessment: Criterion is defined as "a property or characteristic by which the quality of something may be judged, but makes no statement or assumptions about actual quality" [8]. It differs from standards, which are about definite levels of performance. Generally, an assessment is "a representational technique" rather than a literal one. A construct, refers to concepts that cannot be directly observed, but may be evaluated through a collection of items. It is a particular body of content that an assessment is designed to measure [9].

1.2 Theory and research

Various academic articles were reviewed to form a concise framework of issues to be investigated. Several search terms such as resilience, psychological resilience, social well-being, family-resilience, vulnerable population and assessment criteria, were explored in this review. The information pertaining to the purpose and findings of this paper is summarized.

Folke et al., by exploring the social dimensions of disaster governance, define resilience as the capacity of a system to absorb disturbance and reorganize while undergoing change, so as to retain essentially the same function, structure, identity, and feedbacks [10].

Fletcher et al., in their study exploring the definitions of resilience in psychology literature, outline it as a positive adaption in response to adversities of varying nature, ranging from daily hassles to major life events. Psychological resilience to flood would therefore imply an individual's capacity to return to normalcy in life, both physically and mentally [11]. Jenkins and Meltzer [12] in their report on mental health impacts of the Indian Ocean Tsunami of 2004, explain that the survivors displayed symptoms of anxiety, depression and Post-Traumatic Stress Disorder (PTSD). Displaced victims reported the symptoms to a greater extent as compared to non-displaced victims. A study exploring the psychological impact of natural disasters states that, extended exposure to stress, as experienced in relief camps, can result in a complete breakdown of significant physiological processes [13].

Meadows et al. [14] in their research, identify associations between neighbourhood characteristics and health and well-being of residents. The study describes social well-being through theories in social cohesion and interaction. The theory is vital to a relief camp environment since it functions similar to a close-knit neighbourhood. Three mechanisms linking neighbourhood, individual health and well-being are discussed.

1. Social cohesion: neighborhood's ability to create a sense of community among people.
2. Linking mechanism: physical environment, infrastructure and resources in the precinct.
3. Stress mechanism: neighborhoods with social disorder, few recreation options and limited access to healthy food builds the stress of living in the neighborhood.

Hackbarth et al. [15] underline the significance of assessing family resilience, as a building block of community resilience. Evacuating and keeping families together, providing them with appropriate shelter, support and mental health services are enlisted in Relief Camp Guidelines of India as interventions to help restore social cohesion and infrastructure [16]. These were found to be similar in principle to the three mechanisms discussed by Meadows et al. [14].

Maheen and Hoban [17] in their study concentrating on the plight of rural women in relief camps, emphasize the need to ensure social well-being of vulnerable population. Women were found to be more vulnerable than men in the same natural-disaster setting. Pre-existing gender inequality and socio-cultural community dynamics put women at greater risk. Pregnant women are particularly vulnerable because of their limited access to prenatal and obstetric care during disasters. The psychological effects of disaster are drastic among children, women and dependent elderly population [18].

The Guidelines for Assessment in Emergencies [19] provides a list of constructs for post-disaster assessment. The application of construct-item approach in existing disaster assessments is critical to this paper as the proposed index relies on the suitability of this framework. The framework suggests drawing assessment items to validate the enactment of policies relating to assessment constructs. Table 1 represents the assessment criteria and items prompted by the guidelines for this purpose.

Table 1: Assessment criteria based on *Guidelines for Assessment in Emergencies* [19].

Assessment constructs	Assessment items
Livelihood	Livestock; yield; tools and equipment; means of production of crafts, etc.; inputs for production (seeds, etc.); market access
Shelter and household	Essential everyday items; proximity and availability of services
Transportation	Access to transport services; access to places thru transport networks
Communication	Availability of communication resources (telephones, etc.)
Natural resources	Right to availability of usable water; energy availability; recreational spaces; pollution and hazard-free environment
Human capacities	Labour availability; skills; training
Financial means	Wages and salaries; self-employment; prices and rates; access to credit; access to savings; insurance

2 RESEARCH METHODOLOGY

Fig. 1 outlines the research methodology commencing from review of existing studies on relief camps and well-being to establish the research gap, and culminating at development of a comprehensive assessment index. The mechanism illustrated was employed to filter out issues affecting social well-being in relief camps through a series of steps.

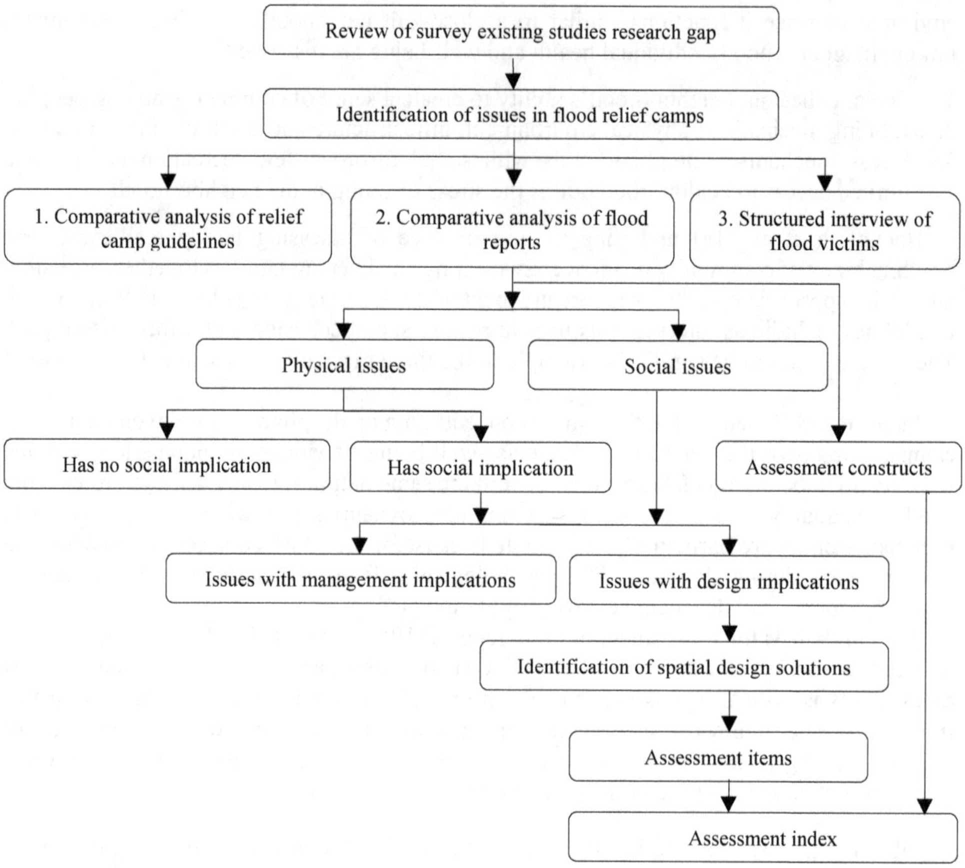

Figure 1: Research methodology.

Issues were identified primarily through three methods:

1. Comparative analyses of four relief camp guidelines of state, national and international scope (elaborated in Section 3).
2. Review of flood assessment reports of two prominent cases of floods in India; Kerala floods of 2018 [20] and Bihar floods of 2008 [21].
3. Structured interview of a focus group of 255 victims, sheltered in 10 relief camps, during the floods in Kerala in 2018.

The first two methods help outline the issues and identify constructs. The interview helps determine the weights associated with specific constructs based on regional factors and helps in customizing region-specific indices.

3 DATA COLLECTION AND ANALYSIS

For the preliminary identification of the issues and solutions, a comparative study of the following four relief camp guidelines of varying scopes was carried out to understand key requirements for well-being (Table 2):

1. Minimum Standards of Relief, Kerala [22] (KSDMA) and,

2. Guidelines for Relief Camp Management, Assam [23] (RDMD) for state-level guidelines;
3. Guidelines for Minimum Standards of Relief in Camps, India [16] (NDMA) for national-level guidelines;
4. Camp Planning Standards [24] (UNHCR) for international-level guidelines.

Table 2: Comparative study of existing relief camp guidelines.

		KSDMA	RDMD	NDMA	UNHCR
1	**Area/person (sqm)**	3.5	3.5–4.5	3.5	Cold/urban areas: 3.5; others: 4.5–5.5
2	**Special provision for diff.-abled, old, medically unfit**	✓	✓	✓	–
3	**Dairy products for mothers and children**	✓	–	✓	–
4	**Precautions against malnutrition**	✓	–	✓	–
5	**Special consideration for women**	✓	✓	Women police officers to be present for assistance	–
6	**Identification of vulnerable groups**	✓	✓	✓	–
7	**Provision of keeping domestic animals**	✓	–	–	–
8	**Provision of medical assistance**	Mobile medical teams	1 doctor and paramedic per camp	Mobile medical teams	1 health center per 20,000 people
9	**Psychological counselling**	✓	Families are to be housed together	–	–
10	**Help desks for grievances**	✓	✓	–	–
11	**Provision of communication devices**	–	✓	–	–
12	**Leisure/entertainment**	Toys and games for children	TV, newspapers, books, temp. schools	–	–
13	**Whether inhabitants are part of camp management**	Community kitchens	Camp management and security	–	–
14	**Livelihood support**	–	Awareness about NREGA[1], etc.; active SIRD[2] and DRDA[3]	–	–

[1]NREGA = National Rural Employment Guarantee Act; [2]SIRD = States Institute of Rural Development; [3]DRDA = District Rural Development Agencies; – no mention.

Despite extensive guidelines, relief camps face multiple issues upon post-disaster activation. Therefore, a comprehensive study of issues faced in flood relief camps from two distinct geographical and cultural contexts in India was carried out.

Table 3: Comparative study of flood assessment reports.

Construct	Kerala	U.P. and Bihar
Health	Spread of communicable diseases due to crowding; malnutrition in children; water-borne diseases	Medical teams deployed in relief camps with essential equipment
Gender	Fear of trafficking, teasing; women lack repair-skills	–
Education	Loss of study material, school days, school used as relief camp; school premises-suffered damages	Alternate learning spaces and study material provided in camps
Children	Violence and abuse; fear of return of floods and damage from shelter; need for mental support	Child psychological care and counselling support provided
Old and differently-abled	Difficulty in using washrooms, other services; access within building difficult; no special support; seeing, hearing, physical, mental health issues	–
Nutrition	Quantity of consumption of food remained unaffected after the floods; although food was contaminated	Nutritional supplements; special care of malnourished children
Water	Water borne diseases reported; reduced availability of potable water	Disinfection and regular testing of water to confirm potability
Social inclusion	–	Community kitchens; alternate learning; maternal health centres; *anganwadis* (preschools)
Sanitation	Lack of sanitary systems, open defecation reported	Family hygiene kits were provided
Shelter	Damage to houses – proportional to duration of stay at relief camp; legal property papers lost	Temporary shelters in camps with tarpaulin sheets
Mass awareness	–	Awareness for school dropout kids, their parents; mothers of malnourished children; hygiene drills
Livelihood	Harvests, cattle, cattle sheds lost; labour affected (due to outstation labour's return to respective home states); tourism, hospitality affected	–
Privacy	Lack of private bathing spaces and latrines. Separate spaces provided for families, lactating mothers; emergency rooms for pregnant women	Installation of temporary toilets and bathing cubicles

Joint Detailed Needs Assessment Report [20] cites the issues faced by relief camps in Kerala, a state in southern India, following the floods of 2018. Flood Emergency Response in Bihar and Uttar Pradesh [21] documents the details of the relief operations undertaken in the states Bihar and Uttar Pradesh in northern India during the overwhelming 2008 floods.

The former disaster being an unprecedented one, the state was underprepared and the report states the problems of inadequacy faced by relief camps. The latter being a recurring event, was tackled proficiently, with the report highlighting their solution-oriented approach. The findings and observations from this comparative study are given in Table 3.

This comparative study, along with the study of camp guidelines helps find the multiple primary identifiers that can translate as assessment constructs to evaluate relief camps.

To gain a primary understanding of the weights of the issues falling under specific identifiers or constructs, structured interviews were conducted on a focus group of 255 flood victims from 10 selected flood relief camps in Kerala.

The selection criteria involved a minimum estimated capacities of over 500 individuals with an equitable representation from across the state in order to capture regions with varying severity of flooding as depicted in Fig. 2. Among these 10 camps, nine are academic institutions and one is town-hall. A close-ended questionnaire, binding on 20 constructs identified through studies and open-ended discussions with relief camp occupants was conducted as part of the interview. The questions were designed to cover the maximum scope of each construct. Each question was presented with a response scale denoting their satisfaction in each construct they represented, ranging from 1 to 10.

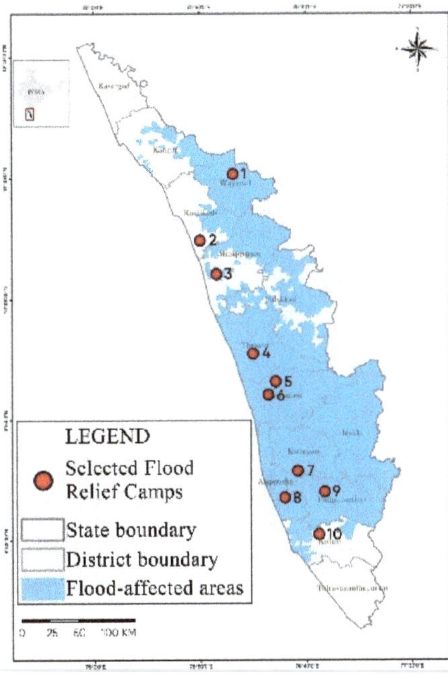

S.No.	Relief Camp	Capacity (in persons)
1	Govt. Higher Secondary School, Neervaram, Wayanad	600
2	Govt. Arts and Science College, Kozhikode	2500
3	Govt. Higher Secondary School, Tirurangadi, Malappuram	600
4	Govt. Higher Secondary School, Kodakara, Thrissur	1000
5	Govt. Higher Secondary School, Puliyanam, Ernakulam	1000
6	U.C College, Aluva, Ernakulam	1000
7	Town Hall, Changanassery, Kottayam	1600
8	Govt. Girls' High School, Haripad, Alappuzha	800
9	S.N.D.P High School, Edapariyaram, Pathanamthitta	700
10	Govt. Higher Secondary School, Puthur, Kollam	600

Figure 2: Selected flood relief camps in Kerala for primary data collection.

Various guidelines recommend culture-specific diet and clothing for the relief camp inhabitants. The regional factors, thus, needs to be identified and used to guide the development of assessment index. The questionnaire identifies the strengths and weaknesses of each construct and can help determine the weights to be assigned to the respective constructs for specific regional cases.

4 FINDINGS AND DISCUSSION

Based on the data, two measures of central tendency were calculated for each construct as depicted in Table 4 and Fig. 3.

Table 4: Percentage distribution of responses received from inhabitants of the ten selected flood relief camps (relative occurrences of modal values marked in bold).

Code	Percentage distribution of response (%)									
	1	2	3	4	5	6	7	8	9	10
HEA-01	0.4	1.6	15.3	20.4	**20.8**	18.4	18.4	2.4	1.6	0.8
GEN-02	0.8	17.6	18.8	**20.4**	18.8	18.4	3.9	0.4	0.4	0.4
PRI-03	0.4	20.0	20.8	**22.0**	20.0	13.7	1.2	0.8	0.8	0.4
EDU-04	0.8	15.7	15.7	17.6	**23.9**	7.1	6.7	6.7	5.1	0.8
CHI-05	0.4	1.6	3.1	22.7	**26.3**	22.7	21.6	0.8	0.4	0.4
ODA-06	8.6	20.4	**23.5**	22.7	21.6	1.2	0.4	0.8	0.4	0.4
VUG-07	0.4	24.7	23.9	**26.3**	22.4	0.4	0.8	0.4	0.4	0.4
NUT-08	0.8	0.4	0.4	1.6	31.0	**33.3**	30.2	0.4	0.8	1.2
WAT-09	0.4	0.4	1.2	0.8	25.1	**26.7**	24.3	19.6	0.4	1.2
SHE-10	0.4	0.4	22.4	23.1	**25.1**	16.1	11.4	0.4	0.4	0.4
SAF-11	20.4	24.7	**25.9**	24.3	1.2	0.4	0.4	0.8	1.2	0.8
SEC-12	0.8	23.5	24.7	**25.1**	23.5	0.4	0.8	0.4	0.4	0.4
HES-13	0.4	29.4	**30.2**	24.7	12.2	0.8	0.4	0.8	0.4	0.8
SIN-14	0.4	0.8	0.4	0.8	24.3	**25.5**	24.7	22.4	0.4	0.4
LIV-15	0.8	19.6	19.2	**20.8**	18.4	19.6	0.4	0.4	0.4	0.4
TRA-16	0.8	0.4	0.4	0.4	22.7	**28.2**	23.5	22.7	0.4	0.4
COM-17	0.4	0.8	0.4	0.8	4.3	8.6	**30.6**	26.7	25.5	2.0
MAW-18	0.4	0.4	0.4	0.8	23.9	**25.1**	23.1	24.3	1.2	0.4
LEI-19	0.4	0.8	20.4	23.5	**25.5**	13.7	14.1	0.8	0.4	0.4
LST-20	1.2	25.1	30.2	**31.0**	4.3	3.1	2.0	2.0	0.4	0.4

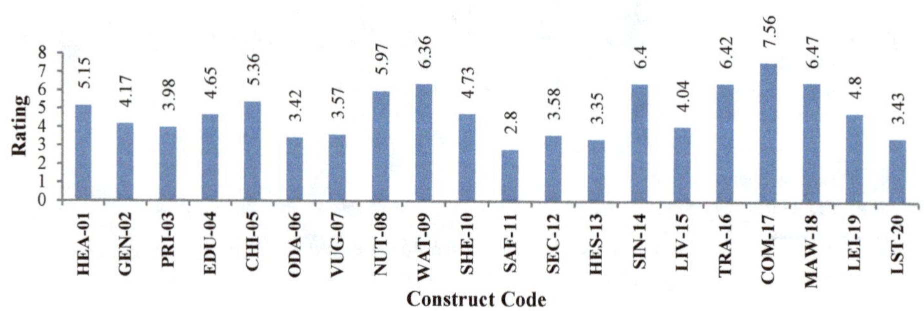

Figure 3: Graphical representation of mean ratings given by the inhabitants of 10 selected flood relief camps on the 20 indicators.

a. Health *(Mean: 5.15, Mode: 5):* Communicable diseases, water-borne diseases and malnutrition were serious issues in relief camps. Proper physical distancing and provision of clean flood and water need to be ensured.
b. Gender *(Mean: 4.17, Mode: 4):* Lack of gender segregation, especially in the event of being unaccompanied by family, and fear of teasing created a hostile environment for girls. Separate toilets for females, although recommended, were not provided in all cases.

c. Privacy *(Mean: 3.98, Mode: 4):* Privacy was compromised in most relief camps due to overpopulation. Lack of private bathing spaces and latrine were critical issues. Families could not have private spaces as recommended, due to lack of partitions and enclosures.
d. Education (Mean: 4.65, Mode: 5): Education was affected for both the inhabitants and the regular students at the schools and colleges operating as relief camps. Alternate learning spaces were necessary, and access to these spaces needed to be ensured.
e. Children *(Mean: 5.36, Mode: 5):* Children underwent psychological trauma caused by fear of return of floods. Shutting down of schools and adversely affected their life patterns. Transforming the relief camp premises through wall paintings is recommended. Protection from abuse through passive surveillance can be ensured in planning.
f. Old and Differently-Abled *(Mean: 3.42, Mode: 3):* Although most of the camps had differently-abled-friendly fixtures, inadequate access and lack of help led to mental stress over helplessness in the elderly and differently-abled. Vision and hearing impairments was common in the elderly, causing them difficulties in habituating to a new environment.
g. Other Vulnerable Groups *(Mean: 3.57, Mode: 4):* Lack of sufficient care, privacy, nutritional diet and medical care creates a hostile environment for pregnant women, lactating mothers, patients or malnourished people. Unsanitary conditions caused risk of spread of diseases to infants.
h. Nutrition *(Mean: 5.97, Mode: 6):* Primary survey showed that most inhabitants were satisfied with the food provided in relief camps. Though, in some cases, food was reported to be contaminated by floods due to improper storage environment. Dry storage areas are a necessity during floods.
i. Water *(Mean: 6.36, Mode: 6*): Supply of clean water was an issue due to contamination of water sources. Stagnation of water in the premises led to spread of diseases. Clean water storage facilities are a necessity. Also, the water points need to be accessible for the differently-abled and elderly.
j. Shelter *(Mean: 4.73, Mode: 5):* Lack of a shelter to return to lengthens the stay of victims in relief camps. Temporary shelter arrangements were made to help the inhabitants gradually relocate away from the relief camp buildings. Alternately, temporary shelters served as classrooms. The quality of life in these makeshift structures needs to be assessed.
k. Security *(Mean: 2.80, Mode: 3*): Loss of belongings caused by absence of storage spaces for valuables such as cash, documents and jewelry are a matter of concern. Relief camps without adequate number of vaults or depositories spread discomfort due to fear of theft and loss of valuables among the residents, who have already suffered from loss of property and livelihood in floods.
l. Safety *(Mean: 3.58, Mode: 4):* Safety, in this context refers to human safety. Two priority areas here are safety for women and children which are already covered. Safety breach occurs when vulnerable groups find it difficult to thrive in an environment. A majority of the inhabitants felt unsafe as per the primary survey.
m. Hygiene and Sanitation *(Mean: 3.35, Mode: 3):* Lack of hygienic living conditions, due to moisture on surfaces, overcrowding, poor waste management and intermingling of sick and the healthy was reported. Latrine facilities were in shortage due to which residents had to engage in open defecation in many areas. Primary survey suggests that a majority of inhabitants found the camp unhygienic and faced issues. Lack of ownership and accountability among the inhabitants has contributed to contamination of precincts.
n. Social Inclusion *(Mean: 6.40, Mode: 6):* It is important to make the inhabitants comfortable in the social environment. Isolated cases of caste-based discrimination were

reported from Kerala as per the JDNA report. Initiatives such as community kitchen (engaging women and children), alternate learning systems, with volunteers for schools, maternal health centers and preschool restoration were certain interventions practiced in relief camps that has helped build an inclusive environment.

o. Livelihood *(Mean: 4.04, Mode: 4):* Livelihood destruction is a primary contributor to mental agony among victims. Congregation spaces in relief camps can be used to deliver awareness regarding welfare schemes and employment opportunities. Space can be assigned for skill training to help generate self-employment. Women could use this opportunity to initiate self-help groups to support families and bolster the economy. Very few camp inhabitants had income generation during their stay in relief camps.
p. Transportation *(Mean: 6.42, Mode: 6):* Access to relief camps is a necessity. Disruption of transportation results in mental stress among victims. It is crucial to have access to transport and to places on transport network.
q. Communication *(Mean: 7.56, Mode: 7):* Having functional channels for communication is critical during floods. Emergency situations can erupt and helplines need to be set up. Since phone lines could be damaged, alternate methods need to be looked for. Provision of phone-charging points is important.
r. Mass Awareness *(Mean: 6.47, Mode: 6):* At multiple occasions, when people need to be addressed at large about issues like hygiene, livelihood, malnourishment, education, etc., a public-address system, and/or a gathering space for announcements is necessary.
s. Leisure *(Mean: 4.80, Mode: 5):* Leisure activities suitable for all age groups should be organized as they play a critical role in reviving emotional well-being. Kids need a safe and open area to play. Spaces designated for television and radio promote public engagement. Community gathering spaces have been used for celebrating events.
t. Livestock *(Mean: 3.43, Mode: 4):* Livestock, albeit suffering habitat loss and trauma similar to humans, is often overlooked in relief camps. Only few of the rural relief camps equipped with space for livestock, which is important to the rural population.

The discourse on assessment constructs and factors affecting each, points towards assessment items. The assessment items, here, refer to the possible spatial implications that could satisfy the issues mentioned under each construct. Table 5 identifies 50 assessment items, identified from multiple sources, enumerated under the 20 constructs, which together present an index for assessing social well-being in flood relief camps. The assessment items mentioned against the corresponding constructs, are inferred from the relief camp guidelines and flood reports, and address the vital issues undermining social well-being, identified through the mixed-review of reports and questionnaires. Together, they put forth an all-encompassing compilation of proposals to be enacted in relief camps that could significantly ensure social well-being of its residents. Conversely, relief camps could be assessed on the basis of the given attributes to verify their relief standards on the basis of social well-being.

5 CONCLUSION

Indifference towards social well-being, a significant determinant of personal health as per the World Health Organization, impedes the post-traumatic recovery of the victims, and proves detrimental to their health and overall resilience of the community. The need for a criterion-referenced assessment index to evaluate social well-being in flood relief camps was thus established and substantiated with shortcomings observed in mixed-review of relief camp guidelines and flood assessment reports.

Table 5: Index of assessment constructs and corresponding assessment items.

Assessment construct	Solution references		Spatial implications/Assessment item
Health	NDMA Guidelines	1	Light and ventilation
		2	Secluded area for the sick
		3	Heaters to counter dampness caused by floods
		4	Dedicated spaces for medical teams, equipment, medicine
Gender	NDMA Guidelines	5	Partitions for gender segregation in living area
		6	Separate toilets for male and female
Privacy	KSDMA Guidelines	7	Segregation of spaces for families/ individuals
		8	Private bathing spaces
		9	Private latrine facility
Education	NDMA Guidelines	10	Provision of space for informal classes
		11	Safe access to nearby school
Children	JDNA Report, Kerala Flood 2018	12	Wall paintings to make camp child-friendly
		13	Enclosure to protect children
		14	Play areas with public surveillance
Old and differently-abled	NDMA Guidelines	15	Universal access to relief camp building
		16	Universal access to toilets
		17	Well-lit access to toilets at all times
Other vulnerable groups	NDMA Guidelines	18	Emergency room for pregnant women
		19	Private room for lactating mothers
Nutrition	UNICEF Report, 2008	20	Cool, dry kitchen and storage space for food supplies
Water	JDNA Report, Kerala Flood 2018	21	Universal access to water points
		22	No stagnation is to be ensured
Shelter	JDNA Report, Kerala Flood 2018	23	Shelter provision for long duration of stay
		24	Sleeping area segregated from activities spaces
		25	Personal spaces of comfort for each family/ individual
Security	KSDMA	26	Lockable doors and windows
		27	Provision for storing valuables
Safety	NDMA Guidelines, Red Cross Guidelines	28	Safe living condition of residents
		29	Well-lit circulation spaces
		30	Female police officers in the camp
Hygiene and sanitation	JDNA Report, Kerala Flood 2018 KSDMA Guidelines	31	Cleaning of toilets at regular intervals
		32	Clearing of debris at the earliest
		33	Dustbins placed all across the camp
		34	Space for storage of waste until collection

Table 5: Continued.

Assessment construct	Solution references		Spatial implications/Assessment item
Social inclusion	KSDMA Guidelines	35	Inclusive arrangement of inhabitants' living spaces
		36	Alternate *anganwadis* (preschool facility)
		37	Community kitchens
Livelihood	KSDMA Guidelines	38	Provision of space for workshops
		39	Designated space for awareness programs
Transportation	Red Cross Guidelines	40	Functioning access path to relief camp
		41	Parking space for rescue vehicle/ ambulance
		42	Gate of the camp enabling movement of lar vehicles/boats
Communication	NDMA Guidelines	43	Provision of communication devices, newspapers, radios
		44	Provision of charging points for cellular phones
		45	Designated desk for reporting grievances and complaints
Mass awareness	KSDMA Guidelines	46	Provision of a gathering space for public address
		47	Public announcement equipment like speakers
Leisure	KSDMA Guidelines	48	Provision of common areas for community gatherings
Livestock	KSDMA Guidelines	49	Allocated space for livestock
		50	Storage of water, food and shelter for pets, livestock

General theories on social well-being were explored and direct parallels were found between well-being mechanisms and the inadequacies reported in flood relief camps. Construct-item approach of criterion-referenced assessment structure was adopted as a reliable gauge as it presents an operative index to evaluate social well-being in relief camps.

The research, based on comparative analyses of the mentioned sources, helped gain perspective about the current gaps at achieving social well-being. It further led to the identification of assessment constructs and corresponding issues. Close-ended survey of relief camp inhabitants revealed lack of well-being in the relief camps and further emphasized the role of regional factors in achieving it. Furthermore, the rating scale employed proved to be an effective mechanism to determine the weights of each construct.

The issues and the corresponding spatial recommendations were identified and translated as assessment items. The study concluded with the enumeration of 50 assessment items under 20 constructs. Unlike the assessment index, which is region-specific due to inherent cultural biases, the method is universally applicable. The output of this research can serve as an operative index of assessment criteria to ensure effectiveness in spatial design solutions to ensure social well-being in flood relief camps.

ACKNOWLEDGEMENTS

This research was developed under the UKIERI-UGC flagship project between School of Planning and Architecture, Delhi (SPA-D) and University of the West of England, Bristol, UK (UWE Bristol) as part of institutionalizing disaster management in undergraduate curriculum. The opinions and findings expressed here, however, are those of the authors alone, and are not necessarily the views or positions of UKIERI-UGC. The authors thank colleagues from university, and the numerous relief camp volunteers for assisting in dissemination of the questionnaire. The authors are thankful to Dr Sudha T and Dr Jaya Kumar for assistance and guidance towards completion of this manuscript.

REFERENCES

[1] Crabtree, A., *Questioning Psychosocial Resilience After Flooding*, Springer Science+Business Media Dordrecht, 2013. DOI: 10.1007/s11205-013-0297-8.

[2] Sendai Framework for Disaster Risk Reduction 2015–2030, United Nations, 2015.

[3] Intergovernmental Panel on Climate Change, Fifth Assessment Report, 2014.

[4] Liao, K.-H., A theory on urban resilience to floods: A basis for alternative planning practices. *Ecology and Society,* **17**(4), 2012.

[5] Bashawria, A., Moodleya, L. & Garritya, S., *An Overview of the Design of Disaster Relief Shelters*, 2014. DOI: 10.1016/S2212-5671(14)01019-3.

[6] World Health Organization, Constitution of the World Health Organization, 71st World Health Assembly, Geneva, 2018.

[7] Plodinec, M.J., Edwards, W.C. & Whitea, R.K., Applications of a "whole community" framework for enhancing community or campus resilience. Salford Quays, UK, 2014. DOI: 10.1016/S2212-5671(14)00907-1.

[8] Sadler, D.R., Formative assessment and the design of instructional systems. *Instructional Science,* 1987. DOI: 10.1007/BF00117714.

[9] Hanson, F.A., *Testing Testing: Social Consequences of the Examined Life*, University of California Press: Berkeley, CA, 1993.

[10] Folke, C., Hahn, P.O.T. & Norberg, J., Adaptive governance of social-ecological systems. *Annual Review of Environment Resources,* 2005. DOI: 10.1146/annurev.energy.30.050504.144511.

[11] Fletcher, D. & Sarkar, M., Psychological resilience: A review and critique of definitions, concepts, and theory. *European Psychologist*, **18**(1), 2013.

[12] Jenkins, R. & Meltzer, H., *The Mental Health Impacts of Disasters*, Government Office of Science: UK, 2012.

[13] Freedy, J., Shaw, D., Jarrell, M. & Masters, C., Towards an understanding of the psychological impact of disasters: An application of the conservation resources stress model. *Journal of Traumatic Stress*, 1992. DOI: 10.1007/BF00977238.

[14] Meadows, S.O., Miller, L.L. & Miles, J.N.V., Associations between neighborhood social and economic characteristics and resident health and well-being, RAND, 2014.

[15] Hackbarth, M., Pavkov, T., Wetchler, J. & Flannery, M., Natural disasters: An assessment of family resiliency following Hurricane Katrina. *Journal of Marital and Family Therapy*, **38**(2), 2012. DOI: 10.1111/j.1752-0606.2011.00227.x.

[16] National Disaster Management Authority, Guidelines for Minimum Standards of Relief in Camps, Government of India, 2017.

[17] Maheen, H. & Hoban, E., Rural women's experience of living and giving birth in relief camps in Pakistan, 2017. DOI: 10.1371/currents.dis.7285361a16eefbeddacc8599f326a1dd.

[18] Makwana, N., Disaster and its impact on mental health: A narrative review. *J. Family Med. Prim. Care*, **8**(10), 2019. DOI: 10.4103/jfmpc.jfmpc_893_19.
[19] International Committee of the Red Cross, Guidelines for Assessment in Emergencies, International Federation of Red Cross and Red Crescent Societies: Switzerland, 2014.
[20] Sphere India, Kerala Floods: Joint Detailed Needs Assessment Report, Sphere India, National Coalition of Humanitarian Agencies in India: New Delhi, 2018.
[21] UNICEF, Evaluation of UNICEF's Flood Emergency Response in Bihar and Uttar Pradesh, TARU, 2008.
[22] Kerala State Disaster Management Authority, Minimum Standards of Relief, Kerala, Department of Revenue and Disaster Management, Government of Kerala, 2017.
[23] Revenue and Disaster Management Department, Guideline for Relief Camp Management, Assam, Government of Assam, 2015.
[24] UN High Commissioner for Refugees, Camp Planning Standards, United Nations, 2018.

DEVELOPMENT OF A DISASTER RISK PROFILE IN THE PUBLIC HEALTHCARE SYSTEM DURING FLOOD SITUATION: A CASE STUDY OF NAKHON SAWAN CITY MUNICIPALITY, NAKHON SAWAN PROVINCE, THAILAND

KODCHAKORN KRUTPHONG[1], JING TANG[1,2,3] & NATT LEELAWAT[1,3,4]
[1]Risk and Disaster Management Program, Chulalongkorn University, Thailand
[2]International School of Engineering, Chulalongkorn University, Thailand
[3]Disaster and Risk Management Information Systems Research Group, Chulalongkorn University, Thailand
[4]Department of Industrial Engineering, Chulalongkorn University, Thailand

ABSTRACT

The United Nations Office implemented the disaster risk profile idea for disaster risk reduction as one of the tools in the development and country's disaster risk reduction. Due to the worldwide agreement, the Sendai Framework of Disaster Risk Reduction 2015–2030 has one of the priority actions to understand the risk before taking further steps to plan and operate. Furthermore, the knowledge of the emergency management has the topic of the critical infrastructure that needs to be protected to make it operate as usual during the emergency to prevent the worst case. The public healthcare system is one of these critical infrastructures. This research followed this concept and focused on the disaster problem in Thailand, which its major disaster type is a flood. The data from the Department of Disaster Prevention and Mitigation, Ministry of Interior, reported that during the fiscal year of 2010–2019, 48,543 THB of government expenditure on the emergency help cost was used to ease the flood disasters. Thus, this research paid attention to the development of the disaster risk profile in the public healthcare system during the flood situation. The study area was Nakhon Sawan Municipality, Nakhon Sawan Province, Thailand, one of the significantly affected areas from the 2011 Thailand floods. The main target referred to the operation during the 2011 floods in the study area. This research applied the BPMN 2.0 to understand the operation process and the cooperation during the flood situation. The model could explain the function and responsibility of each organization and the disaster risk profile, by focusing on the operational level during the flood situation. The outcome of this was the risk profile and the suggestion to conduct the policy in the next step.

Keywords: risk profile, flood, public health, emergency operation, Thailand.

1 INTRODUCTION

The United Nations office implemented the disaster risk profile idea for disaster risk reduction (UNDRR) as one of the tools in the development of country's disaster risk reduction [1]. From the beginning, of the vision of the sustainable development of the international agreements on a lead of the United Nations, in the field of disaster risk reduction, the international community has the decision to create the shared framework in the name of Sendai Framework of Disaster Risk Reduction 2015–2030 [2]. This framework has the priority of the Disaster Risk Management in each country in the government policy for the relation in the wider area.

At the global level, the World Health Organization (WHO) Regional Office for Europe states that "floods are the most common natural disaster" [3]. And the data from the same organization also states that

www.witpress.com, ISSN 1743-3509 (on-line)
doi:10.2495/DMAN210151

> "Floods can cause widespread devastation, resulting in loss of life and damages to personal property and critical public health infrastructure. Between 1998 and 2017, floods affected more than 2 billion people worldwide".

According to the recent research, not only countries in Europe were affected by floods, but also those in Asia. It was stated that "analysis of flood trends in Asia shows that flood disasters in most of the Asian countries are increasing significantly, especially in the last few years Asia witnessed a rapid increase of flood disasters" [4].

In Thailand, the data from the Department of Disaster Prevention and Mitigation (DDPM), showed that in 10 years, between 2001–2012, there were 189 times of flood disasters. The other source collecting data between 1989–2009 (20 years) revealed that Thailand faced 213 flood disasters which resulted in 115,786.67 million Baht of damage cost. In particular, according to "2011 Great Flood: the lesson from the experience" [5] affected the area of 57,600 square kilometers in 65 provinces. There were approximately 12 million victims and 815 deaths from the incident. The calculated damage cost in the economic sector was 1.425 trillion Baht, seven times more than the government budget in that year. This great flood disaster did not affect only Thai people or business, but also the supply chain because of the flood in Rojana Industrial Park, Ayutthaya, and other industrial parks. The information from the Bank of Thailand showed that the damage cost from seven industrial parks in Ayutthaya and Pathum-thani was approximately 24 billion Baht. The data from Thailand's Land development department also showed that from 2009 to 2019, there were three levels of the chronic flood area in Thailand: 8–10 times, 6–7 times, and less than three times. Most of the areas with 8–10 times of the chronic flood were in the lower northern and the central parts of Thailand [6]. According to the budget report in the section of government cash advance from DDPM in 10 years between 2010–2019 (Fig. 1), the governors had submitted the request for the advance payment during the emergency situation in the flood disaster section every year, with the highest in 2012, the year after the great flood. Thus, this information showed the minimum damage cost that came from flood disasters each year. It also displayed the delay of the budget and the disaster period because the local government needed to pay first during the situation before submitting the request for the related organization's budget and the number of the provinces affected by the flood disasters in the past decade.

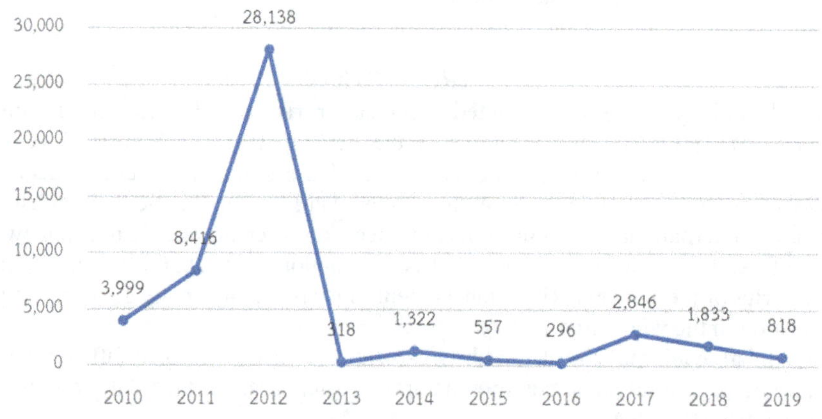

Figure 1: Show the amount of flood situation emergency compensation cost (million Baht) in Thailand from DDPM budget report in 10 years (2010–2019).

The idea of the critical infrastructures during the emergency made this operating system more important. With regard to this field of knowledge, during the emergency the critical infrastructures needed to be protected to prevent the worse effect from the disaster [7]. The damage to the public healthcare system from the synopsis of Thai cabinet in 2010 flood disaster [8] stated that the damage cost of hospitals was 50.3 million Baht, with number of deaths and injuries. During the 2011 great flood, the effect was more severe due to the wider flooded areas. The learning from disaster management and emergency management mentioned the critical infrastructures and key assets during the disaster. These critical infrastructures are based on basic human needs. If the government, both at the local and the national levels cannot manage these infrastructures properly, they could cause the disaster. Moreover, if the infrastructures could not operate as usual during the disaster, the situation could get worse. Healthcare and Public Health are one of the most crucial factors because the "health sector plays a significant role in response and recovery across all other sectors in the event of a natural or manmade disaster" [7]. Thus, from the perspective of disaster management, it is important to maintain the critical infrastructures before the disaster happens in the preparation phase.

The disaster risk profile is one of the tools of disaster risk management. This tool has been widely implemented with the support of UNDRR. It is seen as an important tools to use in the disaster risk assessment process in several countries and in different dimensions such as economics, social, environment, etc. The risk assessment is inside the whole process of disaster management that can be conducted at any level, either a small or a big scale. The Sendai Framework for Disaster Risk Reduction 2015–2030 has the aspect of the "understanding disaster risk is the priority for action" [2]. However, with the complexity of the guideline the level that frequently implements is at the national level or wider, after the process with a national disaster risk profile as the product of the assessment. The national government would get the resources of the policy planning process in the future along with scientific information of the disaster risk that has the expected benefit for the policy management. !

2 RESEARCH METHODOLOGY

This research started with the review the operation system review during the flood situation by examining and conducting the interviews regarding to the great flood in 2011. The next step was to visualize the operation system details by creating a diagram of Business Process Modelling Notation (BPMN 2.0). Finally, the stakeholder analysis was conducted to find the relations of the public healthcare system and related organizations during the flood.

2.1 Operation system visualization

This part of the research was conducted by reviewing and interviewing about the 2011 great flood operation in Nakhon Sawan city municipality. It covered both the municipality and the hospital (Sawan Pracharak hospital). It visualized the operation system and assessment for the interconnection between the numerous organizations inside this area, especially the public healthcare system related organizations. The data collection to create the diagram focused on the operation of Sawan Pracharak hospital, as this hospital is the main hospital in the municipality and other parts of the province. The situation resulted in the increasing number of patients in the hospital and caused the disruption to operational activities at the same time. Therefore, the complexity of the operation during this specific situation caused by the separation of work by the related organizations, the separated commanders, and the command systems conducted. It was a common situation that happened during the management that involved numerous stakeholders [9].

The visualization concept and model were conducted using Business Process Modelling Notation (BPMN 2.0) to show the details of the operation and the cooperation between related organizations during the flood situation. BPMN 2.0 is "the graphical notation will facilitate the understanding of the performance collaborations and business transactions between the organizations" [10]. Thus, this model showed the details of the operational activities that were essential for the public healthcare system during the flood situation or other disasters.

2.2 Stakeholders' analysis

After reviewing the details of the operation from BPMN 2.0 diagram, this part conducted the stakeholders' analysis which its concept "is the key to understanding the complex relationships among all the parties involved" [11]. The next step was to conclude all of the previous processes and show the responsible organizations and their ministries. The outcome was to map the interrelation between the organization both at the local and national levels. Afterwards, the previous diagram was brought in to break into the figure and the factor to show the responsible organizations by the risk equation [12]:

$$Risk = \frac{\text{Hazard} \times \text{Vulnerability}}{\text{Capacity}}$$

This equation shows the factors which create the risk and their effects with the level of risk. The multiplier factors, hazard, and vulnerability are the factors that increase the risk by its number/value. However, the factor with the inverse effect to the risk level is the capacity, as the higher capacity will decrease the risk value. These factors are also able to display the healthcare system risk by calculating the vulnerability of the operation, activity, and the capacity of the operation to show how well it could cope with the hazard and vulnerability.

3 FINDINGS

From the research methodology, the findings would be present, starting with the operation system diagram referred to the BPMN 2.0 concept, to show the details of the operation during the flood situation, following by the stakeholders' analysis and the risk factors developed from the first diagram and the analysis.

3.1 Flood operation system details diagram

After the data collection process regarding the information profile of the study area, the diagram by the concept of BPMN 2.0 was created and divided into two areas, i.e. Nakhon Sawan municipality and the public healthcare system (Sawan Pracharak hospital and other hospitals) as shown in Figs 2 and 3. After that, the stakeholder's analysis from this cooperation process during the flood situation was done.

First, Fig. 2 showed the operation of the Nakhon Sawan municipality during the 2011 great flood. The event diagram started with the flood risk season, in Fig. 2(a) that was highlighted. The municipality management board or the mayor ordered the engineering department to monitor the situation and to report the situation back to the management sector of the municipality to make a public announcement. Meanwhile, the operation and management team of Sawan Pracharak hospital prepared the responses to operate during the flood situation. Then, the triggered event occurred, the dam failure, resulting in Nakhon

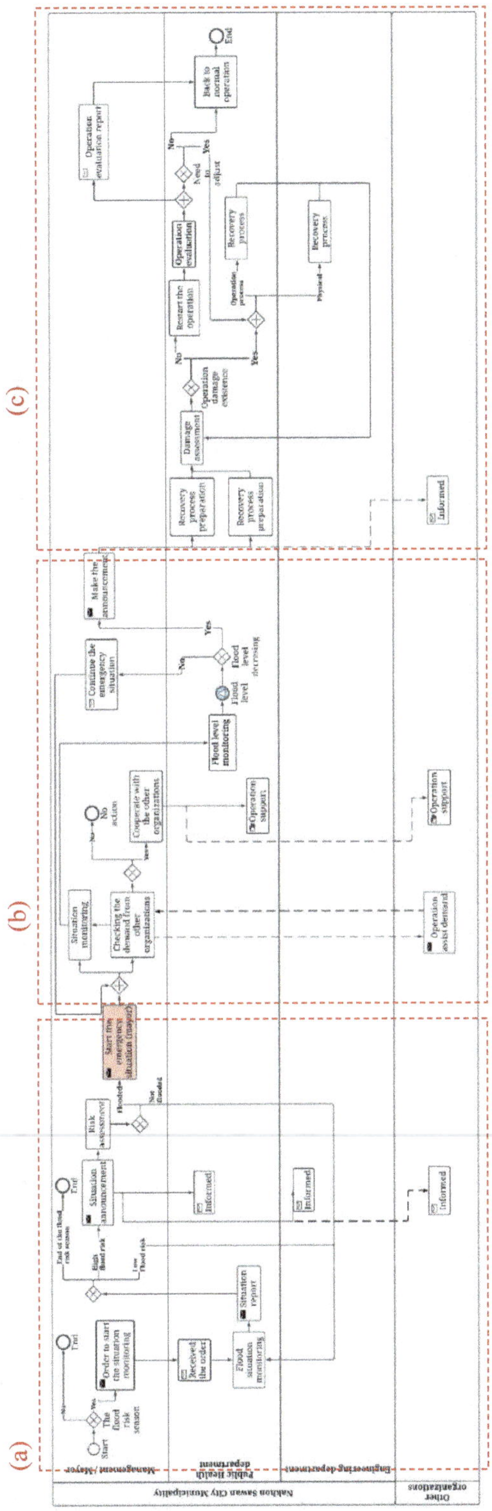

Figure 2: BPMN 2.0 diagram of the operation of the Nakhon Sawan city municipality during the 2011 great flood.

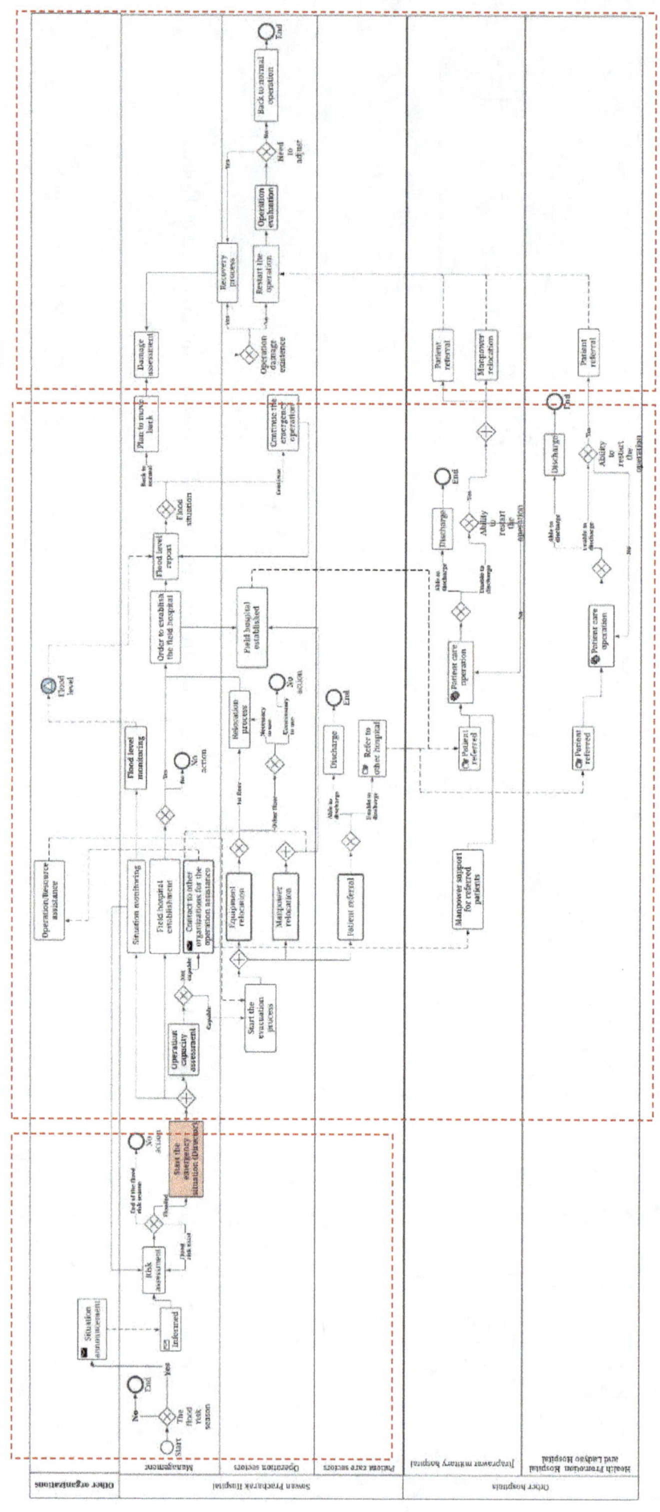

Figure 3: BPMN 2.0 diagram of the operation of the public healthcare system (Sawan Pracharak hospital and other hospitals) during the 2011 great flood.

Sawan city flood. Thus, the management section, with the mayor as the commander, needed to announce to make a public announcement to declare the emergency. The declaration of emergency would be done only if the flood occur.

As shown Fig. 2(b), it showed the operating system after the emergency situation in the municipality. This operation started with the parallel monitoring and the cooperation with the other organizations. As the support role during the flooded period, they also monitored the flood level until it decreased to the proper level to conduct the recovery phase, then they made another announcement. The last part of the operation was in Fig. 2(c), the recovery operation. They started with the damage assessment. The damage were separated into two kinds. The first one is the physical damage which the engineering department was responsible. The latter is the operation damage in the public healthcare service which the public health department was in charge. Both of the recoveries needed to redo the assessment until every section could work as usual before the end of the operation.

The public healthcare service operation system diagram was shown in Fig. 3. The start of the operation was in Fig. 3(a), similar to the municipality's activities. This part was the hospital director's decision, as the emergency commander, to make the emergency announcement. In Fig. 3(b), the top priorities during the emergency were the patient referral and the establishment of field hospitals. Firstly, they separated the patients into two groups, dischargeable patients and non-dischargeable patients at that time. The patients who were not well enough to discharge needed to be referred to the other hospitals; Jiraprawat military hospital, Health promotion hospital, and Ladyao hospital. The relocation process was assisted by the municipality and the military camp. During the operation, at Jiraprawat military, the IPD (in-patient departments) of the field hospital had been set up. It is still operating after the flood situation. During the whole process, the management sector ordered involved departments to monitor the flood level and the situation then made the announcement and took the next step of the operation planning.

After the flood level decreased to the normal condition, as seen in part (c) the hospital also conducted the damage assessment to do the recovery process and prepared for the usual operation. For the referred patients and staff, the relocation operation back to the Sawan Pracharak hospital was conducted by the commander. However, before the normal operation restart, the hospital and the municipality had to assess the damage both operational and physical damage from the flood. After the recovery and the assessment were accomplished, the operation went back to normal again.

3.2 Stakeholders' analysis and list of risk infrastructure

As shown in Figs 2 and 3, the operation in two parts of the organization was separated by the BPMN 2.0 diagram with different commands and orders. However, in reality, they collected data showed that these two operation systems had interdependency during the flood situation. Therefore, this research needed to conduct the stakeholder's analysis to link these operation details together.

After showing the details of the operation during the flood situation, the government played the main role during the disaster to conduct the cooperation. Fig. 4 displayed the stakeholders' analysis diagram divided by the responsible ministries i.e., Ministry of Interior, Ministry of Public Health, and Ministry of Defense. As shown in Fig. 4 The Public Healthcare Service is the organization under the Ministry of Public Health, while the other organizations assisted the operation such as the Provincial Administrative Organizations (PAO) which is under the Ministry of Interior. They distributed the public area under their authority to use for the field hospital. In Fig. 5, existing resources that could support the operation of the public healthcare system were shown. From the government sector, a column of hazard were the ability to monitor the natural hazard and the ability to forecasting. The vulnerability column was to show the facility which could be disrupted during the disaster. Flooding was the government and public healthcare sector's co-responsibility. It could be concluded that the operational activity of the healthcare sector needs to be supported by the government sector (city municipality, PAO, etc.).

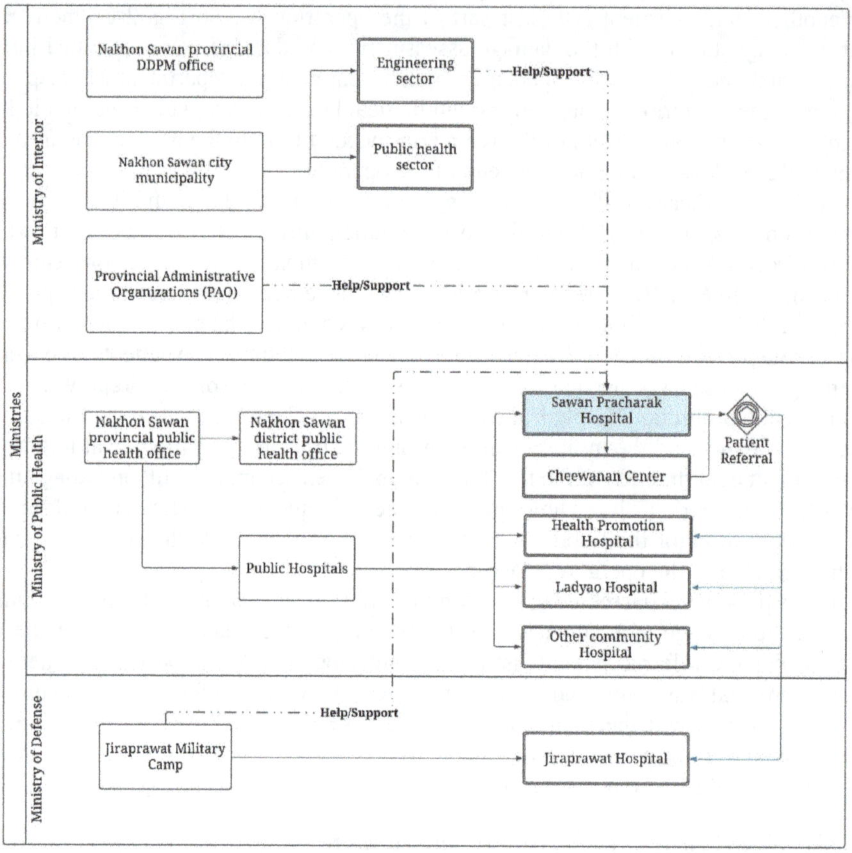

Figure 4: Diagram of stakeholders' analysis during the flood situation.

4 CONCLUSION

The conclusion of the research from the risk profile data assessment found the interconnection and the cooperation pattern of the organizations inside the area of Nakhon Sawan city municipality during the flood. The model was visualized from the previous disaster, the 2011 great flood in Thailand, and the operation during that time. As shown in previous figures, the operation condition was altered during the disaster and emergency as the capability of the organizations and the hospitals was exceeded. Thus, the interconnection to assist the operation needed to be conducted.

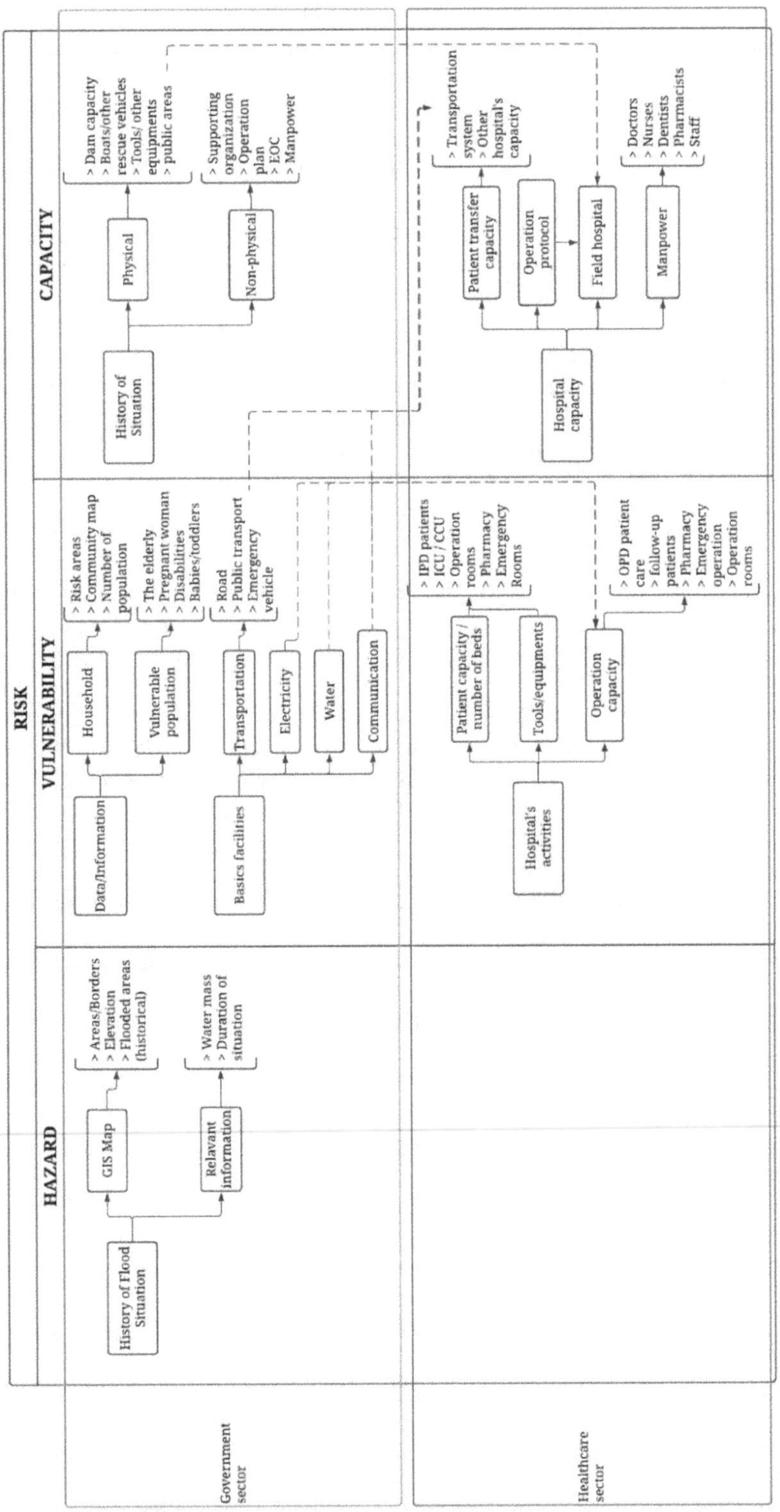

Figure 5: Stakeholder analysis, list of risk infrastructure, separated by the risk equation factors.

In this research, which focused on the operation of the public healthcare system after the assessment, the details of the basics facilities responsible by the government sector needed to be secured for the continuity of the public healthcare system, as included in the critical infrastructures [7]. When the assessment showed the over operation capacity and limited ability of the hospitals during the emergency, further assistance would be requested from the other government organization, both the government sector (municipality, PAO) and the military sector. The risk profile was created to gather the data and information from the responsible organizations and separated the data by the risk equation factors. The relation diagram would help those in the policy management level and the commanders to have the idea to contact the right organizations during the incident.

For further research, after creating the risk profile, the interviews with the specialists in the public healthcare system, the operation staff and the senior government officers should be conducted to gain more insights both in the operation side and the management side. The preference of information during disaster is also interesting to be taken into consideration with the operation analysis [13], [14]. The result of this can support the other area. Also, the outcome of this research will be fundamental to the risk assessment process and the conclusion can support the planning process in the emergency.

ACKNOWLEDGEMENTS

This research was supported by the Collaborative Research with External Partners (CU CRX2101), Japan International Cooperative Agency (JICA) Project for ASEAN University Network/Southeast Asia Engineering Education Development Network (AUN/SEED-Net).

REFERENCES

[1] United Nations International Strategy for Destruction Reduction (UNISDR), Words into action guidelines: National disaster risk assessment, special topic: Public communication for disaster risk reduction. United Nations Office for Disaster Risk Reduction, 2017.

[2] United Nations Office for Disaster Risk Reduction (UNDRR), What is the Sendai Framework? https://www.undrr.org/implementing-sendai-framework/what-sendai-framework. Accessed on: 31 Mar. 2021.

[3] World Health Organization (WHO), How flooding affects health. https://www.euro.who.int/en/health-topics/environment-and-health/Climate-change/news/news/2013/05/how-flooding-affects-health. Accessed on: 31 Mar. 2021.

[4] Dutta, D. & Herath, S., Trend of floods in Asia and flood risk management with integrated river basin approach. *Proceedings of the 2nd International Conference of Asia-Pacific Hydrology and Water Resources Association*, **1**, pp. 55–63, 2004.

[5] Suriyawongsepaisan, P. (ed.), *The Great Flood 2011: Lesson Learn from the Situation.* 1st ed., Health System Research Institute (HSRI): Nonthaburi, 2012.

[6] Land Development Department, Map of chronic flood between 10 Years. Land Development Department: Bangkok, 2019.

[7] Fagel, M.J., *Principles of Emergency Management and Emergency Operations Centers (EOC)*, 1st ed., CRC Press: Boca Raton, FL, 2010.

[8] Ministry of Public Health, Report of damage from flood disaster in medical and public health section, 2010.

[9] Kodaka, A., Ono, T., Watanabe, K., Leelawat, N., Chinranapakdee, C., Tang, J., Ino, E. & Kohtake, N., A dependent activities elicitation method for designing area business continuity management. *2020 IEEE International Symposium on Systems Engineering (ISSE)*, 2020.

[10] Object Management Group, Home: BPMN 2.0 charter. https://www.bpmn.org/. Accessed on: 25 May 2021.
[11] Sapapthai, S., Leelawat, N., Tang, J., Kodaka, A., Chintanapakdee, C., Ino, E. & Watanabe, K., A stakeholder analysis approach for area business continuity management: A systematic review. *Journal of Disaster Research*, **15**(5), pp. 588–598, 2020.
[12] United Nations International Strategy for Destruction Reduction (UNISDR), Hyogo framework for action 2005–2015: Building the resilience of nations and communities to disasters 2005, UNISDR., 2005.
[13] Leelawat, N., Muhari, A., Srivichai, M., Imamura, F. & Bricker, J.D., Preference for information during flood disasters: A study of Thailand and Indonesia. *Sustainable Future for Human Security*, ed. B. McLellan, Springer: Singapore, pp. 335–349, 2018.
[14] Leelawat, N., Pee, L.G. & Iijima, J., Mobile apps in flood disasters: What information do users prefer? *Proceedings of the 12th International Conference on Mobile Business*, 2013.

PROPOSED FLOOD REACTION MODEL FOLLOWED BY THE EVACUATION OF THE POPULATION IN V4 COUNTRIES

DANIEL BREZINA[1] & KATARÍNA HOLLÁ[2]
[1]Armed Forces Academy of General Milan Rastislav Štefánik, Slovakia
[2]Faculty of Security Engineering, University of Žilina, Slovakia

ABSTRACT

In the last few years, the number and severity of natural disasters is growing, especially floods. This phenomenon does not bypass V4 countries. Lower government levels of crisis management are confronted with many requirements. Their objective is to ensure the desired level of prevention of crisis events, as well as the ability to effectively respond to real threats to residents and property. This is associated with the need to optimally decide and subsequently effectively use the available resources needed to solve natural disasters. The main aim of this paper is to proposals flood reaction model followed by the evacuation of the population. The authors deal with the issue of optimizing decision-making processes, while emphasizing its practical application for lower government levels of V4 country (Slovak Republic, Czech Republic, Republic of Poland, and Republic of Hungary). The main reason for the focus of the paper is due to the constant increase in the number of floods in the countries of Central and Eastern Europe as well as the absence of the model in V4 countries. The paper is divided into a theoretical and practical part. The theoretical part contains a comparison of crisis management systems in V4 countries. The practical part consists of the processes and proposal of the flood reaction model followed by the evacuation of the population in the conditions of V4 countries. The model is shown in the form of an algorithm with a detailed defined analysis of the competence and tasks of lower government levels on a section of crisis management. For a better view, the model is divided into three logically and methodically follow-up phases.

Keywords: crisis management, floods, evacuation, decision-making process, authorities, V4.

1 INTRODUCTION

Deciding is a significant function of crisis management. Its importance is growing in the solving of crisis events. It is a difficult and challenging process that affects the nature of the crisis events itself. Crisis managers, individual bodies and public authorities, participating in the solutions of natural disasters are exposed to the media and inhabitants. These require a quick and effective response to extraordinary events that minimize damage and losses. Tasks of public administration, with an emphasis on local government level, in the solution of crisis events are extensive and complex. The main motivation for the selection of the chosen theme was the fact that there are many uncertainties and problematic areas that were not solved in the past. If we want to talk about the real optimization of crisis management decision-making processes at operational level, it is necessary to clearly define their roles and competences and propose effective measures applicable to practical conditions.

2 COMPARING CRISIS MANAGEMENT SYSTEMS IN V4 COUNTRIES

Proposing a flood reaction model followed by the evacuation of the population in the V4 countries requires knowledge and relations that apply to V4 crisis management systems. Individual V4 countries were compared based on predetermined criteria. Crisis management systems in each country were influenced by historical assumptions as well as the species and characteristics of a crisis event with the most extensive impact on life, health, property, and

WIT Transactions on The Built Environment, Vol 207, © 2021 WIT Press
www.witpress.com, ISSN 1743-3509 (on-line)
doi:10.2495/DMAN210161

the environment. This has led this to changes to the legal environment and adjustment of security documents in each country.

2.1 Slovak Republic

The Government of the Slovak Republic is in the highest position in crisis management system in the Slovak Republic. The Ministry of the Interior of Slovak Republic is the central authority of crisis management. The executive authority of the crisis management is the crisis staff. The crisis staff is not a permanent authority. It is activated only when there is a danger of a crisis situation or when a state of crisis is declared. Local government level (District Office) has a very important role in the process of solution of crisis events. The executive components of crisis management are Regional/District of the Police Force and Regional/ District Directorate of the Fire and Rescue Corps. The most competent institutions for solution of crisis events are the Integrated Rescue System. According to Šimák [1] the crisis management system uses mainly staff organizational structures. They form a special public authority, t. j. state administration (central and local) and self-government (regional and local). Powerful crisis management components must be able to intervene immediately at a particular place with the necessary capacity of human and technical means, including adequate technology. It is essential that all powerful crisis management components have clearly defined competencies allowing intervention in any environment.

Local government level authorities, in particular District Offices, manage selected segments of state administration, including crisis management section. Effective reaction to crisis events, especially on natural disasters, requires unambiguous determination of responsibility and tasks at lower levels of management. Neubauerová and Dubrovina [2] characterizes local government level as a state budgetary organization. It is linked to the Ministry of Interior of the Slovak Republic through financial relations. Local government authorities are a hierarchical lower level of state administration.

There are several legislations in the prevention and resolution of crisis events. These include two constitutional acts, 23 acts and more than 10 decrees issued by various ministries. In the Slovak Republic, the Integrated Rescue System was set up in 2002 because of the European Union's experiences. The main purpose was to increase the quality of rescue activities. According to Act no. 129/2002 Coll. about the integrated rescue system [3], as amended, the Integrated Rescue System ensures rapid use and coordination of rescue and rescue entities in the event of a risk of crisis event or during its solution.

The effective crisis management process requires the determination of tasks and determine the responsibility of authorities and institutions in the crisis management system at all government levels, especially at lower government levels. Constitutional Act no. 227/2002 Coll. on state security at war, state of war, state of exceptional and state of emergency, as amended [4], defines states of crisis resulting from the name of the constitutional act, including the authority responsible for its declaration. This Constitutional Act defines and provides for the competence, responsibility, and action of constitutional authorities during the declaration of a particular state of crisis, including the implementation of the tasks of the Security Council established in the relevant territorial unit (district, county, state). The purpose of Act no. 42/1994 Coll. on civil protection of the population, as amended [5], is to establish tasks and scope of state administration authorities, self-governing regions, and municipalities in the field of civil protection of population, including law and obligations of legal persons in ensuring civil protection of the population. This act contains details of the announcement of the emergency in terms of its nature and scope of the negative consequences of crisis event within different territorial districts (municipality, district, region, state). This act also describes the principles of governance, organization, and implementation of rescue work. A substantial part of Act no. 387/2002 Coll. on the governance of the state in crisis situations outside the time of war and state of war, as amended [6], is dedicated to issue of public authorities is devoted to public authorities. This act defines crisis management authorities and sets out tasks within their scope. The position of the forces and means in the crisis management system does not occur in any of the three-above legislation in issue of crisis management. The District Office is responsible for several segments of state administration, including a section of crisis management. They are governed by acts and lower legislation in its activities. These create space for an effective and effective reaction to crisis events, with an emphasis on natural disasters.

2.2 Czech Republic

Extensive floods in the Czech Republic in 1997, resulted in a change in the legal environment in crisis management, including the perception of crisis management by not only competent authorities, but also by the public. The impact of these floods led to the creation of a specific legal framework, t. j. Act no. 240/2000 Coll., on Crisis Management and amending certain acts, as amended (the so-called Crisis act). The basic aim of this act was the definition of all entities, authorities and institutions that are directly or indirectly involved in the prevention, progress and disposal of the impacts caused by extraordinary events [7], [8]. According to Vaníček [9] a legal environment in the field of crisis management in the Czech Republic does not specify the differences between the term crisis, an extraordinary situation, and an extraordinary event. However, on closer examination, it is possible to find differences mainly based on the possibility of threats for the existence of the state and life of the nation.

In the case of a threat or creation of a crisis event and in direct dependence on its character, a crisis state is declared. Its fundamental aim is a legal confirmation of changes to the competences of individual crisis management authorities. The right to declare state of crisis only have crisis management authorities. The government of the Czech Republic may declare a state of emergency in the case of natural disasters, ecological accidents, industrial accidents, and other dangers. Parliament of the Czech Republic on a proposal for the Government of the Czech Republic may declare a state of threats to the state if the sovereignty or territorial integrity of the state is endangered [10]. It is possible for crisis management authorities in the Czech Republic to consider all the authorities that ensure the analysis and evaluation of possible endangerment of its security, planning, implementation, and control of activities carried out in the context of the prepared measures as well as solution of crisis situations [11], [12]. Security councils and crisis staffs belong to crisis management. Security councils is to coordination instruments in the process of solution crisis events. Crisis staffs can be as advisory employment authorities established by the government, region, or municipalities [10].

The main role of the Government of the Czech Republic in the crisis management system in the Czech Republic is to correction for various crisis events through other crisis management authorities. It has certain powers depending on the nature of the declared state of crisis. For example, the Government of the Czech Republic may set out to what extent and for which some civil rights will be limited. The Government of the Czech Republic implements specific economic or regulatory measures [13].

Territorial self-government authorities have its justification in the crisis management system in the Czech Republic based on the delegated competence. County authorities (Regional president and Regional Office) ensure the preparedness of the crisis, the preparation of rescue work and the protection of the population. Region Security Council is the working authority of regional president. At the same time, it is also a coordinating authority in preparing the region on crisis events. The Regional Office organizes and ensures synergies and cooperation between the Municipality Office and other competent authorities and municipalities in the region.

Municipality authorities (Mayor of Municipal and Municipal Office) ensure the preparedness of the municipality to address crisis situations and perform activities to mitigate their impacts, within its territorial scope. The Mayor of Municipal sets up and manages crisis staff (municipal work authority). The Municipal Office provides synergies to the Regional's fire-rescue service in the processing of the crisis plan of the region and the municipal crisis plan [11].

2.3 Republic of Poland

According to Dworzecki [14] the crisis management system in the Republic of Poland is relatively complicated and demanding. History of the formulation of this type of structure in the Republic of Poland is the best proof. The shortcomings of systemic solutions allowing the coordination of activities of all subjects that form various rescue structures were found during large floods in July 1997. As Lizak [15] writes the crisis management system in the conditions of Republic of Poland is the civilian component of the national security system. It is a response to various internal and external threats such as natural disasters and terrorism. Individual competences and tasks are regulated and defined in the Constitution of the Republic of Poland [16] some of the acts and internal procedures of governing authorities at different government levels.

The Government of the Republic of Poland is a crucial element and is responsible for maintaining public order and internal security at national level. Within the Government of the Republic of Poland, the Governmental Crisis Management Team is established by the Minister of Interior and Administration of the Republic of Poland. Most of the rescue components is subordinate to the Ministry of the Interior and Administration of the Republic of Poland, which is responsible for maintaining public order, protection of population in crisis situations, including disaster prevention and crisis situations. Other rescue components are subordinated to the Government Centre for Security. The structure of fire components and civil protection components are united into one unit [17].

The 11th Chapter of the Constitution of the Republic of Poland [16] deals with the issue of extraordinary measures. State of exceptional, state of emergency or state of natural disasters can be declared in situations of special danger and where constitutional measures are insufficient (Article 228). The Government of the Republic of Poland may declare the state of natural disasters in a particular area or throughout the Republic of Poland longest for 30 days. During this time, it is necessary to minimize the impacts of natural disasters or technological accident (Article 232). The issue of crisis management system in the Republic of Poland is described in more detail in Act of 26 April 2007 on crisis management [18]. The authorities responsible for crisis management, their roles and competences, as well as the rules for the financing of crisis management tasks are contained in this act (Article 1).

According to Zdzymira and Rozwadowski [19] a structure of temporary crisis management authorities (teams and centres) obliges the competent authorities to make crisis management teams in suitable regime. It is a principle that in great territorial units (for example, in larger cities, especially in the voivodes), crisis management centres operate as a whole. These work continuously at relevant levels of crisis management.

2.4 Republic of Hungary

The last characterized system of crisis management is a crisis management system in the Republic of Hungary. Frequent disasters occurring in recent years in the territory of the Republic of Hungary caused the crisis management system in the Republic of Hungary to be exposed to great pressure. Petkov and Tagarev [20], for one of the most famous and the most tragic exceptional events, it considers the red sludge leak from the ruptured dam of the aluminium sludge pond near the village of Ajka in 2010. Following this event, the new Act was adopted in 2011 Act no. 128 on disaster management [21]. Its main goal was to increase the efficiency of the National Disasters Management System in the Republic of Hungary and improving protection measures against risks caused by nature or human. This act laid the basis of the cooperation system of crisis management authorities with the structures of compulsory and voluntary components of civil protection [17].

National Directorate General for Disaster Management (NDGDM) obligations and competences (NDGDM) are clearly defined in Act no. 128 on disaster management. The basic function of NDGDM is the protection of lives and property of the population living in the Republic of Hungary, including the protection of critical infrastructure elements. Its main mission is the prevention of disasters and the implementation of rescue actions in civilian emergency situations. The new organizational structure is based on three pillars, specifically civil protection, fire protection and industrial security [20], [21].

On 25 April 2011, a new basic act was adopted by the Republic of Hungary, in which various crisis conditions are described. The Government of the Republic of Hungary may declare the state of danger in the case of a natural disaster or industrial accident that threatens the life and property of the population, or to mitigate its impacts (Article 53). All activities and tasks are coordinated at the level of the region in case of threat to extensive territory or where local forces and means designed to address crisis events are insufficient. The Government of the Republic of Hungary has responsibility for managing during crisis events that negatively affect multiple districts or which cannot be controlled at the region level [21].

The Mayor of the Municipality is responsible for preparing for the protection against disasters and guiding activities in the process of solution of crisis events at local government level within its territorial scope. The Committee on Local Defence coordinates and manages all measures and activities in the event of a threat to the exceeding territory of individual parts within the municipality. The main responsible person is the mayor and other members of this committee are mayors of city parts. They carry out their activities with the support of District Directorate of Disaster Protection. The leaders of regional and local governments (Presidents of the General Assembly of Regional and Mayors) have responsibility for planning, organizing, managing, and implementing elements of critical infrastructure and assistance to the population for extraordinary events [17].

3 DATA AND METHODOLOGY

The main aim of this paper is the processing and proposing a flood reaction model followed by the evacuation of the population for V4 countries. The main methods include analysis, comparison and algorithmic procedures. The content of the methodology is an analysis of individual crisis management systems in V4 countries and their subsequent comparison through the selected characteristics of these systems. Currently, there exist many scientific and research methods, techniques and procedures. Models or algorithms used to optimize decision-making processes can be created through them. Algorithmic procedures were used in the processing of the flood reaction model followed by the evacuation of the population for V4 countries.

4 PROPOSED OF DECISION-MAKING PROCESSES OF THE FLOOD REACTION MODEL FOLLOWED BY THE EVACUATION OF THE POPULATION FOR V4 COUNTRIES

Ensuring effective evacuation of the population due to flood is a specific and problematic area, especially at lower levels of crisis management. Individual municipalities and cities plan to declare and manage evacuation in their territory through developed evacuation plans. At first, the expected progress of the flood development needs to be assessed. It consists of five steps:

1. Cause of origin of flood (long-lasting rainfall, intensive rainfall, melting snow or current occurrence of these causes);
2. Gradual rise of the river level to the maximum capacity of the water flow river;
3. Exceeding the capacity of the watercourse bed by the water flow;
4. Drop in water level below to the maximum capacity of the water flow river;
5. Normal flow of the river bed without the possibility of flood risk.

In terms of individual phases of the flood reaction model followed by the evacuation of the population for V4 countries, the progress of the flood development is shown in Fig. 1.

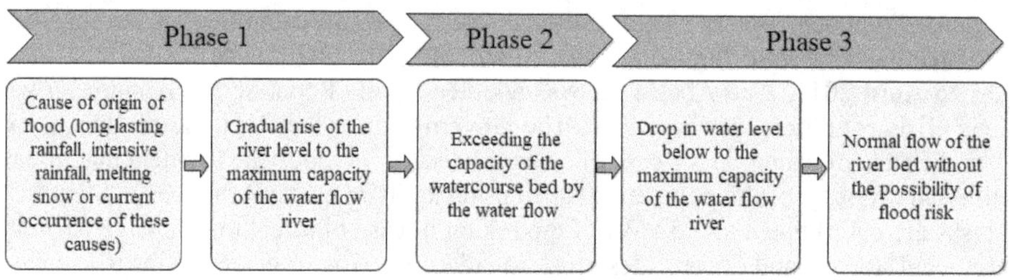

Figure 1: Expected course of flood development.

Decision-making process of phase 1 of the flood reaction model followed by the evacuation of the population for V4 countries is illustrated in Fig. 2. Phase 1 of the model begins at the lowest level of crisis management. The municipal receives information about the possibility of developing floods or its inception based on a gradual rise of the river level to the maximum capacity of the water flow river. This is preceded by an assessment the cause of origin of flood. Individual municipalities commission needed for a solution floods are activated depending on the intensity of flood effects. Their activation will already be carried out in the activation of the warning and notification network, or after the assessment and evaluation of the situation at the municipal crisis staff meeting. The mayor of the city/village envoys an exploration-monitoring group endangered area or location of the occurrence of flood. The municipality declares a state of crisis based on information from the exploration-monitoring group or the results of the decision of the members of the municipality crisis staff. In the case of flood, it is necessary to distinguish three degrees of flood activity with a subsequent state of crisis statement in the local government. The municipality declares or recalls of individual degrees of flood activity in the case of flood in its territory. The municipality shall inform about declaration or recall of the state of crisis and individual degrees of flood activity to the relevant higher level of crisis management (in the figures marked of abbreviation of HLCM). All the necessary authorities and crisis management subjects, including human and material resources needed to solution of floods, is activated after informing about state of crisis.

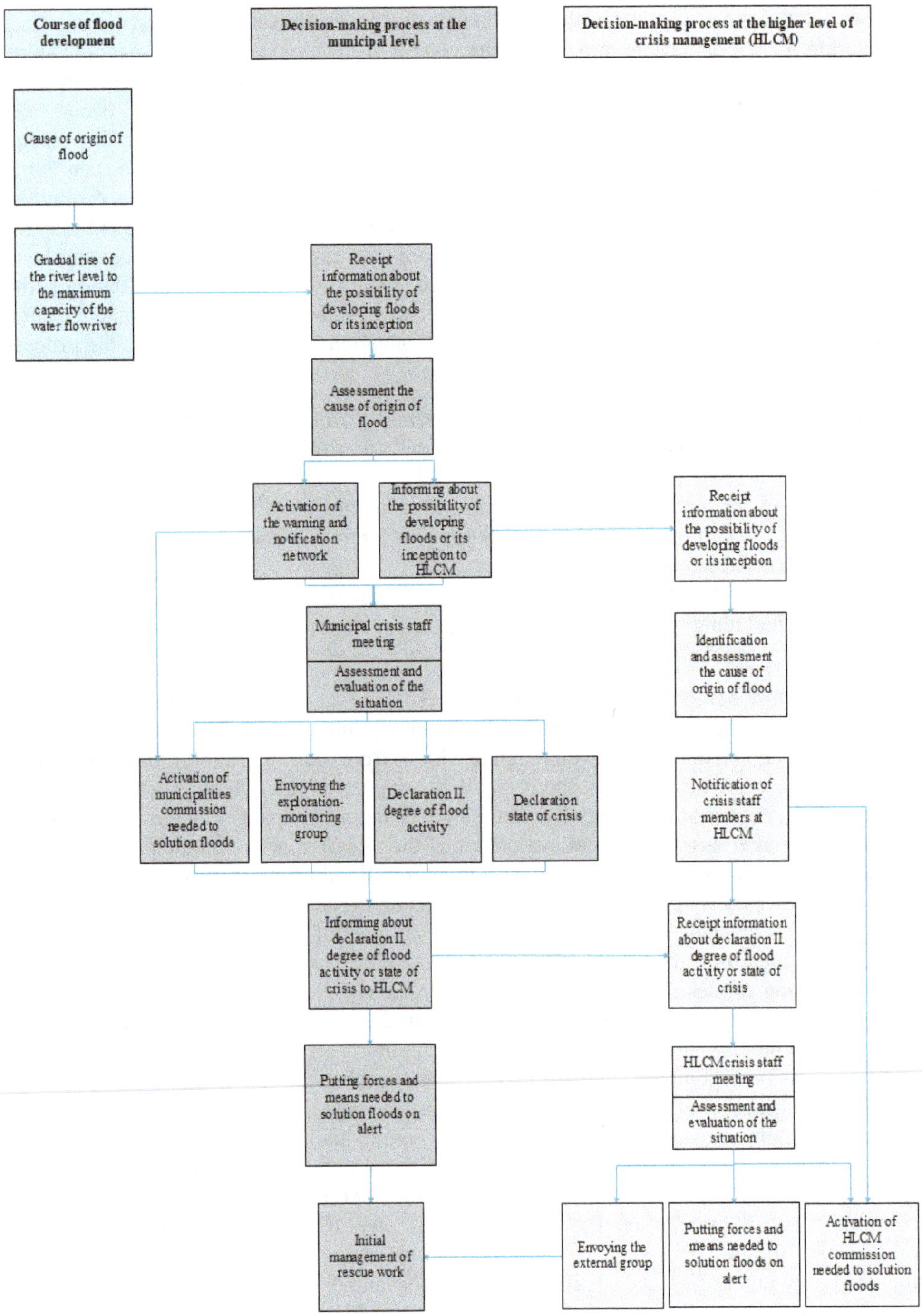

Figure 2: Decision-making process of phase 1 of the flood reaction model followed by the evacuation of the population for V4 countries.

Fig. 3 detail shows phase 2 of the flood reaction model followed by the evacuation of the population for V4 countries. In terms of the course of developing floods, the model starts exceeding the capacity of the watercourse bed by the water flow. Following the assessment and evaluation of the situation at the municipal crisis staff, it is necessary to declare III. degree of flood activity regarding the course of the flood development. The receipt information about declare III. degree of flood activity from the municipality is the beginning of decision-making process at a higher level of crisis management. In case of exceeding the territorial unit of self-government due to the negative effects of floods, responsibility for crisis management in the endangered or affected area assumes the authority at a higher level of crisis management. The decision to ensure and perform the population evacuation is the role of crisis manager, respectively the manager at a higher level of crisis management. Organization and management of the course of evacuation of the population in the process of a flood solution consists of three gradual steps:

- Managing and guidance of evacuants movement to the evacuation centre, including informing of the evacuation centre about the status and progress of the evacuation of population;
- Organizing and managing the of evacuants movement from the evacuation centre to the emergency accommodation place, including informing of the evacuation commission about the activities of the evacuation centre;
- Organizing and ensuring the involvement of evacuants into the emergency supply system, including informing of the evacuation commission about the activities of emergency accommodation place.

The last phase of the model is conditioned by reducing the risk of flood for inhabitants. In this case, it is a drop in water level below to the maximum capacity of the water flow river (Fig. 4). Information about the development of floods from the exploration-monitoring group are a decisive basis of decisions about recall III. degree of flood activity and subsequent declaration II. degree of flood activity. The mayor of the municipality changes the degree of flood activity after assessment and evaluation of the situation at the municipal crisis staff. Upon receipt of information on the change in degree flood of activity, the crisis staff will meet at a higher level of crisis management. Based on the assessment and evaluation of the situation, crisis manager (or the manager at a higher level of crisis management) decides to prepare the measures that will allow the fastest return to normal stabilized state. The implementation of these measures is part of the recovery phase after the effects of floods. The preparation for these measures is carried out in the reaction phase on the flood. In this step decision process, the main task of the municipality is a supportive activity for a higher level of crisis management. They are involved in checking the functionality of equipment providing distribution of energy, water, and gas. After updates to other information from the exploration-monitoring group, the mayor of the municipality recalls II. degree of flood activity and declare I. degree of flood activity. If the situation is allowed this, I. degree of flood activity does not declare, but directly recalls state of crisis. After withdrawal forces and resources needed to solution the flood, the last step of the decision-making process municipality follows. This step is to inform about the change degree of flood activity of all the necessary authorities and institutions, including authorities at a higher level of crisis management. After receipt of the information about recall II. degree of flood activity or recall of state of crisis in the territory of the municipality, it follows of the issuing an order to ending work in the area affected by the flood.

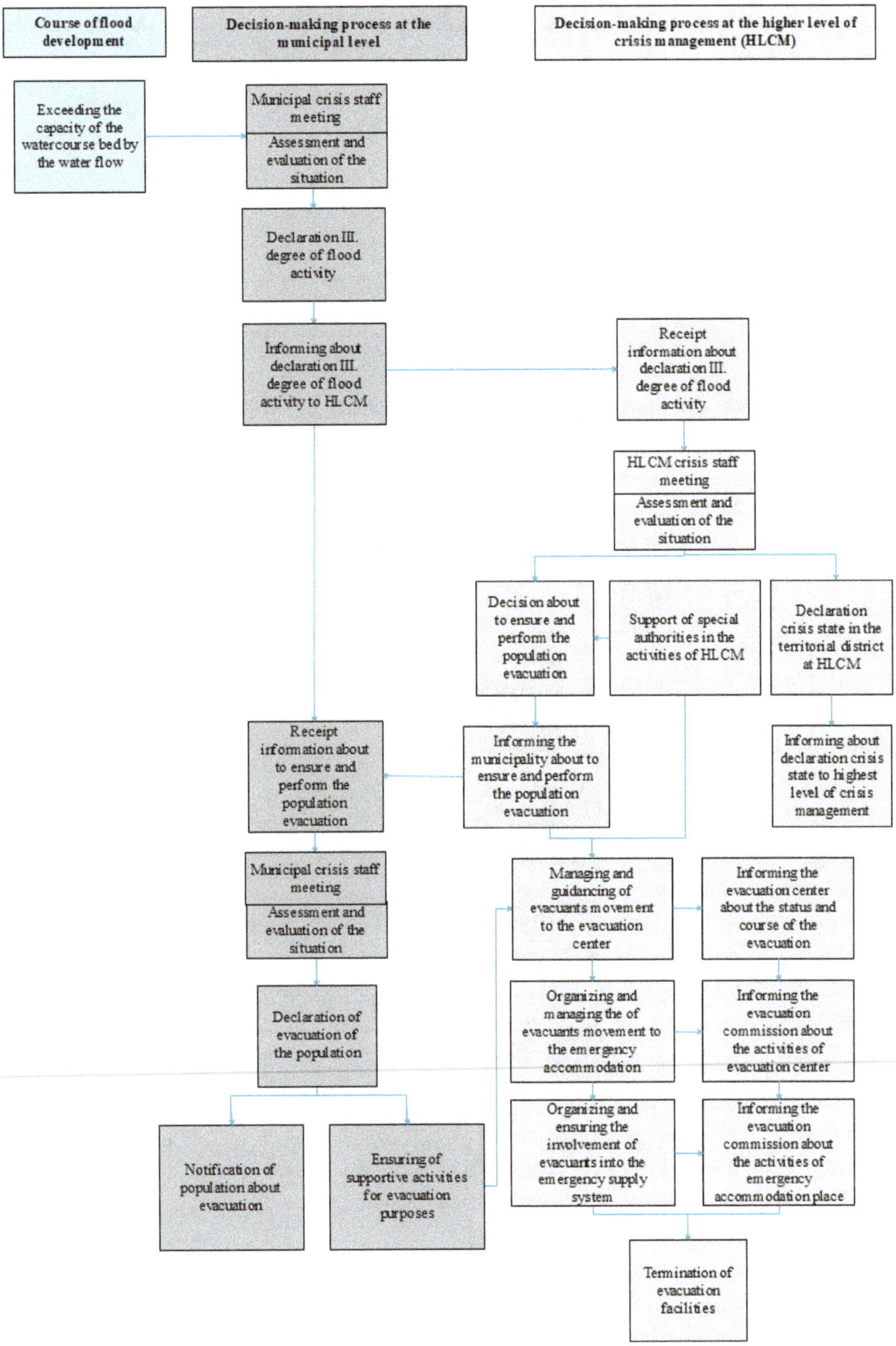

Figure 3: Decision-making process of phase 2 of the flood reaction model followed by the evacuation of the population for V4 countries.

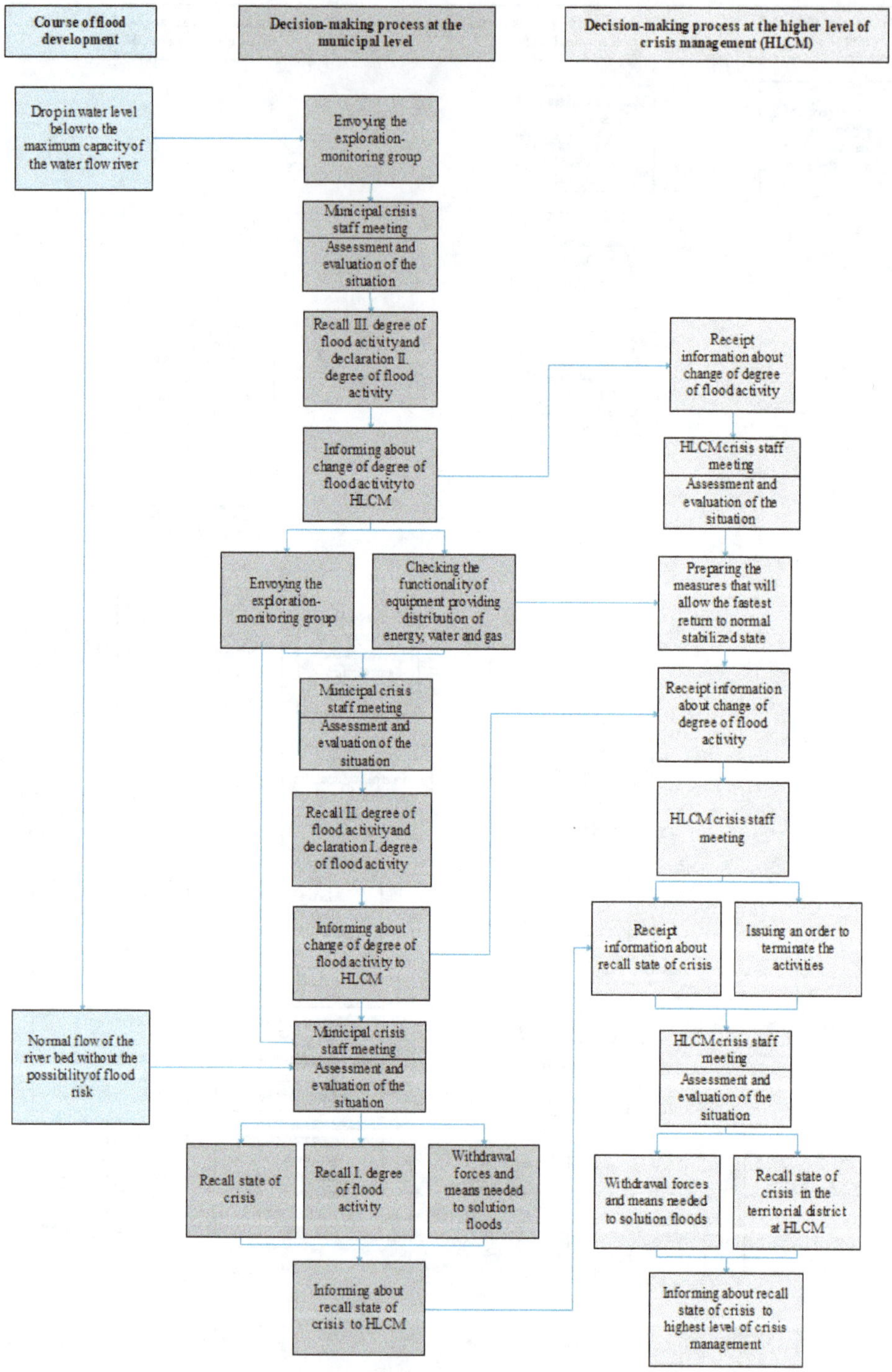

Figure 4: Decision-making process of phase 3 of the flood reaction model followed by the evacuation of the population for V4 countries.

5 CONCLUSION

Based on the understanding of the current state of crisis management systems in V4 countries, it was possible to propose an effective and efficient flood reaction model followed by the evacuation of the population for V4 countries. This model is adapted to the requirements determined by the flood, including the nature of rescue work, with the subsequent necessary process of evacuating the population from the affected area. Crisis management cross-border cooperation between V4 countries must continue to be strengthened and streamlined. Crisis management authorities at all government levels, in particular in the area of border between countries, must be aware of crisis situations risks even beyond those countries whose negative effects and impacts can endanger their own populations and territory.

In addition to the faculty experts of the Faculty of Security Engineering of the University of Žilina and Department of Safety and Defence of Armed Forces Academy of General Milan Rastislav Štefánik, the authors had the opportunity to communicate with many experts from various domestic and foreign institutions dealing with crisis management issues. An essential contribution to the creation of theoretical and practical part of the paper was participation at foreign stages, namely in Tadeusz Kościuszko Land Forces Military Academy in Wroclaw and the Occupational Safety Research Institute in Prague. Other institutions with authors cooperated, they also include the Academy of Police Corps in Bratislava and the 5th Special Purpose Regiment in Žilina.

ACKNOWLEDGEMENT

This research was funded by Research and Education Grant Agency in Slovak Republic, grant number VEGA 1/0581/19 Stating the Social and Individual Risk of Employees and Public Resulting from Impacts of the Domino Effects Caused by Industrial Accidents and Methods of Reducing their Possible Escalation.

REFERENCES

[1] Šimák, L., *Krízový manažment vo verejnej správe [Crisis Management in the Public Sector]*. Druhé prepracované vydanie [2nd revised ed.]. EDIS: Žilina, 2016.

[2] Neubauerová, E. & Dubrovina, N., Verejná správa v slovenskej republike – aktuálne problémy [Public administration in the Slovak Republic – current problems]. *Public Administration*, 2010.

[3] Act no. 129/2002 Coll. about the Integrated rescue system, as amended (Slovak Republic).

[4] The Constitutional Act no. 227/2002 about state security in time of war, state of war, state of exceptional and state emergency (Slovak Republic).

[5] Act no. 42/1994 Coll. about civil protection of the population, as amended (Slovak Republic).

[6] Act no. 387/2002 Coll. about state management in crisis situations outside the time of war and state of war, as amended (Slovak Republic).

[7] Veishar, A., *Krajina, lidé a povodne v povodí řeky Moravy spoločnosti [Landscape, People and Floods in the Morava River Basin of Society]*. Regiograph: Brno, 131 pp., 2002.

[8] Brázdil, R., *Historické a současné povodně v České republice [Historical and Current Floods in the Czech Republic]*. Masaryk University in cooperation with Czech Hydrometeorological Institute in Prague: Brno, 369 pp., 2005.

[9] Vaníček, J., *Právní úprava krizového řízení v ČR: vybrané problémy právní teorie i praxe [Legal Regulation of Crisis Management in the Czech Republic: Selected Problems of Legal Theory and Practice],* 1. vydanie [1st ed.], Eurolex Bohemia: Prague, 402 pp., 2006.

[10] Constitutional Act no. 300/2000 Coll., amending Constitutional Act no. 1/1993 Coll., The Constitution of the Czech Republic, as amended by Constitutional Act no. 347/1997 Coll., and Constitutional Act No. 110/1998 Coll., about the security of the Czech Republic (Czech Republic).

[11] Act no. 430/2010 Coll. about crisis management – Crisis Act, as amended (Czech Republic).

[12] Horák, R., Krč, M., Ondruš, R. & Danielová, L., *Průvodce krizovým řízením pro veřejnou správu [Crisis Management Guide for Public Administration],* 1. vydanie [1st ed.], Linde: Prague, 407 pp., 2004.

[13] Antušák, E. & Kopecký, Z., *Úvod do teorie krizového managementu [Introduction to the Theory of Crisis Management],* 2. vydanie [2nd ed.], University of Economics in Prague, Faculty of Business Administration, Department of Management, Crisis Management Section: Prague, 95 pp., 2003.

[14] Dworzecki, J., Crisis management system in Poland. *The Science for Population Protection,* Vol. 2, 2012. http://www.population-protection.eu/prilohy/casopis/eng/11/52.pdf. Accessed on: 18 Jun. 2021.

[15] Lizak, S., Critical infrastructure protection in Poland. *Population Protection 2007*, Ostrava, pp. 192–197, 2007.

[16] The Constitution of the Republic of Poland of 2nd April 1997. *Dziennik Ustaw [Journal of Laws].* no. 78. item 483 (The Constitution of the Republic of Poland).

[17] International CEP Handbook, *Civil Emergency Planning in the NATO/EAPC Countries*, Swedish Civil Contingencies Agency (MSB), 2009.

[18] Act of 26th April 2007 about Crisis Management (Republic of Poland).

[19] Zdzymira, M. & Rozwadowski, P., Emergency management process at a voivodeship level according to the Emergency Management Act of 26 April 2007. *Przedsiębiorczość I Zarządzanie 2017 [Entrepreneurship and Management 2017].* SAN, **18**(5). pp. 53–72, 2017.

[20] Petkov, V. & Tagarev, T., *Hungary: Capabilities, Organisations, Policies, and Legislation in Crisis Management and Disaster Response,* Institute of Information and Communication Technologies/Centre for Security and Defence Management: Sofia, Number 127, 2015.

[21] Act no. CXXVIII of 2011 about Disaster management and related amendments to certain acts (Republic of Hungary).

REVIEWING SEISMIC VULNERABILITY DIMENSIONS: CURRENT TRENDS AND METHODOLOGICAL CHALLENGES

LUIS IZQUIERDO-HORNA & YUSTIN YEPEZ
Universidad Tecnológica del Perú, Peru

ABSTRACT

Traditionally, seismic vulnerability has been assessed considering, for the most part, infrastructure as the main evaluation component, leaving aside other dimensions of study. This result does not adequately evaluate all the anthropogenic characteristics that make a certain system more susceptible to experiencing economic, patrimonial, and human losses in the face of natural phenomena. In this context, Peru is exposed to many different phenomena of the earth's internal and external geodynamics (i.e., earthquakes, tsunamis, volcanism, mass movements, heavy rains, drought, etc.), earthquakes being the ones that have caused the most damage and repercussions. It is for the latter, considering that seismic hazards are determined by the geographical conditions of the study area, that the main objective of this review is to study and highlight different perspectives of vulnerability analysis (i.e., social, cultural, economic, etc.) when seismic events happen. This review shows the main assessment parameters used to describe each dimension of analysis and in addition, a review of the main existing methodological frameworks is carried out, aimed at showing a comprehensive perspective of the context analyzed in order to improve the conditions and livelihoods of the population exposed to these hazards.

Keywords: seismic vulnerability, vulnerability dimensions, earthquakes, risk and disaster management, Peru.

1 INTRODUCTION

Peru is exposed to many different events of natural origin due to its multiple geographical interactions [1]. In addition, it is located in a demarcation of high seismic movement caused by the subduction process between the Nazca oceanic plate that moves under the South American continental plate, causing these geological phenomena to be severely recurrent [2].

Throughout history, there have been several earthquakes of great magnitude. According to the historian Silgado Ferro [3], one of the oldest and most devastating earthquakes on record occurred in 1604 with a magnitude of 8.4. It was originated in the southern coast of Peru and affected the cities of Arica, Tacna, Moquegua, Arequipa, extending to Ica, being perceived up to 130 km from the coast inland and 1,650 km (approximately) from north to south, causing 23 human losses in the Port of Pisco. Later on, other seismic events, also of great magnitude, would occur in 1784 (magnitude of 8.1 in Arequipa, 54 deaths, 500 injured and countless buildings destroyed), 1868 (magnitude of 8.6 in Moquegua, where 300 aftershocks were counted, 150 deaths), 1942 (magnitude 8.4, in the border region of Ica and Arequipa, 30 people died due to houses collapsing and 25 were injured due to many different reasons) [4], and in 1970 (magnitude 7.8 and a large flood in Callejón de Huaylas, Ancash, 67,000 people died and 150,000 were injured). In recent times, the 2007 earthquake on the central coast and with a magnitude of 7.9 caused significant damage, especially in Pisco (considered the epicenter and where damage was generated to approximately 80% of the structures), minor damage was also caused in towns near the coast [5]; but mainly it caused several –and very sad– human losses. These events not only generate material damages and significant human losses, they also interrupt different business activities (i.e. social, economic, community development, infrastructure and public resources) of a community [6].

In this context, factors associated with sustainable development increase the destructive effects of earthquakes, since they are enhanced by the disorderly growth of cities, it is because of this that it is important to study what makes a certain community susceptible – or not – to experiencing large or few losses; that is, they are more vulnerable compared to their peers. In this sense, Adger [7] states that the study of vulnerability faces important challenges in different human dimensions, in which a solid conceptual framework must be included, addressing perceptions of vulnerability and governance. Currently, the studies carried out on seismic vulnerability should not only focus only on quantitative approaches, but also on qualitative ones, on methods capable of translating the abstract concept of vulnerability into practical tools, classifications and comparative criteria through demographic, social, economic, environmental, and institutional factors [8], [9].

On the other hand, studies are usually based on factors related to infrastructure in Peru; however, there are more elements (i.e., social, economic, cultural, resilience, etc.) that need to be further explored [10], therefore, the purpose of this review is to assess the descriptive potential of the main indicators used for each analysis dimension. In addition, a complete description of the methodological proposals for seismic vulnerability is offered, integrating their assessment. Lastly, this review carries out a systematization of the information, using a hierarchical analysis process of the results obtained from the seismic vulnerability evaluations available in the literature.

2 MAIN PARAMETERS USED FOR THE ASSESSMENT OF SEISMIC VULNERABILITY

The concept of "vulnerability assessment" not only includes quantitative approaches, it also seeks discussing and developing all kinds of methods capable of translating the abstract concept of vulnerability into practical tools that work as support for decision-making [9]. Likewise, determining the physical, social, economic and environmental parameters are the starting point to mitigate risk and promote a culture that is resilient to disasters [11]. At the global level, in order to represent the vulnerability context of a community, different contexts were analyzed to determine relevant indicators. Typically, these indicators are based on the geographic environment of the disaster, pre-existing social, economic and political conditions, hazard characteristics, degree of exposure, impact and response scales, and the disaster phase (i.e. before, during, after) [12].

Due to the multiple nature vulnerabilities and complexities that occur in our country, Peru, parameters (i.e. social, economic and environmental) were proposed in order to be taken into account and integrate them into criteria for prevention and reduction of risks in the various land use planning processes, planning, investment and environment management programs for each specific sector [10].

2.1 Socioeconomic dimension

This dimension is important to assess because it is inversely related to vulnerability, that is, the less privileged sectors suffer the greatest losses due this type of phenomenon occurring and also have a limited capacity for recovery [13]. The socioeconomic dimension can be defined as the probability of being below the poverty line at a time of chaos [9]. In this sense, in China, the percentage of the low-income population and the percentage of the population aged 15 years and over that possess an education degree below a high school diploma were analyzed [6]. Along the same lines, in Indonesia the debt and lack of productive assets in women were analyzed [14]. In Malaysia, parameters such as: gross income and gross domestic product (GDP) were considered [15]. Likewise, in Uttarakhand (India) they were based on the poverty rate, agricultural workers and the unemployed [16]. Similarly, in Colombia and Mexico, certain parameters were studied such as: average economic income, economic loss, minimum or total intervention costs, and interruption costs [17], [18]. In the United States, they used parameters such as: personal wealth, economic dependence on a single sector, and occupation [8], [19]; the percentage of households that could seek temporary shelter after an earthquake and the total economic cost required for the replacement, reconstruction, and recovery of residential buildings were also analyzed [20]. In the case of Italy, the following were considered: occupational separation rate, employed female labor force, employed labor force, and unemployment and employment rate [21], [22]: and in Portugal, parameters such as unemployment rate and social dependency rate were also used [23], [24].

At the national level, the dependent population, daily personal income and permanent disability were studied [25]. In addition, in Pisco parameters such as: unemployment, insufficient income, employment instability, difficulty and total impossibility to accessing formal education, recreation and health services were used [2]. Lastly, CENEPRED [10] proposed a list of economic parameters in order to contextualize aspects related to the person and the built environment.

2.2 Sociocultural dimension

The sociocultural dimension establishes the link between the population and its environments with social and institutional relations, population concentration, level of poverty, level of human development, cultural values, among others. It also includes factors related to social equity, access to basic human rights, systems of collective organization, levels of ability to read and write, education, existence of peace and security, systems of good governance, traditional values, customs, and ideological beliefs [13]. Likewise, Ngo [26], states that variations within the elderly population, such as gender, marital status, race, education and religion, can affect the type of vulnerability of the population when facing a natural disaster. Historically ingrained patterns of discrimination and social relationships, inequitable access to resources and power are important determinants of vulnerability [9]. At the international level, in Southeast Asia they used parameters such as: age, education and ethnicity [27], gender, disability status, illiteracy [28] and migration [29].

In this sense, in Indonesia they considered the following parameters: homeless women, violence against women, widows with dependents, female heads of the household, and sexual abuse to women [14]. Likewise, in Manzanillo (Mexico) parameters such as: level of education, knowledge, perception of the individual, historical knowledge, informality of the neighborhood, informality of housing, existence of prevention plans, and community response were used [18]. Also, the United States, Romania and Turkey used parameters such as: race, ethnicity [30], age, education, inventory and housing ownership, and dependency on infrastructure [8], [31], [32]. In addition, in Italy, Portugal and the Netherlands they used parameters such as: age, gender, family structure [24], education, employment, race and ethnicity [21], [33], population density, foreign residents, and social cohesion [22].

At the national level, the following indicators were proposed: education level, age profile, permanent disability, and family structure [25]. In addition, in Huaraz they used social factors such as: population with primary education, population with higher education, illiterate population, indigenous population, population over 65, women, informal settlement, population with disabilities, population with health insurance, etc. [34].

2.3 Sociodemographic dimension

The demographic dimension is related to the population groups that are affected by similar circumstances that limit their capacities for self-sufficiency [18].

Internationally, indicators such as the percentage of the female population, the percentage of the elderly population, the gross birth rate, and the percentage of the population with physical and mental disabilities were used [6]. In this sense, parameters such as: population density, household density [15] and dependency rate [27], [28] were also used. Similarly, in Iran they based their parameters on the density of the child population, the density of the elderly population, the density of the female population, the population density at the educational level, and the population density of the employment situation [35]–[37]. In Manzanillo (Mexico), they used parameters such as degree of population concentration, dependent population, and installed anthropic activity [18]. Also in the United States, they used parameters such as density of the built environment, children population and elderly population [8], percentage of immigrant population, percentage of institutionalized elderly population, percentage of black population, and percentage of Hispanics [19]. It should be noted that, even in the Netherlands, single-parent households were considered as a parameter [33].

At the national level, the following demographic indicators were proposed: access to basic medical services and anthropogenic activity [25]. On the other hand, in Acobamba (Carhuaz, Ancash) parameters related to urban planning, business planning, natural resource management, risk and emergency assessment, environmental impact, service management, transportation routes, cartography, and sociological and demographic studies were used [38]. Lastly, in Pisco the following were used: children population and elderly population [2], [39].

2.4 Environmental dimension

The United Nations [40] state that the environmental dimension refers to the degradation degree of nature and its resources, and climate change. In addition, they point out that parameters such as decrease in biodiversity, poor sanitation, air and water pollution, water scarcity and soil degradation, increase the vulnerability of a population. Often, in the long term, environmental changes lead to increased morbidity, premature death and a reduction in the quality of life [26]. In this sense, Birkmann and Wisner [9] propose two points of view, the first one is to focus on the susceptibility and fragility of ecosystems and the environmental components themselves, and the second one considers the interruption of environmental services that are essential for human wellbeing. Along these lines, in the mountainous land of Uttarakhand (India), parameters such as: percentage of forest cover area, forest fires, percentage of wasteland and percentage of landslide area out of the total area, were used [16]. On the other hand, in Sanandaj (Iran) environmental parameters were proposed (i.e. road distance, distance from a fault, slope, elevation and geology) concluding that they increase the impact on the earthquake risk assessment [36].

Lastly, in Peru the following environmental parameters were proposed: soil loss, soil geological characteristics, deforestation, species of flora and fauna by geographic area, water loss, exploitation of natural resources, etc. [10].

2.5 Physical dimension

The physical dimension refers to the susceptibility of the constructive or natural environment of the population when facing hazards [13].

Internationally, in Dhaka (Bangladesh) they used parameters such as usage and size of the building, land area, photograph of the building, and construction and elevation plans [28]. Likewise, in Uttarakhand (India) they used: proximity to the river, frequency of floods and average rainfall in the last 50 years [16]. In that sense, in Iran they used parameters such as the age of the building, type of structure, floor number [41]–[43], quality of construction materials, coefficient of seismic resonance [35] and land use [36]. Similarly, in Los Angeles (United States) they were based on criteria such as the area of land that could be burned due to an induced fire and the amount of debris measured in thousands of tons [20]. Along these lines, in Vaslui (Romania) they were based on factors such as: age of the building, construction material, number of floors, number of exits, etc. [44]. Similarly, in Turkey, Italy and Portugal, factors associated with the structure were used, such as the position and structural system of the stairs, the width and natural lighting of the evacuation route, irregularities and interaction, floors and roofs, etc. [23], [24], [32], [45], [46].

At the national level, in Cusco they used factors such as: typology of buildings, type of structure, thickness of facade, among others [47]. Similarly, in the district of Punta Negra (San Bartolo, Lima) they used parameters such as number of floors, predominant structural system, type of building material, etc. [48]. Lastly, in Pisco they were based on: damage to the building, maximum rates of soil movement, and fragility of derivatives [49].

3 METHODOLOGICAL APPROACHES APPLIED TO THE ASSESSMENT OF SEISMIC VULNERABILITY

3.1 Machine learning approaches

Nowadays, many researchers have tried to use machine learning to integrate many objective algorithms to reduce the margin of error. However, this approach is not as common in seismic vulnerability assessment. Thus, in Tehran (Iran), a brand new machine learning was used that evaluated seismic vulnerability based on the analysis method of the relationship between the step-wise weight assessment ratio analysis (SWARA) and the geographic information system (GIS), achieving that users with little knowledge in the area can easily determine seismic vulnerability [35].

3.2 GIS and other assessment approaches

The GIS approach is the most used by researchers because it helps generating a vision of who is vulnerable and in which locations, which allows the implementation of specific impact reduction strategies.

Internationally, in the case of Portugal, seismic vulnerability was analyzed with social and physical factors by analyzing multi-criteria decisions in a GIS environment [23], [24], [46]. Likewise, in Italy, through integrated indices, seismic vulnerability was assessed in a GIS environment [22], [45]. Similarly, in Vaslui (Romania) seismic vulnerability was studied through a hierarchical analysis process in a GIS environment [44]. In addition, in Southeast Asia and the Middle East, to reduce disaster risk, social vulnerability was assessed through multi-criteria analysis integrated in a GIS environment [15], [27], [28], [37]. Also, in Los Angeles (United States) the vulnerability of urban areas towards seismic hazards was assessed using the fuzzy logic approach and multi-criteria spatial analysis in a GIS environment [20].

At the national level, in Peru, micro zoning maps were made to assess physical and social vulnerability towards natural disasters through a multi-criteria analysis in a GIS environment [10], [25], [38], [47], [48], [50].

3.3 AHP approaches

The Analytic Hierarchy Process (AHP) approach allows managing the subjectivity of the judgment associated with the weighting of the estimation of the vulnerability parameters through the construction of a comparison matrix, allowing the reduction of uncertainty and error that exist in the assessing process [20], [23], [24], [28], [29], [35], [36], [42], [43], [51].

3.4 Principal component analysis: PCA

The Principal Component Analysis (PCA) approach describes a data set in terms of new uncorrelated parameters using multivariate analysis. In this sense, in Manzanillo (Mexico) it was proposed to work through principal component analysis with composite indicators of the different levels of vulnerability of the population when natural disasters occur [18]. Likewise, in the United States, indices for social vulnerability due to earthquakes were formulated for different magnitude levels through principal component analysis [19], [30]. Furthermore, in Europe parameters of social vulnerability to natural disasters were identified through principal component analysis [24], [31], [33]. Along the same lines, to mitigate disaster risk, in Central Java (Indonesia) social vulnerability was assessed through principal component analysis [27]. Similarly, in Iran, the ratios of social vulnerability to disasters were adequately classified with the principal component analysis [29]. Also, in Huaraz (Peru), to calculate the response time in evacuations, a social vulnerability index was created through the principal component analysis [34].

4 CONCLUSIONS AND RECOMMENDATIONS

In general terms, the studies analyzed suggest that assessing vulnerability to seismic events requires a broad approach that analyzes and includes all the parties involved, that is, people and the built environment, since a partial approach only composes a vision biased from the problem to be addressed, which causes – in turn – partial solutions in the short, medium and long term. In fact, the studies suggest analyzing many different dimensions that allow the evaluator to understand the situational reality and the context in which the events are taking place. Hence, the results validate the importance of conducting studies with more diverse approaches that collect a greater amount of information so that the exposed population can be segmented much better. For example, dimensions such as those related to the cultural aspect, take on special relevance when it comes to understanding the many conceptions and representations of hazard or, even more so, when trying to understand the link between the worldviews of the population and the events per se. In fact, it is expected that the dimensions described in this document work as support information to generate more exhaustive studies when assessing seismic vulnerability. On the other hand, when comparing the analyzed studies, it was observed that –in some cases– different parameters were obtained during the seismic vulnerability assessment for similar dimensions. These differences are related to the availability of information on the sector and the kind of work technique. Furthermore, most of the studies did not perform any uncertainty or sensitivity analysis. Thusly, comparisons should be made with caution, bearing in mind that the statistical approaches applied will tend to provide better results as better quality of the database is available.

In general, there has been progress in terms of vision and methods to determine the level of seismic vulnerability in a given case study in recent years. For example, Izquierdo-Horna and Kahhat [52] show how the level of vulnerability to seismic events varies for a Peruvian case study when the social dimension is taken into consideration. This research shows the importance not only of the inclusion of a human development approach but also of the importance of the integration approach between various dimensions (i.e., socioeconomic dimension, sociocultural dimension, sociodemographic dimension, environmental dimension and physical dimension). This integrated and articulated approach facilitates decision making, since the affected population takes on a leading role in the subsequent evaluation of seismic risk, becoming a social agent [8], [9]. This scenario allows dealing with social incommensurability [7]. Finally, at the national level, it is important to mention that although the guidelines and regulations provided by CENEPRED have undergone updates in the evaluation of vulnerability, giving more emphasis to the person through its social, cultural and environmental dimensions, there is still a significant gap to overcome in terms of study techniques, sources of information and characterization factors [1]. To conclude, we hope that this review of seismic vulnerability assessment will serve as a basis for future studies to consider including as many dimensions as possible in order to have a more complete view of the area of interest.

ACKNOWLEDGEMENT

This project was funded by Universidad Tecnológica del Perú, within the framework of the "Research Projects I+D+i 2021–1" agreement.

REFERENCES

[1] CENEPRED, Manual para la evaluación de riesgos originados por Fenómenos Naturales v3.0, Lima, 2019. https://cenepred.gob.pe/web/wp-content/uploads/Guia_Manuales/Borrador EVAR V3/MANUAL_EVARv3-PRE.pdf.

[2] Llocclla, E., *Evaluación del riesgo sísmico en los distritos de Ica, Parcona y La Tinguiña*, 2005. http://sinpad.indeci.gob.pe/IntranetOracle/Documentos/Informe_riesgos_sismico_Ica_parte1.pdf.

[3] Silgado Ferro, E., Earthquakes of the XVI and XVII centuries in Peru. *Geogr. Soc. Lima*; XCII, 1974. https://app.ingemmet.gob.pe/biblioteca/pdf/P-226.pdf.

[4] INDECI, Sismos ocurridos en e Perú a traves del tiempo. *Compend Estadístico Prevención y Atención Desastr 2006 Sismos Ocurr en el Perú*, **7**, 2006.

[5] Tavera, H. & Bernal, I., The Pisco (Peru) earthquake of 15 August 2007. *Seismol. Res. Lett.*, **79**, pp. 510–515, 2008.

[6] Lee, Y.-J., Social vulnerability indicators as a sustainable planning tool. *Environ. Impact Assess. Rev.*, **44**, pp. 31–42, 2014.

[7] Adger, W.N., Vulnerability. *Glob. Environ. Chang.*, **16**, pp. 268–281, 2006.

[8] Cutter, S.L., Boruff, B.J. & Shirley, W.L., Social vulnerability to environmental hazards. *Soc. Sci. Q.*, **84**, pp. 242–261, 2003.

[9] Birkmann, J. & Wisner, B., *MEASURING the un-measurable: The Challenge of Vulnerability*, 2006. http://www.ihdp.unu.edu/file/get/3962.pdf.

[10] CENEPRED, *Manual para la evaluación de riesgos originados por Fenómenos Naturales*, 2nd ed., Lima, 2014. https://www.cenepred.gob.pe/web/wp-content/uploads/Guia_Manuales/Manual-Evaluacion-de-Riesgos_v2.pdf.

[11] United Nations, Hyogo Framework for Action 2005–2015: Building the Resilience of Nations and Communities to Disasters. *World Conf Disaster Reduct.*, 2005. https://www.unisdr.org/2005/wcdr/intergover/official-doc/L-docs/Hyogo-framework-for-action-english.pdf.
[12] Rufat, S. et al., Social vulnerability to floods: Review of case studies and implications for measurement. *Int, J, Disaster Risk Reduct.*, **14**, pp. 470–486, 2015.
[13] Ocola, L., Peligro, vulnerabilidad, riesgo y la posibilidad de desastres sísmicos en el Perú. *Rev. Geofísica,* **61**, pp. 81–125, 2005.
[14] Yumarni, T., Amaratunga, D. & Haigh, R., Assessing gender vulnerability within post-earthquake reconstruction: Case study from Indonesia. *Procedia Econ. Financ.,* **18**, pp. 763–771, 2014.
[15] Suhaiza Sauti, N., Effendi Daud, M. & Kaamin, M., Proposed method and framework for evaluating and calculating a seismic vulnerability index of Malaysia. *IOP Conf. Ser. Earth Environ. Sci.*, **616**, 012034, 2020.
[16] Kumar, D. & Kumar Bhattacharjya, R., Study of integrated social vulnerability index (SoVI int) of hilly region of Uttarakhand, India. *Environ. Clim. Technol.*, **24**, pp. 105–122, 2020.
[17] Yamin, L.E. et al., Probabilistic seismic vulnerability assessment of buildings in terms of economic losses. *Eng, Struct.*, **138**, pp. 308–323, 2017.
[18] Thomas, J.E., Evaluación de la vulnerabilidad social ante amenazas naturales en Manzanillo (Colima): Un aporte de método. *Investig. Geográficas*, **81**, pp. 79–93, 2013.
[19] Schmidtlein, M.C. et al., Modeled earthquake losses and social vulnerability in Charleston, South Carolina. *Appl. Geogr.*, **31**, pp. 269–281, 2011.
[20] Rashed, T. & Weeks, J., Assessing vulnerability to earthquake hazards through spatial multicriteria analysis of urban areas. *Int. J. Geogr. Inf. Sci.*, **17**, pp. 547–576, 2003.
[21] Frigerio, I. & De Amicis, M., Mapping social vulnerability to natural hazards in Italy: A suitable tool for risk mitigation strategies. *Environ. Sci. Policy.*,**63**, pp. 187–196, 2016.
[22] Azzimonti, O.L. et al., Combining hazard, social vulnerability and resilience to provide a proposal for seismic risk assessment. *J. Risk Res.*, **23**, pp. 1225–1241, 2020.
[23] Martins, V.N., Silva, D.S. & Cabral, P., Social vulnerability assessment to seismic risk using multicriteria analysis: The case study of Vila Franca do Campo (São Miguel Island, Azores, Portugal). *Nat. Hazards*, **62**, pp. 385–404, 2012.
[24] Fernandez, P., Mourato, S. & Moreira, M., Social vulnerability assessment of flood risk using GIS-based multicriteria decision analysis: A case study of Vila Nova de Gaia (Portugal). *Geomatics, Nat. Hazards Risk*, **7**, pp. 1367–1389, 2016.
[25] Izquierdo-Horna, L. & Kahhat, R., Methodological framework to integrate social and physical vulnerability in the prevention of seismic risk. *WIT Transactions on Engineering Sciences*, pp. 69–79.
[26] Ngo, E.B., When disasters and age collide: Reviewing vulnerability of the elderly. *Nat. Hazards Rev.*, **2**, pp. 80–89, 2001.
[27] Wijaya, A.P. & Hong, J.-H., Quantitative assessment of social vulnerability for landslide disaster risk reduction using gis approach (Case study: Cilacap Regency, Province of Central Java, Indonesia). *Int. Arch. Photogramm. Remote Sens. Spat. Inf. Sci.*, **42**(4), pp. 703–709, 2018.
[28] Rahman, N., Ansary, M.A. & Islam, I., GIS based mapping of vulnerability to earthquake and fire hazard in Dhaka city, Bangladesh. *Int. J. Disaster Risk Reduct.*, **13**, pp. 291–300, 2015.

[29] Fatemi, F. et al., Social vulnerability indicators in disasters: Findings from a systematic review. *Int. J. Disaster Risk Reduct.*, **22**, pp. 219–227, 2017.

[30] Hung, L., Wang, C. & Yarnal, B., Vulnerability of families and households to natural hazards: A case study of storm surge flooding in Sarasota County, Florida. *Appl. Geogr.*, **76**, pp. 184–197, 2016.

[31] Armaş, I. & Gavriş, A., Census-based social vulnerability assessment for Bucharest. *Procedia Environ. Sci.*, **32**, pp. 138–146, 2016.

[32] Yücel, G. & Arun, G., Earthquake and physical and social vulnerability assessment for settlements: Case study Avcılar District. *Megaron.*, **5**, pp. 23–32, 2010.

[33] Koks, E.E. et al., Combining hazard, exposure and social vulnerability to provide lessons for flood risk management. *Environ. Sci. Policy.*, **47**, pp. 42–52, 2015.

[34] Hofflinger, A., Somos-Valenzuela, M.A. & Vallejos-Romero, A., Response time to flood events using a social vulnerability index (ReTSVI). *Nat. Hazards Earth Syst. Sci.*, **19**, pp. 251–267, 2019.

[35] Lee, S. et al., SEVUCAS: A novel GIS-based machine learning software for seismic vulnerability assessment. *Appl. Sci.*, **9**, p. 3495, 2019.

[36] Yariyan, P. et al., Earthquake risk assessment using an integrated fuzzy analytic hierarchy process with artificial neural networks based on GIS: A case study of Sanandaj in Iran. *Int. J. Disaster Risk Reduct.*, **50**. DOI: 10.1016/j.ijdrr.2020.101705.

[37] Ahmad, R.A., Singh, R.P. & Adris, A., Seismic hazard assessment of Syria using seismicity, DEM, slope, active faults and GIS. *Remote Sens. Appl. Soc. Environ.*, **6**, pp. 59–70, 2017.

[38] Lazaro, M.R., *Nálisis de peligros y vulnerabilidades para la gestión del riesgo de desastres, utilizando el sistema de información geográfica (SIG) en la localidad de Acopampa – Carhuaz, Ancash.* Tesis de Titulación, Universidad Nacional Santiago Antunez De Mayolo, 2015. http://repositorio.unasam.edu.pe/handle/UNASAM/1190.

[39] INDECI, Mapa de peligros, plan de usos del suelo y medidas de mitigación de la ciudad de Pisco, 2001. https://sigrid.cenepred.gob.pe/sigridv3/documento/5226/descargar.

[40] United Nations, On better terms: A glance at key climate change and disaster risk reduction concepts, Switzerland, 2006. https://www.preventionweb.net/files/3854_Onbetterterms1.pdf.

[41] Sadrykia, M., Delavar, M. & Zare, M., A GIS-based fuzzy decision making model for seismic vulnerability assessment in areas with incomplete data. *ISPRS Int. J. Geo-Information*, **6**, p. 119, 2017.

[42] Delavar, M.R. & Sadrykia, M., Assessment of enhanced Dempster-Shafer theory for uncertainty modeling in a GIS-based seismic vulnerability assessment model, case study: Tabriz City. *ISPRS Int. J. Geo-Information*, **9**, p. 195, 2020.

[43] Karimzadeh, S. et al., A GIS-based seismic hazard, building vulnerability and human loss assessment for the earthquake scenario in Tabriz. *Soil Dyn. Earthq. Eng.*, **66**, pp. 263–280, 2014.

[44] Albulescu, A.-C., Larion, D. & Grozavu, A., Multi-criteria assessment of the seismic vulnerability of school units in Vaslui city, Romania. *WIT Transactions on Engineering Sciences*, pp. 17–28.

[45] Chieffo, N., Formisano, A. & Miguel Ferreira, T., Damage scenario-based approach and retrofitting strategies for seismic risk mitigation: An application to the historical Centre of Sant'Antimo (Italy). *Eur. J. Environ. Civ. Eng.*, **8189**, pp. 1–20, 2019.

[46] Vicente, R. et al., Seismic vulnerability and risk assessment: Case study of the historic city centre of Coimbra, Portugal. *Bull. Earthq. Eng.*, **9**, pp. 1067–1096, 2011.

[47] Brando, G. et al., Structural survey and empirical seismic vulnerability assessment of dwellings in the historical centre of Cusco, Peru. *Int. J. Archit. Herit.*, **3058**, pp. 1–29, 2019.

[48] CISMID, *Análisis de Riesgo en Zonas Urbanas del Distrito de Punta Negra*, 2017. https://sigrid.cenepred.gob.pe/sigridv3/storage/biblioteca//6994_analisis-de-riesgo-en-zonas-urbanas-del-distrito-de-punta-negra.pdf.

[49] Matsuzaki, S. et al., Evaluation of seismic vulnerability of buildings based on damage survey data from the 2007 Pisco, Peru earthquake. *J. Disaster Res.*, **9**, pp. 1050–1058, 2014.

[50] Santa-Cruz, S. et al., Social sustainability dimensions in the seismic risk reduction of public schools: A case study of Lima, Peru. *Sustain. Sci. Pract. Policy.*, **12**, pp. 34–46, 2016.

[51] Jena, R. et al., Integrated model for earthquake risk assessment using neural network and analytic hierarchy process: Aceh province, Indonesia. *Geosci. Front.*, **11**, pp. 613–634, 2020.

[52] Izquierdo-Horna, L. & Kahhat, R., An interdisciplinary approach to identify zones vulnerable to earthquakes. *Int. J. Disaster Risk Reduct.*, **48**, 101592, 2020.

EXPERIMENTAL STUDY FOR RETAINING WALLS USING PILE COUNTERMEASURE AGAINST EARTHQUAKES

ENAYAT MALLYAR & KAKUTA FUJIWARA
Department of Civil Engineering, Tokai University, Japan

ABSTRACT

In many coastal areas, port facilities such as retaining walls were highly damaged due to earthquakes. Deformation of retaining walls is one of the primary sources of damage in the port facilities. The liquefaction of the backfill causes a large lateral force to the walls, which is triggered by the wall's displacement. The retaining wall has been constructed widely and required to be resistant during earthquake motion. Several effective countermeasures to mitigate liquefaction have been proposed. Using piles as a countermeasure nearby the retaining walls has been proposed that the displacement of the retaining walls could be reduced. However, the high rigid material pile is required high cost. Therefore, the authors conducted a series of shaking model tests in three cases with different high rigid and low rigid types of piles to investigate the effectiveness of piles against the deformation of the retaining wall. The deformation of the retaining wall, caused by liquefied soil, was obtained by the sensors measurement and it is shown that even mix (high rigidity and low rigidity) piles in the backfill could effectively reduce the wall displacement.

Keywords: retaining wall, pile countermeasure, shaking table test, earthquake.

1 INTRODUCTION

Soil liquefaction brought severe damage to structures in many past earthquakes. Port structures have often suffered damages in strong earthquakes. Loose sandy soil behind retaining walls are susceptible to liquefaction. The liquefaction of the backfill causes a large lateral force to the wall which is cause the deformation and extensive damage to the retaining walls. During several destructive earthquakes such as the 1983 Nihonkai–Chubu and the 1995 Hyogoken-Nambu (Kobe) earthquake [1] a large number of retaining walls, including quay walls, were damaged. Although extensive analytical and experimental studies (using a vibration table and centrifuge model) have been conducted on retaining walls behavior undergoing lateral flow of liquefied soil.

Several countermeasure methods against liquefaction have been proposed such as soil improvement techniques represented by sand compaction, stone column and seismic reinforcement using steel piles techniques, can reduced the generated lateral force of liquefied soil. One of the effective methods is the deterrent pile method [2], where steel piles are installed in the backfill, distributive and not connected with retaining walls. It has been verified through centrifuge model tests that these piles prevent soil from flowing laterally and interdict the increasing excess pore water pressure of soil near the piles. However, still it is needed to investigate on different types and numbers of piles because using many steel piles is required high material cost. Steel pile has high rigidity, but it is expensive, other low rigid materials are almost inexpensive. Therefore, the authors conducted series of shaking model tests on different types of piles in the back fill of a retaining wall.

WIT Transactions on The Built Environment, Vol 207, © 2021 WIT Press
www.witpress.com, ISSN 1743-3509 (on-line)
doi:10.2495/DMAN210181

2 TEST CONDITION

2.1 Device and soil

The model was geometrically scaled to approximately 1/20–1/30 of the actual structure, as shown in Fig. 1. The model was built in a rigid container with transparent side wall. Dimension of the container were 0.84 × 0.60 × 0.44 m^3. The rigid container filled in two soil layers; the bottom layer was filled and compacted as a dense layer and one half of the second layer made as a loose layer and the other half was filled with water. The soil for both layers was Toyoura sand (e_{max} = 0.927, e_{min} = 0.625, ρ_s = 2.64g/cm^3). The experiments were conducted in three cases. Case A is no countermeasure case (without pile), as shown in Fig. 1(a). In the Case B ten acryl piles (high rigidity) installed in three rows as shown in Fig. 2(a). In the Case C seven acryl piles (high rigidity) installed in first and second row from the wall and three rubber piles (low rigidity) installed in third row, as shown in Fig. 2(b). The three cases mitigation strategy and the dense and loose layers properties shown in Table 1.

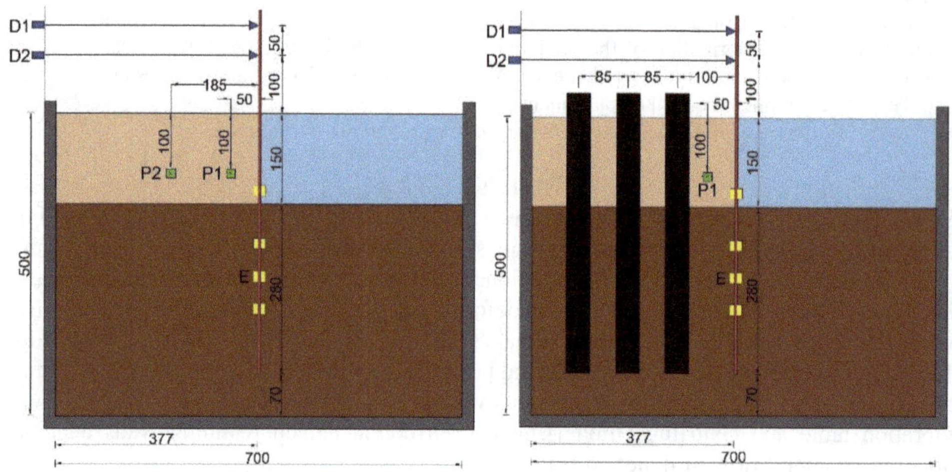

(a) (b)

Figure 1: Test model cross section. (a) Case A; and (b) Cases B and C. Note: D1, D2 = Displacement sensor; P1, P2 = Water pressure sensor; E = Strain gauge sensor.

(a) (b)

Figure 2: The model top view. (a) Case B; and (b) Case C.

Table 1: Summarizes the features of each experiment.

Cases	Dense layer		Loose layer		Mitigation strategy
	Dr(%)	γ_t(kN/m^3)	Dr(%)	γ_t(kN/m^3)	
Case A	94	17.0	40	14.3	No countermeasure
Case B	97	17.1	43	15.2	10 Acryl piles
Case C	91	17.3	44	15.0	7 Acryl and 3 rubber piles

2.2 Materials properties

In this experimental model were used two types of piles, acryl pile represents rigid feature and rubber pile represents low rigidity as shown in Figs 3 and 4. Both types of pile with 40 mm diameter and 450 mm length were installed in backfill of the wall. An acryl plate with 5 mm thickness and 470 mm length as a retaining wall was located inside the dense layer, 70 mm upper than the model box base.

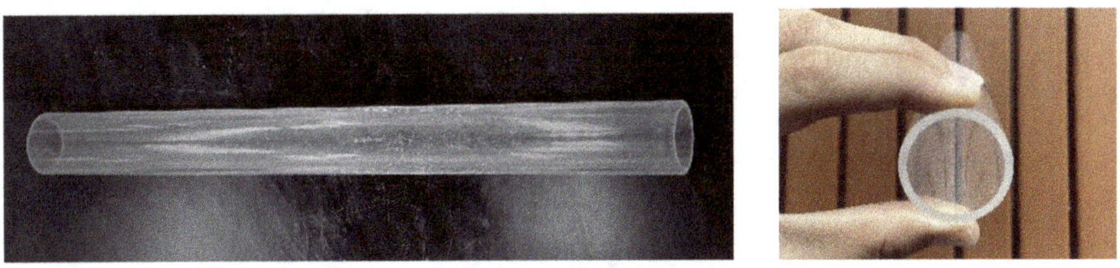

Figure 3: Acryl pile used as a high-rigidity pile.

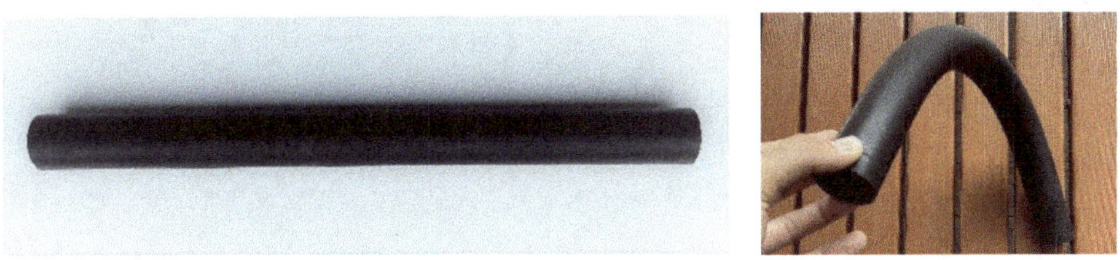

Figure 4: Rubber hose used as a low-rigidity pile.

2.3 Input motion

The input motion was a sine wave of frequency 3Hz for 5 seconds. The sine wave was used as a simple motion. Considering actual field, the characters of the wave; duration, frequency and wave shape should be examined in further study.

Three motions with intervals were given to the device for each case. The maximum acceleration for each motion were 0.5, 1.0 and 2.0 m/s^2. In this paper, the authors regarded 0.5 m/s^2 as a weak motion, 1 m/s^2 as a medium motion and 2 m/s^2 as a strong motion. The acceleration of 0.5 m/s^2 is a weak motion as small as the soil had the excess pore water pressure. The authors increased the acceleration as 1.0 and 2.0 m/s^2 to investigate the effectiveness against the acceleration of earthquake. This model test was equipped with three types of sensors, two displacement sensors sticked on 100 mm and 150 mm upper than top of wall, eight strain gauge sensors sticked on both sides of the wall, and two water pressure sensors that one of them located between wall and piles and another one located between piles in the liquefiable layer as shown in Fig. 1.

3 TEST RESULT

The model was shaken under three input motions. Based on observations, the soil seems to be liquefied, that caused the wall deformations. Detailed information are discussed in this section.

3.1 Displacement

The residual horizontal displacement at the D point (D point presented the top of wall which is derived by two displacement sensors D1 and D2) for all cases shown in Fig. 5(a). The acceleration of zero means before shaking state. The larger the acceleration became, the larger the displacement became, for all cases. Fig. 5(a) shows that the displacement of Case B was smallest among all cases. The displacements for Cases B and C are smaller than that of Case A, which means the pile countermeasures were effective for wall displacement. Additionally, the displacements for Cases B and C are not same, which means the structure of Case B was more effective than that of Case C though the difference was small. The result indicates the displacement for Case C is smaller than Case A, which means even the mix piles countermeasure was effective against wall deformation.

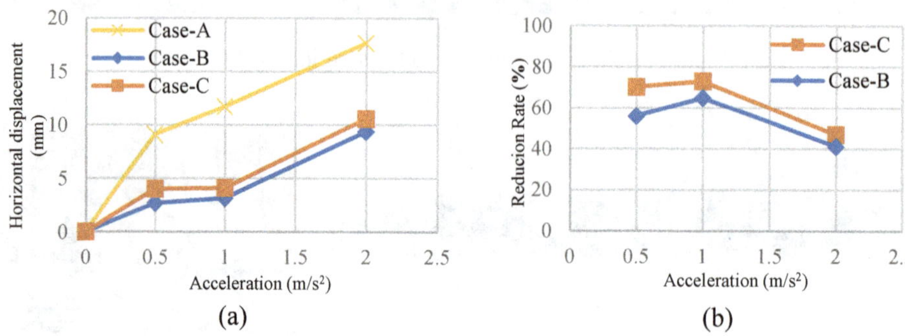

Figure 5: Horizontal displacement of retaining wall. (a) Displacement at point D; and (b) Reduction rate.

The reduction rate means that the displacements of Cases B and C are divided by the displacement of Case A for each acceleration, as shown in Fig. 5(b). The reduction rates of Case B and Case C were lager than 50% for the weak and medium motion (0.5 m/s^2 and 1 m/s^2). On the other hands, the reduction rate of Case B and Case C were less than 50% for the strong motion (2 m/s^2). These results state that the pile countermeasures are more effective for weak and medium motions, even soft pile is effective to reduce the wall displacement. Fig. 5(b) indicates that the reduction rate of Case B was more than that of Case C though the difference was small.

3.2 Bending strain

The residual bending strain along the retaining wall for three motions in all cases are depicted in Fig. 6. The strain gauge sensor at the height of 170 mm for Case A did not work. In these figures the horizontal line at the height of 350 mm shows the border between dense and loose soil layer. Fig. 6 indicates the wall was deformed toward the water side due to the liquefaction of soil. The bigger motion caused the bigger bending strain on the retaining wall among all cases. Through the three motions, the bending strain of Case B was smallest among three cases. The pile countermeasure (Case B) could reduce the wall deformation in spite of the difference of acceleration. This result matches to the result of the displacement indicated in the previous section. On the other hands, the strain for Case C was almost same as that for Case A. This result doesn't match to the result of the displacement indicated in the previous section. As the strain for Case A has a lack of data that mentioned above, the author continues to investigate this result.

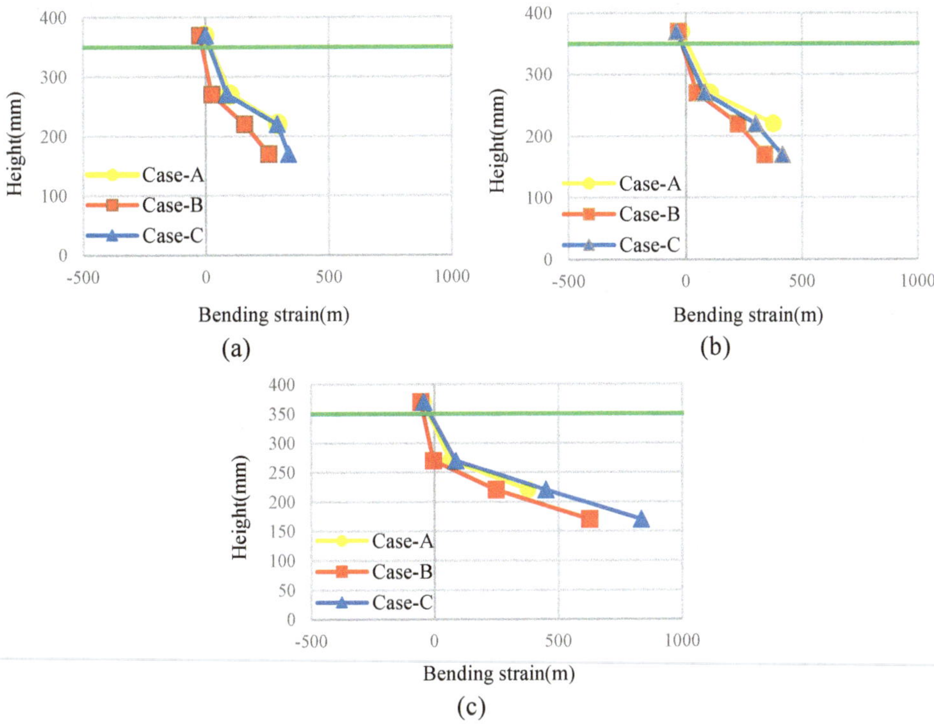

Figure 6: The residual bending strain along the retaining wall. (a) 50 gal; (b) 100 gal; and (c) 200 gal.

3.3 Water pressure

When the earthquake occurs, water pressure increases, and soil gets shear deformation. Shear deformation makes soil liquefied. Figs 7 and 8 show the excess pore water pressure at point P1 and P2 for all cases, respectively. According to the acceleration time histories of excess pore water pressure, high pressure developed at the early stage of shaking. Fig. 7(a) and Fig.

8(a) indicate the excess pore water pressure increased due to the motion that means soil liquefaction. After that, the excess pore water pressure was converged to a constant value. Fig. 7(b), (c) and Fig. 8(b), (c) state the excess pore water pressure started from the remaining value in previous motion, respectively. Compering among 50–200 gal at P1 and P2, it can be seen, as the acceleration got bigger the excess pore water pressure became bigger. The excess pore water pressure were less at P1 than P2. This is because the sand near to the wall was expanded by the deformation of the wall. Fig. 7(a)–(c) and Fig. 8(a)–(c) show that the excess pore water pressure in Case B and Case C were smaller than that in Case A. The result illustrates that the acryl (high rigidity) piles and mix (high rigidity and low rigidity) piles prevented the shear deformation of soil, that caused the prevention of water pressure increment. The reduction of water pressure caused the reduction of the wall displacement.

4 CONCLUSION

For the reduction of the material cost of retaining wall countermeasure, the authors proposed the mix pile countermeasure which consists of high rigidity and low rigidity piles. The authors conducted model tests; three types of motions (weak, medium, and strong) are given continuously with interval to three model cases (no pile, high rigidity piles, mix of high and low rigidity piles). Through the model tests, following conclusions are gained.

1. The sand was liquefied by the motion, which caused wall deformation. The larger the acceleration, the larger the wall displacement became with or without piles.

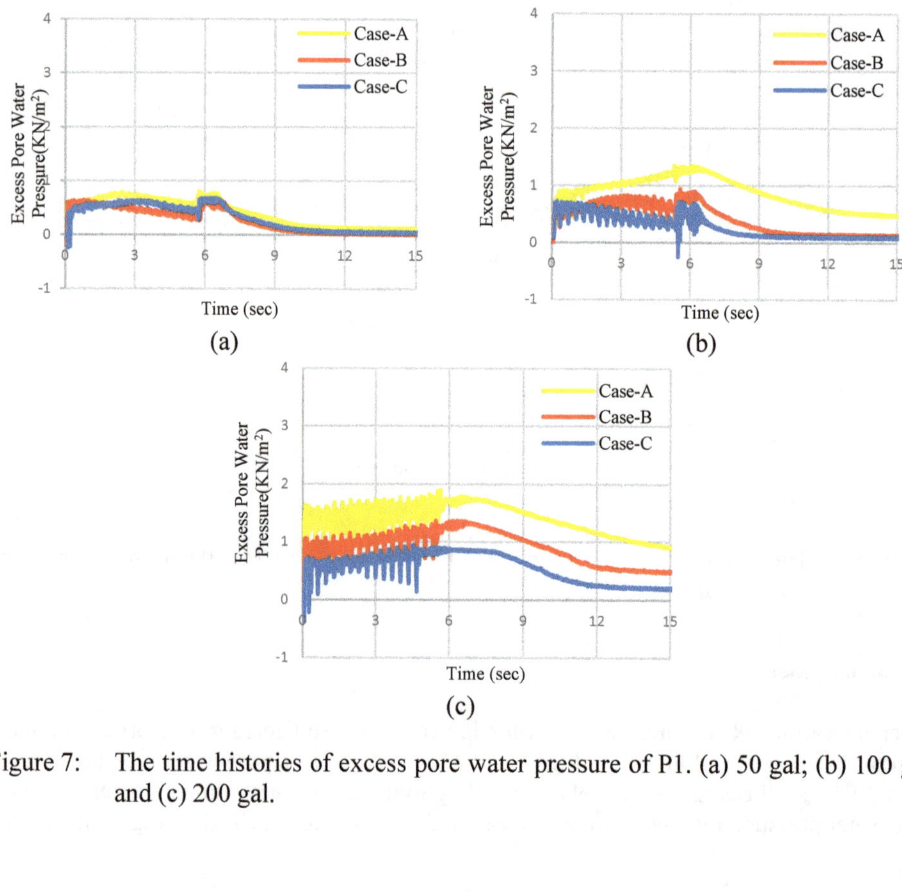

Figure 7: The time histories of excess pore water pressure of P1. (a) 50 gal; (b) 100 gal; and (c) 200 gal.

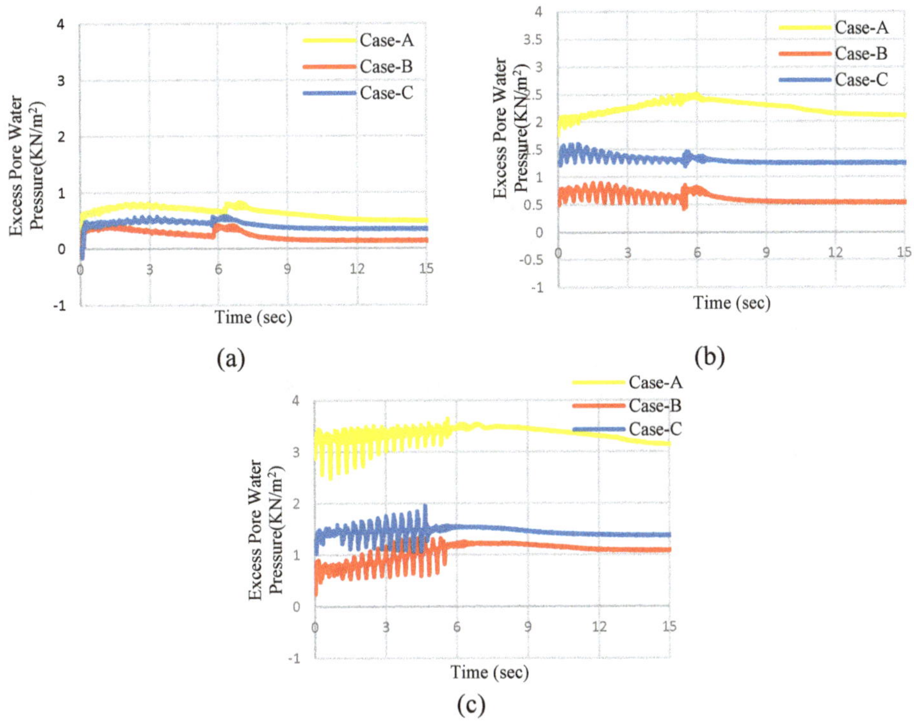

Figure 8: The time histories of excess pore water pressure of P2. (a) 50 gal; (b) 100 gal; and (c) 200 gal.

2. In high rigidity piles case, the wall displacement was 47–73% reduced compared to no pile case. In addition, even in mix pile case, the wall displacement was 40–65% reduced compared to no pile case. There is small difference between high rigidity piles and mix piles countermeasure.
3. Both high rigidity piles and mix piles were more effective against the wall deformation under weak motion than that strong motion.
4. The piles as a countermeasure could prevent water pressure from increasing and prevent the shear deformation of soil.
5. To find out the accurate effectiveness of rubber pile, we need more investigation by considering two rows and one row of piles and changing the number of the rubber piles.

ACKNOWLEDGEMENT

This work was supported by JSPS KAKENHI Grant Number 19K15265. We would like to express our thanks to JSPS.

REFERENCES

[1] Tokimatsu, K. & Asaka, Y., Effects of liquefaction-induced ground displacements on pile performance in the 1995 Hyogoken-Nambu earthquake. *Special Issue of Soils and Foundation*, pp. 163–177, 1998.

[2] Hamada, M. & Higuchi, S., Development of countermeasures for lateral flow of liquefied ground. *Journal of Japan Society of Civil Engineers, Ser. A1 (Structural Engineering and Earthquake Engineering (SE/EE))*, **66**(1), pp. 84–94, 2010.
[3] Fujiwara, K., Numerical study on pile countermeasure against liquefaction behind retaining wall. *29th International Ocean and Polar Engineering Conference.* OnePetro, 2019.
[4] Motamed, R. & Towhata, I., Mitigation measures for pile groups behind quay walls subjected to lateral flow of liquefied soil: Shake table model tests. *Soil Dynamics and Earthquake Engineering,* **30**(10), pp. 1043–1060, 2010.
[5] Ogata, T., Disaster management in Japan. *JMA-PhRMA Joint Symposium Conferences,* **59**(1), pp. 27–30, 2016. https://www.pwri.go.jp/eng/ujnr/joint/41/paper/17_Ikeuchi.pdf.
[6] Miura, K., Kohama, E., Inoue, K., Ohtsuka, N., Sasajima, T. & Yoshida N., Behavior of gravity type quay wall during earthquake regarding dynamic interaction between caisson and backfill during liquefaction. *Proceedings of the 12th World Conference on Earthquake Engineering*, 7, p. 1737, 2000.
[7] Kohama, E., Sugano, T., Shiozaki, Y. & Mitoh, M., Dynamic behaviour of a steel sheet pile quay wall in the full scale experiment with regard to liquefaction. *Proceedings of the 13th World Conference on Earthquake Engineering,* Vancouver, BC, Canada, 2004.

SEMI-QUANTITATIVE METHODOLOGY FOR DISASTER RISK ASSESSMENT: THE CASE STUDY OF PIPUS VILLAGE, AMAZONAS DEPARTMENT, PERU

JHON E. CHAHUA JANAMPA[1*] & JHESSICA M. GUERRERO PINEDO[2]
[1]National Center for Estimation, Prevention and Reduction of Disaster Risk, Ministry of Defense, Peru
[2]Provincial Municipality of Chachapoyas, Regional Government of Amazonas, Peru

ABSTRACT

The objective of this case study was to know the risk levels of Pipus village in Amazonas-Peru in order to have a technical document for the local government to access financial funds to implement actions for risk control and reduction associated with mass movements and floods. To achieve this objective a semi-quantitative methodology was applied based on the hierarchical analysis process. It allows an easy and fast way to quantify the qualitative variables (conditioning factors, triggering factors, fragility, resilience, etc.). For this, a geological, geomorphological and geodynamic mapping of the study area was carried out and the vulnerability index card filling for all the exposed elements as well, which were processed in a geographic information system. This methodology allowed characterizing the hazard, analyzing the vulnerability and calculating the risk of the elements exposed in Pipus village. Thirty seven homes were identified at very high risk of debris flow hazard; a landslide hazard in road; and river flood hazard in a bridge, sanitary system and communal premise, adding a total of six million PEN between damages and losses.

Keywords: semi-quantitative methodology, risk levels, landslide hazard, debris flow hazard, river flood hazard.

1 INTRODUCTION

The Peruvian territory is exposed to various natural events, due to the interaction between the physical conditions that a specific geographic area presents and the factors that trigger it. These events become natural hazards when they cause significant impacts on the population and their livelihoods.

In northern Peru, events due to mass movements and floods are frequent due mainly to rainfall, in which one of the affected places is the department of Amazonas, in which Pipus village is located. These events are recurrent in this populated center from December to March, in which two emergency reports due to landslides were registered [1], [2] this year.

Because of Pipus village does not have infrastructure, measures or actions to reduce the effects of these hazards, it is necessary to determine the risk levels to which the population and their livelihoods are exposed, in order to have a technical document that allows the local government to support the priority of taking risk control measures by accessing financial funds for the execution of these measures.

2 METHODOLOGY

To characterize hazards, analyze vulnerability and calculate risk, the methodology proposed by the National Center for Estimation, Prevention and Reduction of Disaster Risk (Cenepred, by its acronym in Spanish), which is the leading entity in disaster risk assessments, was used. This semi-quantitative methodology [3] is based on the hierarchical analysis process using

ORCID: https://orcid.org/0000-0002-9922-0032

WIT Transactions on The Built Environment, Vol 207, © 2021 WIT Press
www.witpress.com, ISSN 1743-3509 (on-line)
doi:10.2495/DMAN210191

the Saaty matrices [4] for the weighting between quantitative and qualitative variables, facilitating the stratification of hazard, vulnerability and risk levels, which are related to the following equation:

$$R_{ie}|_t = f(P_i, V_e)|_t, \tag{1}$$

where R = risk; P_i = hazard with intensity greater than or equal to i during an exposure period t; and V_e = vulnerability of an exposed element e.

To characterize the hazard, a multicriteria analysis is carried out identifying the factors to which the territory is conditioned, the factor that triggers the hazard and the hazard evaluation parameter (Fig. 1). Each of these factors is assigned a weighted weight, relating them to each other with the hierarchical analysis process.

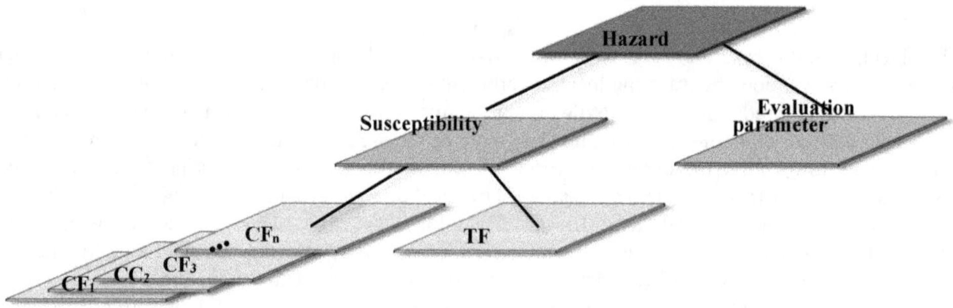

Figure 1: Hierarchical diagram for calculating the hazard. Note: CF_1, CF_2, CF_3, …, CF_n = conditioning factors of the territory; TF = trigger factor.

The vulnerability analysis is obtained by relating its three factors: exposure, fragility and resilience, for each dimension considered in its analysis (social, economic and environmental). These dimensions will depend on the type of exposed element evaluated and a specific analysis must be carried out for each one of them. These dimensions and factors are related to each other with the hierarchical analysis process (Fig. 2).

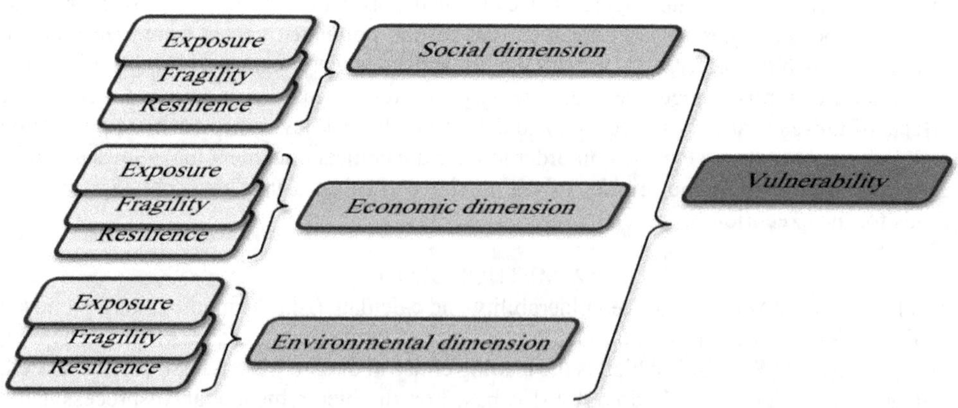

Figure 2: Hierarchical diagram for calculating vulnerability.

With the calculations carried out, the hazard and vulnerability levels are stratified into four levels: Very high, high, medium and low. Each level presents a numerical range which is obtained during the calculation of the variables considered in the study through the hierarchical analysis process.

Finally, each hazard range is multiplied with vulnerability range to obtain risk ranges which are represented in its four levels. This methodology allows to quickly and easily calculate the risk levels to which the population and their livelihoods are exposed, focused on a scale of 1:25,000.

3 CASE STUDY

The study area (Fig. 3) comprises the Pipus village located in the San Francisco de Daguas district, Chachapoyas province and Amazonas department, at the coordinates UTM Zone 18 S: 197721 E and 9311125 N, at an altitude of 1990 masl.

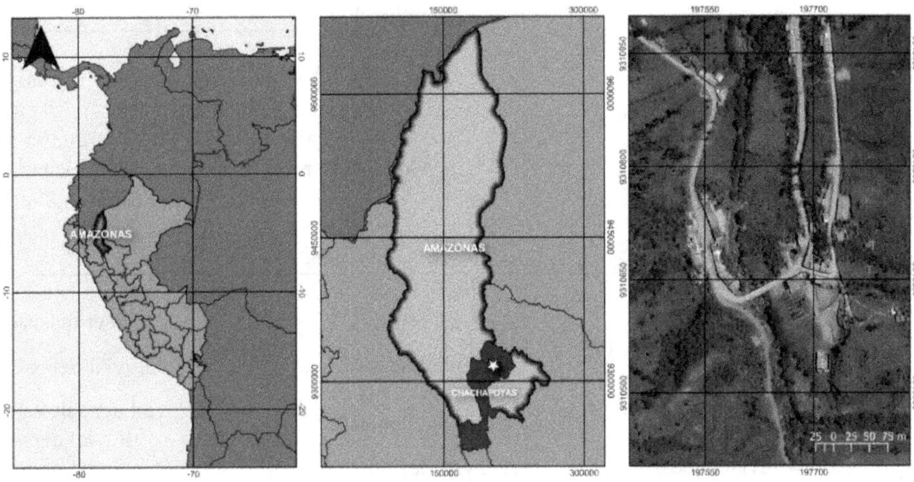

Figure 3: General location of Pipus village.

A field trip was carried out together with the technical team of the Geophysical Institute of Peru (IGP, by its acronym in Spanish) and the Chachapoyas provincial municipality in which a study of the geological, geomorphological and geodynamic characterization of the study area was carried out [5] which were inputs for the characterization of the hazard. Three main hazards were identified: fluvial flood, landslides and debris flow. Likewise, the historical record of the Chachapoyas meteorological station [6] was obtained for a period of 16 years (1996–2010) which were classified in thresholds with percentiles (99p–95p–90p–75p) according to the methodology of the National Service of Meteorology and Hydrology of Peru (Senamhi, by its acronym in Spanish) [7]. For each type of hazard identified, its conditioning factors, triggering factor and its evaluation parameters (Table 1) obtained and previously treated were analyzed.

Table 1 shows the descriptors of each parameter and factor analyzed in a spatial manner for the three types of hazard identified in the study area, which will be used to obtain the hazard levels ranges. For the debris flow hazard, two ravines were identified: the Chilchos and Malcamal ravine, which were analyzed separately; however, they present the same factors and parameters for their calculation.

Table 1: Factors and parameters analyzed for each type of hazard identified.

	Landslide	Debris flow	Fluvial flood
	Unstable areas	Flow affected areas	Recurrence of flooded areas
Evaluation parameter	> 2800 m^2	Greater than 6800 m^2	Very frequent
	2050–2800 m^2	From 5101 m^2 to 6800 m^2	Fairly frequent
	1500–2050 m^2	From 3401 m^2 to 5100 m^2	Frequent
	750–1500 m^2	From 1701 m^2 to 3400 m^2	Infrequent
	< 750 m^2 and non-unstable areas	Less than 1700 m^2	Rare
		Precipitation thresholds	
Triggering factor	Extremely rainy (RR > 35.5 mm)		
	Very rainy (20.0 mm < RR ≤ 35.5 mm)		
	Rainy (13.9 mm < RR ≤ 20.0 mm)		
	Moderately rainy (6.8 mm < RR ≤ 13.9 mm)		
	Slightly rainy (RR < 6.8 mm)		
		Surface geological units	
Conditioning factor 1	Colluvial deposit	Alluvial deposit 2	River reservoir
	Sarayaquillo formation	Alluvial deposit	Alluvial deposit 1
	Goyllarizquizga group	Colluvial deposit	Alluvial deposit 2 and alluvial deposit
	Alluvial deposit	Alluvial deposit 1	Colluvial deposit
	Alluvial and fluvial deposits	Goyllarizquizga Group and Sarayaquillo Formation	Goyllarizquizga group and Sarayaquillo formation
		Geomorphological units	
Conditioning factor 2	Mountainside	Alluvial channel	River bed
	Plateau	Alluvial fan	Flood plain
	Alluvial fan	Alluvial terrace	Alluvial terrace and alluvial channel
	Alluvial channel	Floodplain and riverbed	Alluvial fan
	Terrace, plain and riverbed	Plateau and mountainside	Plateau and mountainside
		Slope of the terrain	
Conditioning factor 3	Greater than 35°	From 0° to 5°	From 0° to 5°
	From 25° to 35°	From 5° to 15°	From 5° to 15°
	From 15° to 25°	From 15° to 30°	From 15° to 25°
	From 5° to 15°	From 30° to 45°	From 25° to 35°
	From 0° to 5°	Greater than 45°	Greater than 35°

Table 2: Variables of the vulnerability factors for each element exposed to the hazard.

ti	Exposure	Fragility	Resilience
		Social dimension	
Houses (debris flow)	Number of people per house	Age group	Educational level achieved Type of insurance Training on DRM issues
		Economic dimension	
	Close to hazard	Predominant wall material State of conservation Antiquity	Average family income Labor regime
		Economic dimension	
Road (landslide)	Close to hazard	State of conservation	Compliance with construction processes in accordance with current regulations
		Economic dimension	
Bridge (f. flood)	Close to hazard	State of conservation	Compliance with construction processes in accordance with current regulations
		Economic dimension	
Slaughterhouse (f. flood)	Close to hazard	State of conservation	Compliance with construction processes in accordance with current regulations
		Economic dimension	
Parking (f. flood)	Close to hazard	State of conservation	Compliance with construction processes in accordance with current regulations
		Economic dimension	
Sewage system (f. flood)	Close to hazard	State of conservation	Compliance with construction processes in accordance with current regulations
		Environmental dimension	
	Close to water sources	Exploitation of natural resources on the banks of the river	Knowledge and compliance with environmental regulations
	Close to the landfill		Training of operators in environmental conservation

Table 3: Values of the hazard, vulnerability and risk levels range.

	Level	Hazard	Vulnerability	Risk
			Number of houses in Chilchos ravine	
Debris flow	Very high	$0.261 < P \leq 0.490$	$0.264 < V \leq 0.452$	$0.069 < R \leq 0.221$
	High	$0.146 < P \leq 0.261$	$0.161 < V \leq 0.264$	$0.024 < R \leq 0.069$
	Medium	$0.069 < P \leq 0.146$	$0.084 < V \leq 0.161$	$0.006 < R \leq 0.024$
	Low	$0.034 \leq P \leq 0.069$	$0.039 \leq V \leq 0.084$	$0.001 \leq R \leq 0.006$
			Number of houses in Malcamal ravine	
	Very high	$0.261 < P \leq 0.490$	$0.262 < V \leq 0.448$	$0.068 < R \leq 0.219$
	High	$0.146 < P \leq 0.261$	$0.157 < V \leq 0.262$	$0.023 < R \leq 0.068$
	Medium	$0.069 < P \leq 0.146$	$0.089 < V \leq 0.157$	$0.006 < R \leq 0.023$
	Low	$0.034 \leq P \leq 0.069$	$0.044 \leq V \leq 0.089$	$0.001 \leq R \leq 0.006$
			Road	
Landslide	Very high	$0.266 < P \leq 0.487$	$0.261 < V \leq 0.463$	$0.069 < R \leq 0.226$
	High	$0.144 < P \leq 0.266$	$0.148 < V \leq 0.261$	$0.021 < R \leq 0.069$
	Medium	$0.068 < P \leq 0.144$	$0.083 < V \leq 0.148$	$0.006 < R \leq 0.021$
	Low	$0.034 \leq P \leq 0.068$	$0.046 \leq V \leq 0.083$	$0.001 \leq R \leq 0.006$
			Bridge	
Fluvial flood	Very high	$0.264 < P \leq 0.490$	$0.265 < V \leq 0.454$	$0.070 < R \leq 0.222$
	High	$0.142 < P \leq 0.264$	$0.150 < V \leq 0.265$	$0.021 < R \leq 0.070$
	Medium	$0.070 < P \leq 0.142$	$0.085 < V \leq 0.150$	$0.006 < R \leq 0.021$
	Low	$0.035 \leq P \leq 0.070$	$0.046 \leq V \leq 0.085$	$0.001 \leq R \leq 0.006$
			Municipal slaughterhouse	
	Very high	$0.264 < P \leq 0.490$	$0.263 < V \leq 0.445$	$0.070 < R \leq 0.218$
	High	$0.142 < P \leq 0.264$	$0.160 < V \leq 0.263$	$0.023 < R \leq 0.070$
	Medium	$0.070 < P \leq 0.142$	$0.085 < V \leq 0.160$	$0.006 < R \leq 0.023$
	Low	$0.035 \leq P \leq 0.070$	$0.047 \leq V \leq 0.085$	$0.001 \leq R \leq 0.006$
			Municipal parking	
	Very high	$0.264 < P \leq 0.490$	$0.263 < V \leq 0.460$	$0.069 < R \leq 0.225$
	High	$0.142 < P \leq 0.264$	$0.154 < V \leq 0.263$	$0.022 < R \leq 0.069$
	Medium	$0.070 < P \leq 0.142$	$0.084 < V \leq 0.154$	$0.006 < R \leq 0.022$
	Low	$0.035 \leq P \leq 0.070$	$0.040 \leq V \leq 0.084$	$0.001 \leq R \leq 0.006$
			Sewer system component	
	Very high	$0.264 < P \leq 0.490$	$0.270 < V \leq 0.464$	$0.071 < R \leq 0.227$
	High	$0.142 < P \leq 0.264$	$0.148 < V \leq 0.270$	$0.021 < R \leq 0.071$
	Medium	$0.070 < P \leq 0.142$	$0.080 < V \leq 0.148$	$0.006 < R \leq 0.021$
	Low	$0.035 \leq P \leq 0.070$	$0.039 \leq V \leq 0.080$	$0.001 \leq R \leq 0.006$

Table 4: Quantity of elements exposed to hazard, vulnerability and risk levels.

	Level	Hazard	Vulnerability	Risk
			Number of houses in Chilchos ravine	
Debris flow	Very high	–	15	9
	High	13	5	8
	Medium	7	1	4
	Low	1	–	–
			Number of houses in Malcamal ravine	
	Very high	–	7	1
	High	1	9	14
	Medium	15	–	1
			Road section (m)	
Landslide	Very high	77.79	77.50	77.13
	High	236.26	268.05	150.33
	Medium	616.75	677.52	795.61
	Low	92.27	–	–
			Bridge	
Fluvial flood	Very high	1	–	1
	High	–	1	–
			Municipal slaughterhouse	
	Very high	–	1	–
	High	1	–	1
			Municipal parking	
	Very high	1	1	1
			Sewer system component	
	Very high	1	–	1
	High	–	1	–

For the vulnerability analysis, files were generated with specific data on exposure, fragility and resilience for each type of element exposed to these hazards. Houses were identified as being exposed to debris flow hazard; the local road to landslide hazard; and the bridge, municipal parking, municipal slaughterhouse and the sewage system component to fluvial flood hazard (Table 2).

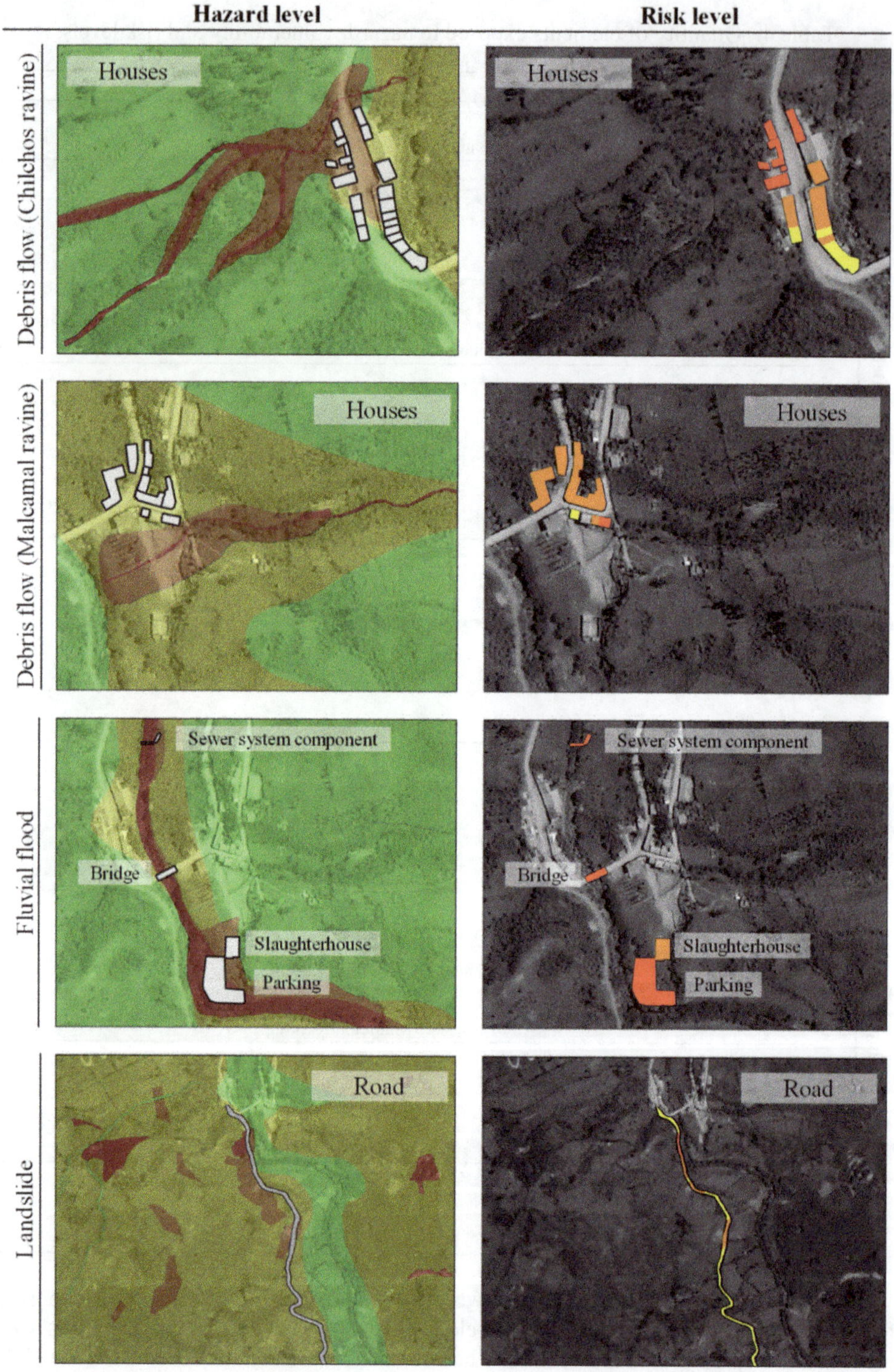

Figure 4: Hazard and risk maps according to the type of element exposed in Pipus village.

Table 2 shows the variables used for each factor and dimension of vulnerability used to calculate the vulnerability levels range.

The variables presented in Tables 1 and 2 were worked with the hierarchical analysis process to obtain the weighted weights of each of the variables.

4 DISCUSSION AND RESULTS

The weighted weights were calculated with the hierarchical analysis process of the variables presented in Tables 1 and 2, in which the hazard, vulnerability and risk levels range were obtained for each type of danger and exposed element. (Table 3).

The ranges of values in Table 3 were used to classify the hazard, vulnerability and risk levels presented by the different elements exposed for each type of hazard, which are shown in Table 4.

In Table 4, it can be seen that the exposed elements are mostly with high and very high risk levels, in which only one house presents a medium risk level to debris flow hazard and a 795 meter stretch of The road presents a medium risk level to landslide hazard. Hazard and risk levels can be seen spatially in Fig. 4, according to type of hazard and type of exposed elements.

Likewise, a quantification was made of the probable effects that these three hazard may cause in Pipus village (Table 5), classified according to damage value, determined by the value of the physical assets destroyed; and losses, determined with the value of changes in the production flows of goods, services and the provision of additional services (emergency care, humanitarian aid, temporary tents, etc.).

Table 5: Probable effects of hazards in Pipus village according to type of hazard.

	Probable effects		
Hazard	Damage	Losses	Total
Debris flow in Malcamal ravine	S/ 2,590,150.00	S/ 223,000.00	S/ 2,813,150.00
Debris flow in Chilchos ravine	S/ 2,144,300.00	S/ 193,300.00	S/ 2,337,600.00
Landslide	S/ 155,250.00	S/ 35,500.00	S/ 190,750.00
Fluvial flood	S/ 680,000.00	S/ 320,000.00	S/ 1,000,000.00
	S/ 5,569,700.00	S/ 771,800.00	S/ 6,341,500.00

Note: The values are expressed in PEN national currency with the symbol S /.

In Fig. 4, the location of the different elements exposed to the evaluated hazard is observed, in which it is appreciated that the risk levels oscillate between very high and medium, which indicates that immediate and priority activities must be developed for control and risk reduction.

Likewise, in Table 5, it is possible to observe the different probable effects that these hazards may cause, which amount to a total sum of S /. 6’341,500.00 between damages and losses in the Pipus village.

5 CONCLUSIONS AND RECOMMENDATIONS

The semi-quantitative methodology for the evaluation of the risk disasters caused by natural hazards allowed to quickly know the different hazard, vulnerability and risk levels to which the different types of exposed elements are found.

The risk due to the debris flow in the houses of the Chilchos ravine is dominated by the very high level, while in the Malcamal ravine the high level predominates, causing a total effect of S /. 5’150,750.00 which leads to an unacceptable risk, which can be reduced with the construction of flow diversion channels along the ravine and reforestation with pine trees in the upper part of the ravine to stabilize the slopes, in which the volumes of solids that contribute to the formation of debris flows are present.

On the road evaluated, the landslide risk that predominates is the medium level, causing a total effect of S /. 190,750.00, therefore, the construction of the channeling of surface and underground water is recommended to reduce the water table, retaining walls for protection on the road and reforestation with pine trees in the upper part of the slopes to stabilize the susceptible slope to landslides.

The fluvial flood risk would cause a total effect of S /. 1'000,000.00 in which the very high level predominates. To reduce this risk, the construction of riverine defenses with gabion and reinforced concrete walls is recommended on both banks of the Olia River, which will reduce the effects of erosion and contain the volume of water in times of flooding of the river.

The non-structural measures that are recommended for risk reduction and control is to propose processes to strengthen organizational capacities in Pipus village, train the population in compliance with technical construction standards, conduct awareness talks on management of the disaster risk and promote environmental practices for the preservation of ecosystems.

Finally, the results of the risk assessment reports will allow local authorities to technically support the risk levels present in the Pipus village in order to gain access to financing for risk reduction and control works through the different financial programs in Peru.

ACKNOWLEDGEMENT

The collaboration of the technical team of the Geophysical Institute of Peru for the elaboration of the geological, geomorphological and geodynamic characterization in Pipus village, the technical team of Chachapoyas Provincial Municipality for the preparation of the risk assessment reports and the National Center for Estimation, Prevention and Reduction of Disaster Risk for technical assistance in the preparation of reports.

REFERENCES

[1] Instituto Nacional de Defensa Civil, Centro de Operaciones de Emergencia Nacional, Deslizamiento en el distrito de San Francisco de Daguas – Amazonas, 2021. https://www.indeci.gob.pe/wp-content/uploads/2021/03/INFORME-DE-EMERGENCIA-N%C2%BA-290-6MAR2021-DESLIZAMIENTO-EN-EL-DISTRITO-DE-SAN-FRANCISCO-DE-DAGUAS-AMAZONAS-1.pdf.

[2] Instituto Nacional de Defensa Civil, Centro de Operaciones de Emergencia Nacional, Deslizamiento en el distrito de San Francisco de Daguas – Amazonas, 2021. https://www.indeci.gob.pe/wp-content/uploads/2021/03/REPORTE-PRELIMINAR-N%C2%BA-973-25MAR2021-DESLIZAMIENTO-EN-EL-DISTRITO-DE-SAN-FRANCISCO-DE-DAGUAS-AMAZONAS.pdf.

[3] Centro Nacional de Estimación, Prevención y Reducción del Riesgo de Desastres, Manual para la evaluación de Riesgos originados por fenómenos naturales, 2014. http://www.cenepred.gob.pe/web/wp-content/uploads/Guia_Manuales/Manual-Evaluacion-de-Riesgos_v2.pdf.

[4] Saaty, T.L., Decision making with the analytic hierarchy process. *Services Science*, **1**(1), p. 86, 2008. https://doi.org/10.1504/IJSSci.2008.01759.

[5] Instituto Geofísico del Perú, Caracterización geológica, geomorfológica y geodinámica del poblado de Pipus, distrito de San Francisco de Daguas, provincia de Chachapoyas y departamento de Amazonas, 2021.

[6] Servicio Nacional de Meteorología e Hidrología del Perú, Datos hidrometeorológicos a nivel nacional, 2021. https://www.senamhi.gob.pe/?&p=estaciones.
[7] Servicio Nacional de Meteorología e Hidrología del Perú, Estimación de umbrales de precipitaciones extremas para la emisión de avisos meteorológicos, 2014. https://cdn.www.gob.pe/uploads/document/file/811068/01402SENA-6.pdf.

Author index

www.ingramcontent.com/pod-product-compliance
Lightning Source LLC
LaVergne TN
LVHW062229200926
848484LV00006B/1505
9781784664459